BIOLOGY: UNDERSTANDING LIFE

STUDENT STUDY GUIDE

BIOLOGY: UNDERSTANDING LIFE

STUDENT STUDY GUIDE

A Manual to Assist Students in
the Development of Biological
Science Concepts

Carolyn A. Dennehy, Ph.D.

Assistant Professor of Biological Sciences
Department of Biological Sciences
University of Northern Colorado - Greeley

St. Louis Baltimore Boston Carlsbad Chicago Naples New York Philadelphia Portland
London Madrid Mexico City Singapore Sydney Tokyo Toronto Wiesbaden

A Times Mirror
Company

Vice President & Publisher: James M. Smith
Editor: Ronald E. Worthington, Ph.D.
Developmental Editor: Jean Sims Fornango
Project Manager: Gayle Morris
Manufacturing Supervisor: Theresa Fuchs

Printed in the United States of America.

Mosby-Year Book, Inc.
11830 Westline Industrial Drive
St Louis, Missouri 63146

International Standard Book Number: 0-8151-0837-0

96 97 98 99 00 / 987654321

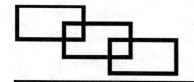

PREFACE

The opportunity to write this book seemed just too good to pass up. The chance to offer a hand to more students than I could ever fit into my classrooms made the reply easy. For me, this experience has been gratifying and reaffirming. It has afforded me the time and the focus to review, ponder, wonder about, and question the science I love and have devoted my professional teaching career to sharing with others.

This guide is written to help you realize your own potential as a learner and to demonstrate that biology is not just something that gets pressed into the pages of a textbook. You and all that surrounds you are this biology. Hopefully, your journey through this book will help you realize how your present knowledge and the addition of important new knowledge can increase your understanding of the all the nuances of life. Knowledge, and all the cognitive processes that accompany its acquisition and use, should become integrated into what and how you think, act, and are.

Each chapter in the study guide begins with a *Key Concept Overview* which discusses the main topics presented in the text. It is written to show how broad concepts are supported by clearly articulated explanations. Biology often requires visual ways to express processes or sequences of events, so I have included many of the diagrams from the text into this section for you to observe and think about. Hopefully, the overview information, coupled with the pictures and diagrams, will help you see and develop a better understanding of the relationships being presented.

The next section in each chapter is called *Chapter Review Activities*. This offers you a variety of ways to acquire, challenge, evaluate, and utilize your understanding of the key concepts. The first of these activities is *Concept Mapping*. It is the first activity in this section because I believe it will help you structure what you know and allow you to see how you have constructed this knowledge in a meaningful context. The real value of doing these maps is the experience of thinking and constructing meaningful connection. A concept map is a visual display of how you have organized and related information from the chapter. There is no right answer in concept mapping, but it does take practice to learn how to construct them. The maps are hierarchically organized concepts delineated by boxes or circles that connect to each other with descriptively labelled lines explaining the relationship between concepts. This activity requires that you evaluate the big ideas within each chapter and not just memorize disjointed facts. I have provided you with an example following this preface that hopefully shows you the structure of a concept map and may help you in getting started. I strongly encourage you to take the time the do this activity. You may want to compare your map with others to see the similarities and differences in the way individuals think about the same information.

Each chapter provides a *Matching* activity. This section reviews the relevant content presented in the text. You should pay attention to those that you were not successful at answering and review that information in the text. A terms list is provided for you and the statements are written to mirror the way they are presented in each of the chapters.

Some chapters in the study guide have an activity called *If.... Then.....* This section asks you to apply what you know about about a concept and justify how it relates to the given situation. There are guiding statements to help you reflect on the most important parts of a concept necessary to supply an appropriate response.

Because observation and identification are important skills, the *Labelling* sections require you to provide information that has been presented for you in the text. As you study the text, try not to skip over the labels in the diagrams. Challenge yourself to recognize those specific parts that have been selected and identified for you. See what you can label on your own before referring back to the text or answer section of the study guide.

There is a *Multiple Choice Questions* section in each chapter. These questions are designed to challenge your recall, ability to compare and contrast, level of content knowledge, ability to analyze statements thoroughly for accuracy and completeness, and to provide you with one form of assessment that tests these things. I strongly suggest that you review the information assoicated with any of these questions you answer incorrectly.

The *Short Answer* section of the study guide, provides a way for you to use your new knowledge at several levels. Sometimes the questions simply ask you to list items, sequence events, or define terms. Others require more thought and encourage you to develop an answer mentally and then write the answer in concise, but accurate detail.

The *Critical Questions* are provided to allow you to apply the knowledge that you have to solving problems or explanation the cause and effects of certain actions. You should think about the questions and organize these thoughts before you attempt to write an answer. Choose the way you express your thoughts carefully so that you say what you mean, but you avoid being repetitious, fragmented, unclear, or off of the subject.

The last activity in each chapter is called *Telling The Story*. Most students have or take little opportunity to express their understanding of scientific principles through writing or dialogue. I have included this section to give you such an opportunity. One of the most powerful ways to demonstrate the level of your knowledge and understanding is to pesent it to someone. Decide on the key concepts you want to discuss. Sequence the parts of the concept so that they build into a complete "story". If you can produce text that is clear, concise, accurate, descriptive, and meaningful to others, it will demonstrate how well you understand the material and will develop a greater confidence for you.

This study guide was prepared with special attention paid to the needs of a diverse population of students. Because most introductory courses have students with varied experiences in studying biological science topics and with different reasons for enrolling in the course, I have tried to offer an array of learning tools which I hope will enhance participation and foster concept development.

Biology is not "just a science." It is perspective about the interactions of living organisms and environments. It has personal, social, political, cultural, and economic significance. And without question, it impacts each of us at many levels. I would challenge each of you to recognize the impact that all life has, has had, and will continue to have on the world. Begin this journey with a different perspective about the value of biology. Become a knowledge consumer, ask questions, ponder and wonder, challenge your own thinking and that of others, and celebrate the opportunity to do so. The rewards are endless.

Carolyn A. Dennehy, Ph.D.
Department of Biological Sciences
University of Northern Colorado
Greeley, Colorado

Example Concept Map

Below is an example of a concept map of the animal kingdom. The map demonstrates how someone might describe their understanding of how the animal kingdom is subdivided. Notice that the "big idea" or concept is the central focus to which all the parts are linked.

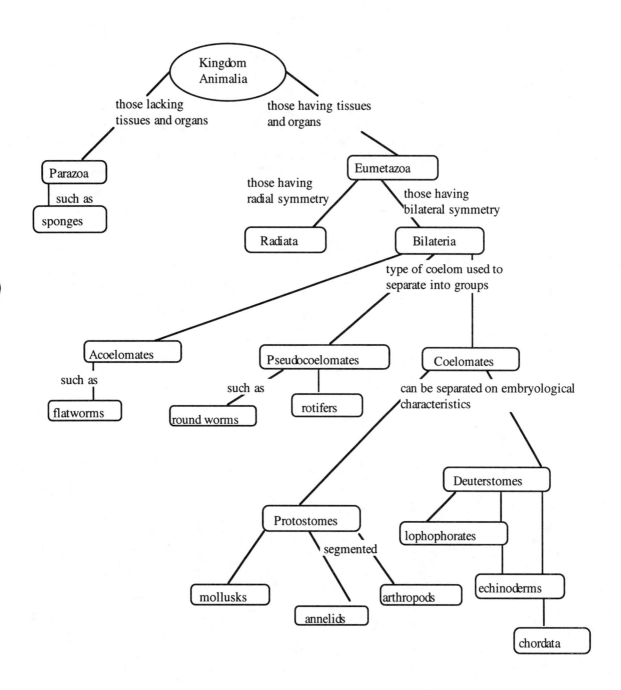

Table of Contents

BIOLOGY: UNDERSTANDING LIFE

STUDENT STUDY GUIDE

BIOLOGY: UNDERSTANDING LIFE

STUDENT STUDY GUIDE

A Manual to Assist Students in
the Development of Biological
Science Concepts

Carolyn A. Dennehy, Ph.D.

Assistant Professor of Biological Sciences
Department of Biological Sciences
University of Northern Colorado - Greeley

 Mosby

St. Louis Baltimore Boston Carlsbad Chicago Naples New York Philadelphia Portland
London Madrid Mexico City Singapore Sydney Tokyo Toronto Wiesbaden

Vice President & Publisher: James M. Smith
Editor: Ronald E. Worthington, Ph.D.
Developmental Editor: Jean Sims Fornango
Project Manager: Gayle Morris
Manufacturing Supervisor: Theresa Fuchs

Printed in the United States of America.

Mosby-Year Book, Inc.
11830 Westline Industrial Drive
St Louis, Missouri 63146

International Standard Book Number: 0-8151-0837-0

96 97 98 99 00 / 987654321

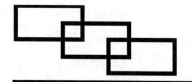

PREFACE

The opportunity to write this book seemed just too good to pass up. The chance to offer a hand to more students than I could ever fit into my classrooms made the reply easy. For me, this experience has been gratifying and reaffirming. It has afforded me the time and the focus to review, ponder, wonder about, and question the science I love and have devoted my professional teaching career to sharing with others.

This guide is written to help you realize your own potential as a learner and to demonstrate that biology is not just something that gets pressed into the pages of a textbook. You and all that surrounds you are this biology. Hopefully, your journey through this book will help you realize how your present knowledge and the addition of important new knowledge can increase your understanding of the all the nuances of life. Knowledge, and all the cognitive processes that accompany its acquisition and use, should become integrated into what and how you think, act, and are.

Each chapter in the study guide begins with a *Key Concept Overview* which discusses the main topics presented in the text. It is written to show how broad concepts are supported by clearly articulated explanations. Biology often requires visual ways to express processes or sequences of events, so I have included many of the diagrams from the text into this section for you to observe and think about. Hopefully, the overview information, coupled with the pictures and diagrams, will help you see and develop a better understanding of the relationships being presented.

The next section in each chapter is called *Chapter Review Activities*. This offers you a variety of ways to acquire, challenge, evaluate, and utilize your understanding of the key concepts. The first of these activities is *Concept Mapping*. It is the first activity in this section because I believe it will help you structure what you know and allow you to see how you have constructed this knowledge in a meaningful context. The real value of doing these maps is the experience of thinking and constructing meaningful connection. A concept map is a visual display of how you have organized and related information from the chapter. There is no right answer in concept mapping, but it does take practice to learn how to construct them. The maps are hierarchically organized concepts delineated by boxes or circles that connect to each other with descriptively labelled lines explaining the relationship between concepts. This activity requires that you evaluate the big ideas within each chapter and not just memorize disjointed facts. I have provided you with an example following this preface that hopefully shows you the structure of a concept map and may help you in getting started. I strongly encourage you to take the time the do this activity. You may want to compare your map with others to see the similarities and differences in the way individuals think about the same information.

Each chapter provides a *Matching* activity. This section reviews the relevant content presented in the text. You should pay attention to those that you were not successful at answering and review that information in the text. A terms list is provided for you and the statements are written to mirror the way they are presented in each of the chapters.

Some chapters in the study guide have an activity called *If.... Then.....* This section asks you to apply what you know about about a concept and justify how it relates to the given situation. There are guiding statements to help you reflect on the most important parts of a concept necessary to supply an appropriate response.

Because observation and identification are important skills, the *Labelling* sections require you to provide information that has been presented for you in the text. As you study the text, try not to skip over the labels in the diagrams. Challenge yourself to recognize those specific parts that have been selected and identified for you. See what you can label on your own before referring back to the text or answer section of the study guide.

There is a *Multiple Choice Questions* section in each chapter. These questions are designed to challenge your recall, ability to compare and contrast, level of content knowledge, ability to analyze statements thoroughly for accuracy and completeness, and to provide you with one form of assessment that tests these things. I strongly suggest that you review the information assoicated with any of these questions you answer incorrectly.

The *Short Answer* section of the study guide, provides a way for you to use your new knowledge at several levels. Sometimes the questions simply ask you to list items, sequence events, or define terms. Others require more thought and encourage you to develop an answer mentally and then write the answer in concise, but accurate detail.

The *Critical Questions* are provided to allow you to apply the knowledge that you have to solving problems or explanation the cause and effects of certain actions. You should think about the questions and organize these thoughts before you attempt to write an answer. Choose the way you express your thoughts carefully so that you say what you mean, but you avoid being repetitious, fragmented, unclear, or off of the subject.

The last activity in each chapter is called *Telling The Story*. Most students have or take little opportunity to express their understanding of scientific principles through writing or dialogue. I have included this section to give you such an opportunity. One of the most powerful ways to demonstrate the level of your knowledge and understanding is to pesent it to someone. Decide on the key concepts you want to discuss. Sequence the parts of the concept so that they build into a complete "story". If you can produce text that is clear, concise, accurate, descriptive, and meaningful to others, it will demonstrate how well you understand the material and will develop a greater confidence for you.

This study guide was prepared with special attention paid to the needs of a diverse population of students. Because most introductory courses have students with varied experiences in studying biological science topics and with different reasons for enrolling in the course, I have tried to offer an array of learning tools which I hope will enhance participation and foster concept development.

Biology is not "just a science." It is perspective about the interactions of living organisms and environments. It has personal, social, political, cultural, and economic significance. And without question, it impacts each of us at many levels. I would challenge each of you to recognize the impact that all life has, has had, and will continue to have on the world. Begin this journey with a different perspective about the value of biology. Become a knowledge consumer, ask questions, ponder and wonder, challenge your own thinking and that of others, and celebrate the opportunity to do so. The rewards are endless.

Carolyn A. Dennehy, Ph.D.
Department of Biological Sciences
University of Northern Colorado
Greeley, Colorado

Example Concept Map

Below is an example of a concept map of the animal kingdom. The map demonstrates how someone might describe their understanding of how the animal kingdom is subdivided. Notice that the "big idea" or concept is the central focus to which all the parts are linked.

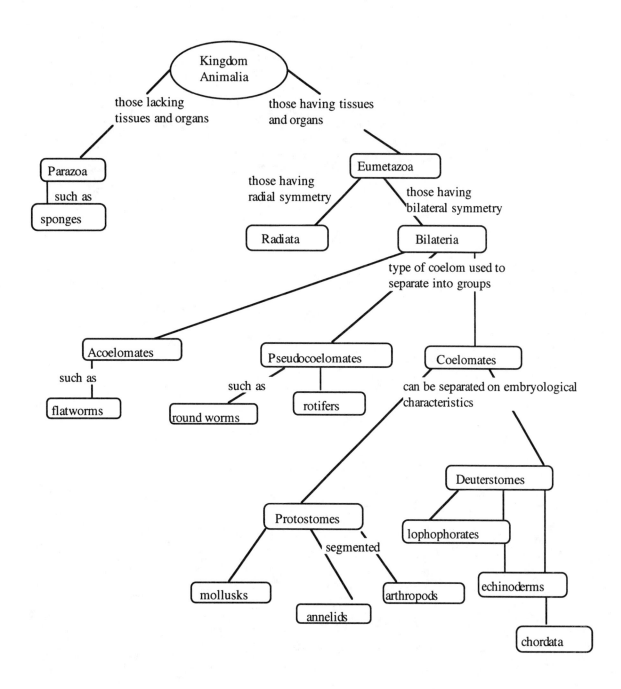

Table of Contents

BIOLOGY: UNDERSTANDING LIFE

STUDENT STUDY GUIDE

CHAPTER 1

THE THEMES OF BIOLOGY TODAY

KEY CONCEPTS OVERVIEW

1.   Scientific Methods

Scientists approach the study of any problem in a systematic way. The word "science," derived from the Latin term *scientia* (meaning to have knowledge or to know), is actually a process of coming to better understand an event in nature. This would imply that science is really focused on *how* knowledge is acquired, rather than simply acquiring facts already known. The systematic ways that scientists, and many others, approach the study of the natural world are referred to as **scientific methods.** These methods, although different in some ways, generally include *observations*, *hypothesis formation*, and *hypothesis testing*.

2.   Theories

Questioning, exploring, and experimenting are processes by which the scientific community gathers supporting evidence to accept or refute hypotheses. When hypotheses can be consistently explained or supported by evidence gathered, scientists begin to formulate what is called a **theory.** Theories are plausible, dependable, scientifically-supported generalizations that come from consistently reported evidence. A scientific theory is described in your text as a synthesis of hypotheses that have withstood the test of time. They are used to explain what scientific laws describe. Theories do not become scientific laws. Unfortunately, the word *theory* is often used differently in our cultural language to refer to an unsubstantiated claim (e.g., diet pills melt fat, salt is bad for you, or vitamin C prevents cancer). Science does not accept any information as valid without substantial evidence.

3.   Themes

Themes are principles, general truths or paradigms which describe a context, a way in which lesser constructs are held and viewed. Accordingly, themes are accepted explanations that transcend time. Specific details related to a theme may change as new or better information is determined. Biologists have catagorized their questions, hypotheses, observations, and data into organizing themes. These themes apply to all living things.

Living things have the following properties:
- Display both diversity and unity
- Are composed of cells and are organized hierarchically
- Interact with each other and with their environments
- Transform energy and maintain a steady internal environment
- Exhibit forms that fit their functions
- Reproduce and pass on biological information to the offspring
- Change over time, or evolve

4.   Taxonomy

Classification is a powerful method of packaging information. Because of the enormous amount of diversity among all living things, scientists require that there be some way to group organisms. Modern **taxonomy** uses evolutionary information concerning ancestry to sort and to group organisms. While this was not the first system devised for classifying organisms, it allows scientists to explain naturalistically why there are inherent groups among living things. Taxonomists start with broad categories and narrow to very specific characteristics that ultimately differentiate each organism from another. The most accepted taxonomic system, and the one used in your text, is the *five kingdom system* (Monerans, Protists, Fungi, Plants, and Animals). These kingdoms of organisms are further subdivided into groupings that reflect an increasing closeness in evolutionary ancestry as determined by a variety of criteria. These subdivisions from more general to more specific are phylum, class, order, family, genus, and species.

CHAPTER REVIEW ACTIVITIES

CONCEPT MAPPING

Construct a concept map that incorporates the ideas in this chapter. The map should reflect your understanding of scientific methodology, theories, themes, and taxonomy. Before beginning, review the examples given in the preface.

MATCHING

Using the list of terms provided, select the term that best relates to the statements below. After answering the questions, check the answers against those provided at the end of this chapter review.

_____ 1. These are the broad categories in which taxonomists group all living things.

_____ 2. These organisms break down organic molecules of dead organisms.

_____ 3. This is the part of the Earth where biological activity exists.

_____ 4. This tentative explanation or prediction guides scientific inquiry.

_____ 5. This population of organisms interbreeds freely in their natural settings and does not interbreed with other populations. It is also a taxonomic subcategory of family and the most narrow and specific taxonomic classification.

_____ 6. This is a pattern of thought in which a person begins with a general statement and proceeds to a specific statement.

_____ 7. This group consists of the individuals of a given species that occur together at one place and at one time.

_____ 8. These living things can be either multicellular or unicellular.

_____ 9. These organisms make their own food using the process of photosynthesis.

_____ 10. These are the molecules that make up genes and constitute the "code of life."

_____ 11. This is the pattern of thought in which a person develops generalizations from specific instances.

_____ 12. These populations of different species interact with one another in a particular place.

_____ 13. This is the factor manipulated during a controlled experiment.

_____ 14. These organisms cannot make their own food and must feed on producers or organisms like themselves.

_____ 15. This method of classification focuses on the common evolutionary history of a diverse array of species.

_____ 16. This is a process by which certain organisms produce their own food by capturing energy from the sun.

_____ 17. This community of plants, animals, and microorganisms interacts with one another and their environments and are interdependent on each other for survival.

_____ 18. This is the factor within a controlled experiment that varies in response to changes in the factor that is manipulated.

_____ 19. This hypothesis of organismal change over time embodies the ideas that organisms alive today are descendants of earlier organisms and that organisms have changed and diverged from one another over billions of years.

_____ 20. This is the standard against which experimental observations or conclusions may be checked to establish their validity.

_____ 21. These are subdivisions of the kingdoms Monera, Plantae, and Fungi and are taxonomically equivalent to phyla.

_____ 22. This describes the natural phenomena in science.

TERMS LIST

A. Consumer	L. Decomposer
B. Control	M. Organisms
C. Evolution	N. Community
D. Kingdom	O. Biosphere
E. Photosynthesis	P. Taxonomy
F. Hypothesis	Q. Population
G. Deoxyribonucleic acid (DNA)	R. Species
H. Independent variable	S. Ecosystem
I. Deductive reasoning	T. Producers
J. Dependent variable	U. Inductive reasoning
K. Divisions	V. Scientific laws

SEQUENCING AND USING TAXONOMIC TERMS

Sequence the following taxonomic terms beginning with the broadest category containing the most organisms and progressing toward the most specific category that identifies individual organisms.

A. Family   E. Phylum

B. Class   F. Order

C. Kingdom   G. Species

D. Genus

1. _____
2. _____
3. _____
4. _____
5. _____
6. _____
7. _____

Classification schemes should reveal certain characteristics about an organism. Organisms classified under these particular groupings will have these characteristics in common. What information could you conclude about an organism from each of the following?

8. Class: Mammalia

_____

_____

9. Species: *domesticus*

_____

_____

10. Phylum: Chordata

_____

_____

11. Genus: *Felis*

_____

_____

12. Develop a statement that clarifies the purpose for and need of a classification scheme in biology.

COMPOSITION OF ORGANISMS AND HIERARCHICAL ORGANIZATION

Differentiate among the following by writing the name of the appropriate individual below the statement that best relates to his contribution.

1.   A German medical microscopist who believed that all cells could arise only from pre-existing cells. In addtion to this, he also made reference to an existing hierarchy within living systems when he wrote: "...the last constant link in the great chain of mutually subordinated formations that form tissues, organs, systems, the individual. Below them is nothing but change."

       _____

2. A German zoologist who emphasized the great difference between the cells of plants and the structures of animals but suggested that they are fundamentally the same. He was one of several nineteenth-century observers to hypothesize that the structure of all organisms is composed solely of cells or the products of cells.

   _____

3. As a botanist among his nineteenth-century contemporaries, he too hypothesized that the unit of structure of all living things is the cell.

   _____

MULTICELLULAR ORGANIZATION

Provide the appropriate level of organization or the term needed to complete the following statements.

4. The simplest nonliving unit of matter that represents the purest form of any element. _____

5. A group of organs that function together to carry out the principal activities of an organism. _____

6. A group of similar cells that work together to perform a function.

   _____

7. A structural and functional unit formed of tissues grouped together.

   _____

8. Single-celled organisms that work and live together as a team.

   _____

MULTIPLE CHOICE QUESTIONS

Place the letter indicating your choice of answers in the space provided by each question. As you read through the selections, mentally note the reasons for your selection and rejections of the choices given.

_____ 1. Of the five accepted kingdoms, which is considered to contain the most primitive organisms?
   a. Animalia
   b. Plantae
   c. Monera
   d. Fungi
   e. Protista

_____ 2. Which of the following would include any nonliving components as well as the living organisms?
   a. Populations
   b. Ecosystems
   c. Individuals
   d. Species
   e. All of the above

_____ 3. An "if .... then ...." statement is characteristic of which of the following?
   a. Principle
   b. Theory
   c. Conclusion
   d. Law
   e. Hypothesis

_____ 4. Which of the following would you LEAST expect to find in a scientific method?
   a. An assumption
   b. Data gathered from experimentation
   c. Conclusions drawn from analysis of data gathered
   d. Experimental testing of hypothesis
   e. A hypothesis

_____ 5. Having observed that your hair is softer after using a certain herbal shampoo for 3 months, you suggest to a friend that this product causes hair to be softer than other shampoos. This would be an example of which of the following?

    a. Prediction

    b. Theory

    c. Inductive reasoning

    d. Law

    e. Deductive reasoning

_____ 6. Scientists use evidence to _____ hypotheses.

    a. Prove

    b. Support

    c. Disprove

    d. All the above

    e. Only B and C

_____ 7. A synthesis of hypotheses that have held over time and assist scientists in making dependable predictions is which of the following?

    a. Principle

    b. Scientific theory

    c. Conclusion

    d. Law

    e. Hypothesis

_____ 8. Biologists' estimation of the number of different species that exist on the Earth varies between which of the following?

    a. 2 to 5 million

    b. 5 to 30 million

    c. 10 to 15 million

    d. .5 to 20 million

    e. Biologists have no way of knowing the number

_____ 9. Populations of different species that interact with one another make up which of the following?
   a. Community
   b. Ecosystem
   c. Niche
   d. Biosphere
   e. Organ system

_____ 10. Consumers differ from producers in that they do which of the following?
   a. Produce their own food through photosynthesis
   b. Break down organic materials from dead organisms
   c. Convert sunlight into chemical energy
   d. Feed on other organisms and release stored energy
   e. None of the above

_____ 11. For a theory to be scientific, which of the following must apply?
   a. Theories are absolutely true.
   b. Theories must be in agreement with previous knowledge.
   c. Theories must be provable beyond all doubt.
   d. Theories must be socially acceptable or economically useful.
   e. Theories must be capable of being faslefied.

CRITICAL QUESTIONS

1. Discuss the relevance of a scientific method.

2. Differentiate between a hypothesis and a theory. Give an example of each.

3. The differences between science and nonscience are often based on the assumptions and processes used to collect, organize, test, and report these assumptions. With this in mind, what should consumers ask themselves about statements like: "Four out of five doctors surveyed recommend the use of Bend Better Topical Ointment for the treatment of arthritis"?

TELLING THE STORY

One way to determine how well you understand new or enhanced knowledge is to write or to talk about it. To help you realize your own understanding of the material in Chapter 1, write about the key concepts (Scientific Methods, Theories, Themes, and Taxonomy). Your writing should reflect the depth and scope of your understanding and should help you identify where you are weak in your conceptualizations.

CHAPTER REVIEW ANSWERS

CONCEPT MAPPING

While there is no right answer for a concept map, you may want to check to see if you were able to link the terms presented in Chapter 1 to the "big ideas." The "linking" statements should connect nouns to action phrases and should result in a "blueprint" of the way you construct the information associated with a concept.

MATCHING

1. D	7. Q	13. H	19. C
2. L	8. M	14. A	20. B
3. O	9. T	15. P	21. K
4. F	10. G	16. E	22. V
5. R	11. U	17. S	
6. I	12. N	18. J	

SEQUENCING AND USING TAXONOMIC TERMS

1. C	2. E	3. B	4. F	5. A	6. D	7. G

8.   The class Mammalia is a subgroup of the phylum Chordata. Mammals are characterized by having skin with hair and by secreting milk from mammary glands to nourish their young.

9.   The species *domesticus* implies that the organism is a domesticated variety. This would separate the organism from other species within a given genus.

10.   The phylum Chordata contains all animals that have vertebrae or a spinal column. This is but one of approximately 19 phyla in the animal kingdom.

11.   The genus *Felis* distinguishes the organism as a cat. All cats are in this genus, but there are various types of cat that would be differentiated from one another only by species.

12.   Statements may vary, but an example of a correct response would be: Classification schemes are essential in differentiating one organism from another based on ancestral history. There is enormous diversity among all living things. Sponges, trees, bacteria, forget-me-nots, and elephants are all

living, but characteristically very different. Biologists must be able to clearly communicate with each other about their findings, their questions, and their hypotheses. Common names are usually regional and would not suffice among the international science community. Therefore a classification system has been designed to show the relationships of organisms by placing the closely related ones together in categories.

COMPOSITION OF ORGANISMS AND HIERARCHICAL ORGANIZATION

1.   Verchow
2.   Schwann
3.   Schleiden

MULTICELLULAR ORGANIZATION

4.   Atom
5.   Organ system
6.   Tissue
7.   Organ
8.   Colonial organism

MULTIPLE CHOICE QUESTIONS

1. C	6. E
2. B	7. B
3. E	8. B
4. A	9. A
5. C	10. D
	11. B

CRITICAL QUESTIONS

1.   Although there are different strategies all scientists use in performing experimentation, there are consistent components to the methods used by them. Scientists ask good questions (pose a problem) that result from observing a situation, formulating a hypothesis, and then testing the hypothesis. From the experimentation, data are gathered and treated statistically to determine if the hypothesis should be accepted or rejected. Then scientists report their findings to increase the body of knowledge available about the subject being studied. While scientific methodology is

common place among legitimate scientists, you probably approach many problems you have in similar ways.

2. A hypothesis is a testable question that results from observations made about a specific situation. It represents a possible answer to a question or an explanation of such observations. A theory is a model that is supported by data from many testable hypotheses. Usually a theory holds up over long periods of time because little evidence (opposing data) has been found to refute it.

3. The general public relies on "truth in advertising" to substantiate claims about marketed products. Often these marketing claims do not provide all the available information and are constructed of very general statements. When the claims are made in language that suggests scientific backing for the product, the public is often mislead. Therefore a consumer might ask themselves questions such as the following:
    - What kind of doctors were these?
    - How many were actually asked?
    - Were they randomly selected or are they biased participants?
    - What does "recommend" mean?
    - What data were used to show the effects of the product (subjects, amount of the product, strength of the product, time used)?

CHAPTER 2

THE CHEMISTRY OF LIFE

KEY CONCEPTS OVERVIEW

1. Chemical Interactions Of Atoms

If you look around you will notice that your surroundings vary in characteristics such as states of matter (solid, liquid, gas), hardness, color, shape, etc. Scientists summarize all the material that exists as matter. The term **matter** is used to refer to everything that has mass (a measurable amount of substance) and takes up space. **Elements** are the basic building blocks from which all matter is composed. They are the purest form of all substances. To understand how elements react with one another, it is important to take a look at their composition. Elements are made up of submicroscopic structures called **atoms,** which are represented by chemical symbols (e.g., H for hydrogen, C for carbon, N for nitrogen, Ca for calcium). In 1913, a Danish physicist named Niels Bohr proposed that every atom was made up of three types of small (subatomic) particles: **protons** (positively charged), **neutrons** (no charge)**,** and **electrons** (negatively charged). The protons and neutrons are found within the nuclear region and constitute the mass of the atom, and the electrons orbit this nuclear region and have little mass to contribute. The chemical behavior or properties of any atom are determined primarily by the arrangement of electrons. Electrons are constantly moving at great speeds and tend to move in specific regions called **orbitals.** The balance of several factors (attractive charges of protons and electrons, the tendency of electrons to move, and opposing charges between electrons themselves) cause electrons to distribute themselves in a predictable pattern around the nucleus. Electron energy increases as the distance from the attractive force of the nucleus increases. It is the electrons in the outermost orbital or energy level that provide a reliable indicator of the chemical activity of an atom, but the entire atomic structure also contributes to this activity. Atoms are most stable when their outer orbital is filled. To accomplish this, electrons are lost, gained, or shared by other atoms. The tendency to have the outer orbital filled with electrons is called the **octet rule**, since the outer orbital of many atoms contain a maximum of eight electrons. The first orbital, filled with only two electrons, is an exception to this rule.

2.   Chemical Bonds

When atoms have the outermost orbital unfilled, they can satisfy the octet rule in one of the following three ways:

- They can gain electrons from another atom.
- They can lose electrons to another atom.
- They can share one or more electron pairs with another atom.

Regardless of which of the above occurs, the resulting reaction forms a **chemical bond**, meaning forces exist that now hold atoms together. If the bond occurs as a result of gaining or losing electrons to another atom, an **ionic** bond occurs. If the bond is formed due to atoms sharing one or more pairs of electrons, a **covalent** bond occurs. When atoms interact in these ways, the resulting structure is called a **molecule**. Molecules can be made up of atoms of the same or different elements. Molecules made up of atoms of two or more elements are called **compounds**.

3.   Water

Water is found in all living things and on three fourths of the Earth's surface. Because of its importance to living systems, some information about its chemistry is necessary.

Molecules that are composed of several atoms have an uneven distribution of charges because the electrons involved in the bonds may be located on one side of the molecule. This causes that side of the molecule to be slightly negative and the other side to be slightly positive. The chemical behavior of water indicates that the electrons are not shared equally by the atoms involved. Oxygen atoms attract electrons more strongly than do hydrogen atoms. If the electrons of a bond are not equally shared by the participating atoms, the bond is called a **polar covalent bond**. When several polar molecules are together, they orient themselves so that the partially positive end of one is near the partially negative end of another. This attraction between two molecules is called a **hydrogen bond**. Biologically, the polar nature of water is very significant because most cells and tissues of organisms in nature contain an average of 70% to 80% water. The water provides a medium in which chemical reactions in organisms occur. Molecules must dissolve in water in order to move easily in and between cells. Polar molecules also interact with one another. Because of this, water forms bonds with other water molecules as well as with those of other polar substances.

Some of the important characteristics of water include the following:

- It is a powerful solvent allowing materials to dissolve in it.
- It organizes nonpolar molecules meaning that nonpolar molecules are excluded from forming hydrogen bonds with water.
- It ionizes resulting in a spontaneous ion formation.

$$H_2O \rightarrow OH^- + H^+$$

Scientists have devised a scale based on the slight degree of spontaneous ionization of water. This scale is called the pH scale. It generally ranges from 0 to 14 and indicates the hydrogen ion concentration in a solution. Pure water has a pH of 7.0. When substances dissociate to form H^+ ions when dissolved in water, an **acid** is formed. The more H^+ ions an acid produces, the stronger it is. This would be indicated as a number lower than 7.0 on the pH scale. Any substance that combines with H^+ ions is said to be a **base** and would be indicated by a number higher than 7.0 on the pH scale. Because the scale is logarithmic, a change of 1 on the scale reflects a tenfold change in pH. An acid with pH 5 is 100 times more acidic than pure water.

4.   Carbohydrates, Lipids, Proteins, Nucleic Acids

In all organisms, the varying numbers and arrangements of carbon atoms form a large variety of organic compounds. These compounds have various functions and are made up of varying numbers of atoms. The greatest majority (96.3%) of atoms found in organisms are nitrogen, oxygen, carbon, or hydrogen. Many large organic molecules, such as starches, are made from small organic compounds that act as building blocks. These small building block molecules are called **monomers**. Molecules formed by linking two or more monomers are called **polymers or macromolecules**. There are four major groups of biologically important micromolecules: complex carbohydrates, lipids, proteins, and nucleic acids. One class of organic molecules, **carbohydrates**, are made of carbon, hydrogen, and oxygen atoms in approximately a 1:2:1 ratio. These atoms link together to form monomers called simple sugars or **monosaccharides**. Glucose, a monosaccharide, is the primary energy-storage molecule used by organisms. Simple sugars like glucose combine with each other by dehydration synthesis to form complex carbohydrates. Organisms use this synthesis process to convert soluble sugars to an insoluble form for storage.

We generally describe **lipids** as organic compounds that include three important categories: (1) **fats, oils, and waxes**; (2) **phospholipids**; and (3) **steroids**. Typically lipids are not soluble in water and therefore this characteristic plays an important role in living organisms. Lipids are primarily composed of carbon, hydrogen, and oxygen atoms, but the ratio is different from that of carbohydrates (fewer oxygen atoms than either hydrogen or carbon atoms). Since lipids are composite molecules, we should be able to identify their subcomponents. Oils and fats are constructed from two different subunits: **glycerol** and **fatty acids**. Waxes have a different molecular backbone from glycerol. Phospholipids are similar to oils except that one of their fatty acids is replaced by a phosphate group attached to a nitrogen-containing group. Steroids are characterized by their arrangement of interlocking rings of carbon and form familiar chemicals such as hormones and cholesterol.

Proteins are made up of monomers called **amino acids.** There are approximately twenty different amino acids. Each one contains an amino group, a carboxyl group, a hydrogen atom, a carbon atom, and a functional group (referred to as an "R" group). A peptide bond is formed following a dehydration synthesis reaction of bonding amino acids together and results in the formation of polypeptides. Individual polypeptide chains, or groups of chains forming a particular configuration, are proteins.

Nucleic acids are complex molecules that store and transfer information within a cell. Because they contain genetic information, nucleic acids are essential to life. They are constructed of monomers known as **nucleotides.** These nucleotides are formed by alternating a phosphate group and a five carbon sugar molecule, either ribose or deoxyribose, bound to one of five nitrogen bases (adenine, guanine, cytosine, thymine, and uracil). Deoxyribonucleic acid (DNA), responsible for storing information regarding the synthesis of proteins, and ribonucleic acid (RNA), responsible for transporting the information from the nucleus to the ribosomes in the cytoplasm and for the actual synthesis of proteins, are essential molecules for organisms to have.

CHAPTER REVIEW ACTIVITIES

CONCEPT MAPPING

Use as many of the terms in the key concept review to construct a concept map that explains your understanding of the relationships among the different kinds of organic compounds in living organisms.

MATCHING

Using the list of terms provided, select the term that best relates to the statements below. If you have difficulty, refer back to the text in Chapter 2.

_____   1.   This term refers to two or more atoms held together by the sharing of electrons.

_____   2.   This is a molecule that is soluble in oil but insoluble in water.

_____   3.   This scale, based on the slight degree of spontaneous ionization of water, indicates the concentration of H^+ in a solution.

_____   4.   These molecules, such as water, have opposite partial charges on different sides of the molecule.

_____   5.   This is one of the major groups of macromolecules composed of subunits called amino acids.

_____   6.   These are able to be dissolved in water.

_____   7.   This is the process by which monomers are combined to form polymers.

_____   8.   This is the term for the volume of space around an atom's nucleus where an electron is most likely to be found.

_____   9.   These differ from other lipids in that one fatty acid group has been replaced by a phosphate group attached to a nitrogen-containing group.

_____   10. This is one of the three factors that influence whether an atom will interact with other atoms, implying that an atom with an unfilled outer orbital has a tendency to interact with another atom or atoms by losing, gaining, or sharing electrons.

_____   11. These forces hold atoms together.

_____ 12. This is a type of chemical bond caused by the electrical attraction created by atoms sharing electrons.

_____ 13. These are short proteins tused as chemical messengers within the brain and throughout the body.

_____ 14. These are polysaccharides in which plants store energy.

_____ 15. This reaction involves an atom losing an electron.

_____ 16. These molecules are made up of atoms of different elements.

_____ 17. These particles carry a negative charge and circle the atom's core.

_____ 18. These are long polymers made of repeating subunits called nucleotides.

_____ 19. These are molecules that contain carbon, hydrogen, and oxygen atoms in a 1:2:1 ratio.

_____ 20. This group of atoms, having definite chemical properties, is attached to the carbon-based core of an organic molecule.

_____ 21. This term refers to the nonpolar molecules that cannot form hydrogen bonds with water.

_____ 22. This is the smallest, sub-microscopic part of an element that retains all the characteristics of that element.

_____ 23. These are weak electric attractions between the H of an NH or OH group, and other oxygen or nitrogen atoms.

_____ 24. These particles are found at the core of an atom and they have no charge.

_____ 25. These charged particles are formed by atoms that have either acquired or lost electrons.

_____ 26. This reaction involves an atom gaining an electron.

_____ 27. This is a fat molecule in which three fatty acids are attached to each of three carbons of a glycerol molecule.

_____ 28. These pure substances are made up of a single kind of atom and cannot be separated into different substances by ordinary chemical methods.

_____ 29. These are highly branched polysaccharides in which animals store glucose.

_____ 30. This is the term for any substance that dissociates to form H^+ ion when dissolved in water.

_____ 31. Found in the nucleus of a cell, these molecules stores information regarding the sequencing of amino acids to produce proteins.

_____ 32. These are subunits of nucleic acids.

_____ 33. These positively charged particles are found at the core of an atom.

_____ 34. This type of chemical bond between atoms is caused by the attraction of oppositely charged particles formed by the gain or loss of electrons.

_____ 35. This is the process by which polymers are disassembled into monomers.

TERMS LIST

A. Electrons	M. Proteins	X. Starches
B. Hydrolysis	N. Functional groups	Y. Atom
C. Acid	O. Orbital	Z. Peptides
D. pH	P. Covalent	AA. Nucleic acids
E. Lipid	Q. Hydrophobic	BB. Polar molecules
F. Oxidation	R. Ions	CC. Reduction
G. Nucleotides	S. Ionic bonds	DD. Protons
H. Soluble	T. Phospholipids	EE. Neutrons
I. Triglycerides	U. DNA	FF. Octet rule
J. Carbohydrates	V. Glycogen	GG. Molecules
K. Elements	W. Chemical bonds	HH. Hydrogen bonds
L. Dehydration synthesis		II. Compounds

TERM RELATIONSHIPS

Each of the groups below have one term that does not belong with the others. Choose the term and explain why you removed it from the group.

1. Enzymes, Lipids, Peptides, Amino Acids, Proteins
2. Carbohydrates, Glycogen, Starch, Nucleotide, Disaccharide
3. Proton, Orbitals, Nucleus, Electron, Nitrogen base
4. Proteins, Lipids, Water, Nucleic Acids, Carbohydrates
5. Fatty Acids, Glycerol, Hydrogen Bonds, Amino Acids, Nucleotides
6. Cellulose, DNA, Protein, Starch, Fatty Acid

COMPLETION

Complete the sentences by supplying the correct term.

1. Enzymes are types of _____.
2. If an atom has an atomic number of 12, it would have _____ protons in the nucleus, and the same number of _____ orbiting the nucleus.
3. Peptide bonds link _____ to form proteins.
4. Organic compounds always contain the element _____.
5. Sex hormones, like the male hormone _____, are examples of chemical compounds called _____. These compounds are types of _____.

6. Carbon has an atomic number of 6. This means that there would be _____ electrons in the first orbital and _____ in the second.

ELEMENTS AND THE PERIODIC CHART

Provide the missing information in the shaded areas in the table below.

COMMON BIOLOGICAL ELEMENTS

ELEMENT	SYMBOL	ATOMIC #	VALENCE	IMPORTANCE
Calcium			+2	
	K		+1	Membrane conduction, nerve function
Oxygen				Cellular respiration, component of water
	I			Component of thyroid hormone
Chlorine				
	N			Component of all proteins and nucleic acids

**To review others that were presented in your text, you may wish to continue to develop this information for other elements. Compare your list with that presented in your text in Table 2-1.

MULTIPLE CHOICE QUESTIONS

Place the letter indicating your choice of answers in the space provided by each question. As you read through the selections, mentally note the reasons for your selection and rejection of the choices given.

_____ 1. The key to the chemical behavior of atoms is based on which of the following?
   a. Ratio of neutrons to protons
   b. The atomic mass
   c. The number and arrangement of their electrons
   d. The level of radioactivity
   e. All of the above

_____ 2. Which of the following gases does not belong with the others?
   a. Argon
   b. Helium
   c. Krypton
   d. Radon
   e. Oxygen

_____ 3. Glucose and fructose are which of the following?
   a. Monosaccharides
   b. Disaccharides
   c. Isomers
   d. Proteins
   e. Polysaccharides

_____ 4. Dehydration syntheses are used to which of the following?
   a. Form polymers
   b. Transport energy
   c. Code for genetic instructions
   d. Produce functional groups
   e. Maintain a narrow pH range

_____ 5. Which of the following nitrogenous bases is found only in RNA?

    a. Adenine

    b. Thymine

    c. Uracil

    d. Cytosine

    e. Guanine

CRITICAL QUESTIONS

1. How would you determine if a food material was composed of carbohydrates, lipids, or proteins? Develop a study that would follow a scientific methodology, and use the information presented in Chapter 2 to design the tests you might use to verify your hypothesis.

2. What are macromolecules and how are they formed?

3. You are given two samples of carbohydrates. The molecular formula for one is $C_6H_{12}O_6$ and the other is $C_{12}H_{22}O_{11}$. Which one is glucose and on what do you base your answer?

4. What is the advantage to animals using adipose cells (fat storage cells) as storage for fat?

5. Differentiate between an isotope and an ion.

TELLING THE STORY

One way to determine how well you understand new or enhanced knowledge is to write or talk about it. To help you realize your own understanding of the material in Chapter 2, write about the key concepts (Chemical Interactions of Atoms, Chemical Bonds, Water, Carbohydrates, Lipids, Proteins, and Nucleic Acids).

CHAPTER REVIEW ANSWERS

MATCHING

1. GG	10. FF	19. J	28. K
2. E	11. W	20. N	29. V
3. D	12. P	21. Q	30. C
4. BB	13. Z	22. Y	31. U
5. M	14. X	23. HH	32. G
6. H	15. F	24. EE	33. DD
7. L	16. II	25. R	34. S
8. O	17. A	26. CC	35. B
9. T	18. AA	27. I	

TERM RELATIONSHIPS

1. <u>Lipids</u> All other words in the group relate to proteins; lipids are another type of macromolecule that have important biological significance.

2. <u>Nucleotide</u> All other words in the group relate to carbohydrates; nucleotides are the molecular subunits of nucleic acids

3. <u>Nitrogen base</u> All other words in the group relate to atomic structure; nitrogen bases are molecular components found in nucleic acids

4. <u>Water</u> All other words in the group are important biological macromolecules; water is a very important molecule but it is very small by comparison to the others

5. <u>Hydrogen bonds</u> All other words are subunits of macromolecules; hydrogen bonds are weak bonds formed when the partial negative charge at one end of a molecule is attracted to the positively charged end of another polar molecule.

6. <u>Fatty Acids</u> All others in the list are polymers.

COMPLETION

1. protein
2. 12, electrons
3. amino acids
4. carbon
5. testosterone, steroids, lipids
6. 2, 4

ELEMENTS AND THE PERIODIC CHART

Ca; 20; trigger for muscle contraction

Potassium; 19;

O; 8; -2

Iodine; 53; -1

Cl; 17; -1; principal negative ion bathing cells

Nitrogen; 7; -3

MULTIPLE CHOICE QUESTIONS

1. C	2. E	3. A	4. A	5. C

CRITICAL QUESTIONS

1. Answers may vary.

2. Macromolecules are large polymers of smaller subunit components called monomers. The important biological macromolecules are carbohydrates, lipids, proteins, and nucleic acids. These molecules form through a process called dehydration synthesis.

3. Glucose is a monosaccharide with a molecular formula of $C_6H_{12}O_6$. All monosaccharides have the same ratio of carbons to hydrogens to oxygens (1:2:1).

4. Animals can store unused energy or fuel in the form of fat, which can later be converted to a usable form or can be used immediately for insulation and protection.

5. Isotopes are atoms of the same element that differ in their number of neutrons. Chemically, however, they behave the same way. Ions are atoms that have gained or lost electrons and now carry a charge. These structures are not stable and tend to react with other ions to form molecules.

CHAPTER 3

CELL STRUCTURE AND FUNCTION

KEY CONCEPT OVERVIEW

1.   Cell Theory

Living things, regardless of their size, shape, or complexity, are made up of small units of life called **cells**. Cells require nutrients, metabolize to produce energy, excrete, reproduce, respond to stimuli, move, grow, and maintain an internal environment different from their surroundings. These characteristics are shared by every type of cell, in every type of organism. Multicellular organisms have cells that work collaboratively to perform these functions and to maintain themselves and the whole organism. Single-celled organisms are independent units that perform all of these necessary functions alone.

Over time, the work of individuals like Robert Hooke (mid 1600's), Anton Van Leeuwenhoek (late 1600's), T. Schwann (1839), M. J. Schleiden (1839), and R. Verchow (mid 1800's) culminated into significant findings that supported the development of a unifying theme about the nature of cells. The **cell theory** is a profound statement regarding the nature of living things. The theory includes three basic tenets about cells:

1. All living things are made up of one or more cells.
2. The smallest living unit of structure and function of all organisms is the cell.
3. All cells arise (come from) preexisting cells.

2.   Cell Size

Not all cells are the same size, nor do they have the same shape. The size of a cell is directly related to its activity level and the speed at which molecules can move across the cell membrane. Because cells require essential materials to be transported through the membrane, moving both in and out of the cell, the size of the cell must be limited to its ability to adequately move materials at a rate sufficient to facilitate vital functions. Therefore, cells must maintain a large **surface area-to-volume ratio** to move substances in and out fast enough to meet their needs (Fig. 3-1). Because of this, cells are usually very small and can accommodate the necessary requirements for life. There are exceptions to this. For example, the hen egg is a very large cell, but the only metabolically active portion is a very tiny area near the surface. The rest of the cell is inactive material that provides storage for nutrients. (See Figure 3-3, P. 55 of the text)

3.   Prokaryotic and Eukaryotic Cells

There are distinctive structures within cells that separate them into two groups, **prokaryotic cells and eukaryotic cells.** This grouping is the fundamental basis for the modern classification scheme in biology and also provides a premise for the evolution of cells and organisms.

Prokaryotes represent the simplest type of living cell (Fig. 3.2). Organisms whose cells never contain a nucleus belong to this group. Despite the fact that these cells contain fewer cellular components, they perform similarly to the more complex eukaryotic cells. Even though they are considered primitive, prokaryotes do have a plasma membrane, cytoplasm, ribosomes, and DNA. Bacteria and related organisms are prokaryotes. They belong to the kingdom Monera. Even though these cells are very primitive, they are everywhere: in the soil, air, water, and on or in every organism. They are able to withstand extreme environmental conditions such as thermal vents on the ocean floor and hot springs like those found in Yellowstone National Park.

Prokaryotes have the following structural features:
- **Cell walls.** The cell walls of prokaryotes contain a substance called peptidoglycan, which is a large molecule that consists partially of sugars and amino acids. This wall protects the cell and prevents entry of harmful substances.

- **Nucleoid.** Although prokaryotes do not have a true nucleus, the DNA of these cells is located in a region called the nucleoid. In addition to the nucleoid, prokaryotes often contain "rings" of DNA called plasmids.

- **Flagella.** The flagella of prokaryotes do not contain microtubules. Instead, the flagella are composed of different protein chains that wrap in corkscrew fashion. Bacteria move by rotating this structure rather than by "whipping" them as do eukaryotes.

(See Figure 3-2, P. 56 in your text)

Organisms within the other four kingdoms (Protista, Fungi, Plantae, and Animalia) are composed of eukaryotic cells. These cells most often contain at least one nucleus and are thought to have evolved from the prokaryotes. As you can well imagine, this group of cells is quite diverse, but they all share a basic structural makeup.

(See Figure 3-22, P. 76 in your text)

All eukaryotes have the following features:

- **Plasma Membrane.** The plasma membrane is composed of a lipid bilayer in which proteins are attached at the surface and embedded. This membrane controls the passage of materials into and out of the cell.

- **Cytoplasm (cytosol).** This is a semifluid material that surrounds the organelles within the cell. It contains water, enzymes that catalyze cellular reactions, structural proteins, and other important molecules.

- **Cytoskeleton.** A network of three different types of fibers, this structure provides support, cellular movement, and assists in stabilizing cellular organelles.

- **Endoplasmic Reticulum (ER).** The ER is a tubular membrane system that compartmentalizes the cytosol. There are two distinctly different types of ER. **Rough ER** is studded with **ribosomes** which are the sites for protein synthesis. Ribosomes can be found on the ER as well as within the cytosol. Proteins made on rough ER are secreted from the cell. **Smooth ER** lacks ribosomes and is associated with carbohydrates and lipid synthesis.

- **Golgi bodies.** Golgi bodies are a system of flattened sacs that refine, package, and distribute macromolecules in vesicles for secretion or delivery to other organelles.

- **Lysosomes.** Lysosomes are the sites for intercellular digestion. These organelles, containing digestive enzymes, are formed by budding from the Golgi complex and function when they fuse with vesicles containing material to be broken down.

- **Vacuoles.** Found mostly in plant cells, these structures serve as storage chambers for water, food, and waste products.

- **Nucleus.** The nucleus is the central "command post" of the cell. It contains most of the genetic information in DNA. The nucleus is enclosed by a nuclear membrane that is studded with nuclear pores that allow molecules to enter and leave the nucleus. The nucleus contains one or more nucleoli, which are the sites for the synthesis and assembly of rRNA and tRNA.

- **Mitochondria.** Mitochondria are sites for cellular respiration. In a complex series of chemical reactions, cellular energy in the form of a molecule called adenosine triphosphate or ATP is produced from nutrient fuel sources . Mitochondria are thought to have originated from **symbiotic bacteria.**

- **Chloroplasts.** Chloroplasts, containing green pigment called **chlorophyll**, are found in plant cells and are the sites of **photosynthesis.** In this process, carbon dioxide and water are transformed into glucose using energy from the sun. Like mitochondria, chloroplasts are thought to have originated from symbiotic bacteria.

- **Cilia and flagella.** Cilia and flagella are structures that permit cell movement. Cilia are short and often cover the entire cell. Flagella are whip-like structures that are present in lesser numbers than cilia. Both of these structures are composed of bundles of **microtubules** covered with the plasma membrane of the cell. A structure called the **basal body** anchors the cilium or flagellum to the cell.

- **Cell walls.** Cell walls are formed by living plant cells. They are made of **cellulose** fibers embedded in a matrix of protein and polysaccharides. The cell wall provides rigidity to plant cells, prevents them from drying out, and allows for development of internal cellular pressure called **turgor pressure.**

4. Movement Of Materials Across the Cell Membrane

Molecules move with constant "jiggling", a random motion called **Brownian movement** which is the visible manifestation of kinetic molecular motion. This results in a net movement of uniformly distributed molecules in a particular direction as they respond to differences in concentration, pressure, or electric charge. Such differences produce what is called a **gradient.**

When molecules move as a result of the gradient difference, no additional cellular energy is required. Movement of this type is called **passive transport.** There are three types of passive transport: diffusion, osmosis, and facilitated diffusion. Diffusion is the movement of molecules due to kinetic molecular motion.

Molecules move with a concentration or pressure gradient from high to low. Charged particles in a cell move toward unlike charges or away from like charges. Movement of substances from regions of high concentration to regions of low concentration is called **diffusion.**
(See Figure 3-22, P. 76 in your text))

Osmosis is a form of diffusion in which water molecules move from areas of high concentration to areas of lower concentration across a semipermeable membrane. If a cell is placed in a solution that is identical in concentration to the internal solution of the cell, there will be no net movement in the water. The surrounding solution is said to be **isotonic** to the cellular solution. If a cell is placed in a solution that has a lower concentration of solutes and a higher concentration of water, the water will enter the cell and cause the cell to swell. This can cause the cell to burst or **lyse.** The surrounding solution is said to be **hypotonic** to the cellular solution. If a cell is placed in a solution that has a higher concentration of solutes and a lower concentration of water, the water will move out of the cell and the cell will shrink. This shrinking process is known as **plasmolysis.** The surrounding solution is said to be **hypertonic** to the cellular solution.

Fig. 3-1 The effects of solute concentration on water movement in red
blood cells.

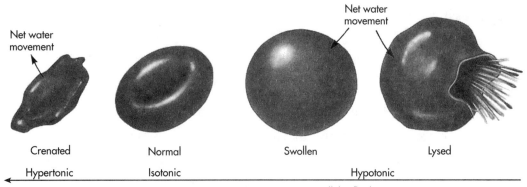

Facilitated diffusion occurs when selected molecules move across the cell membrane through special protein channels. The movement of these molecules occurs with the concentration gradient and therefore does not require additional cellular energy. Review the operational aspects of this process in the diagrams provided in your text.

These passive processes do not facilitate all necessary transport. Sometimes cells require molecules to move against the gradient. These processes are called active transport and they require additional energy from the cell. The sodium-potassium pump is an example of an active transport process.

Movement of large substances from outside to inside the cell is called endocytosis. If the material being brought into the cell is a solid, the process is called phagocytosis. If the material being brought into the cell is a liquid, the process is called pinocytosis.

Movement of materials from inside the cell to the external environment surrounding the cell is called exocytosis. In preparation for this process, the cell packages the material to be discharged in a membrane bound vesicle. The membrane of the vesicle binds to the internal cell membrane and then the contents are released.

Fig. 3-2 Endocytosis.

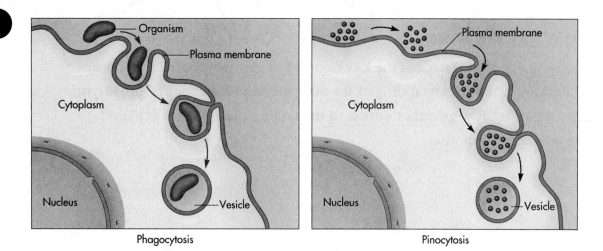

Phagocytosis　　　　　　　　Pinocytosis

Fig. 3-3 Exocytosis.

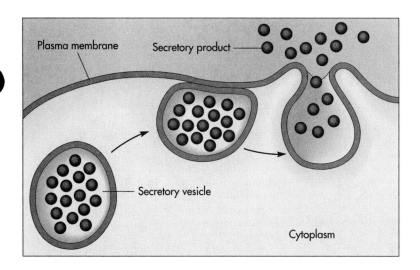

CHAPTER REVIEW ACTIVITIES

CONCEPT MAPPING

Use as many of the terms in the Key Concept Review to construct a concept map that explains your understanding of the structure and functions of prokaryotic and eukaryotic cells. Pay special attention to the linking phrases that explain how you relate one term to another.

IF.... THEN.... CHALLENGE QUESTIONS

Using the "IF" part of the statements below, complete the "THEN" part with guidance from the accompanying question.

1. If a cell shrinks when placed in a solution, then _____. What can you conclude about the solution it was placed in?

2. If cellular structures control for, assist in, and provide a site for protein synthesis, then _____. Which organelles contribute to protein synthesis and where are they located in the cell?

3. If a cell transports materials across its membrane without expending additional energy but requires the assistance of specialized protein carriers, then _____. What transport mechanisms are available to the cell?

4. If hydrophilic portions of a phospholipid molecule in the membrane are attracted to water molecules, then _____. What portions repel water molecules?

5. If enzymes are bound to the inner surface of endoplasmic reticulum and no ribosomes are attached to the outer surface, then _____. What type of ER would this be and what materials would be synthesized there?

6. If cell A has an abundance of mitochondria compared to cell B, then _____. What could be said about the function and requirement differences between the two?

7. If the overall volume (size) of a cell increases from 1 cm^3 to 8 cm^3, then _____. What would happen to its surface area-to-volume ratio and what impact would this have on the cell?

8. If a cell possesses a cell wall, then _____. What conclusions about cell type could be made and what advantage would such cells have?

9. If pinocytosis occurs in a cell, then _____. What materials are being transported and what organelle would be involved?

10. If the net movement of water goes from the external environment into the cell, then _____. What could be said about the concentration of solute inside the cell compared to outside the cell if no additional cellular energy is required?

MATCHING

Using the list of terms provided, select the item that best relates to the statements given. If you have difficulty with any of the statements, review the relevant material in the text.

_____ 1. This region within the nucleus of a cell tcontains a special area of DNA that directs the synthesis of ribosomal RNA.

_____ 2. This thin, flexible structure encloses the contents of a cell.

_____ 3. This movement of a solute across a membrane against the concentration gradient requires the expenditure of cellular energy and the assistance of a specific transport protein.

_____ 4. These organelles, located in different places within the cell, are the sites for protein synthesis.

_____ 5. This form of diffusion occurs when water molecules move from an area of high concentration to an area of low concentration across a semi-permeable membrane.

_____ 6. These membrane-bound storage sacs within cells hold substances such as water, food, and waste products.

_____ 7. This is a process in which cells engulf and package within vesicles large molecules or particles and bring these substances into the cell.

_____ 8. This whip-like extension of some cells is used for movement.

_____ 9. These energy-producing organelles contain a green pigment and are found in the cells of plants and algae.

_____ 10. These shortened, thickened structures consist of DNA coiled tightly around proteins and appear when cells begin to divide.

_____ 11. These differences in concentration, pressure, and electric charge can cause molecules to move randomly and often result in a net movement of molecules in a particular direction.

_____ 12. These membrane-bound organelles contain digestive enzymes that are used to break down materials so that they can be reused or excreted.

_____ 13. This is an extensive system of interconnected membranes that forms flattened channels and tube-like canals within the cytoplasm of a cell.

_____ 14. These cells have a relatively simple structure, were the first cell types on Earth, and are classified as Monerans.

_____ 15. This is an organelle within a cell that collects, modifies, and packages molecules made at different locations within the cell.

_____ 16. This is the term that refers to solutions that have equal solute concentrations to one another.

_____ 17. This viscous fluid within a cell contains all cell organelles except the nucleus.

_____ 18. This is the general term for the transport process that facilitates molecular movement "down" a gradient and across a cell membrane.

_____ 19. This term refers to the discharge of materials from a cell by packaging the substances in a vesicle and moving the vesicle to the cell surface.

_____ 20. This is an organelle thought to have originated as symbiotic bacteria and is the site for cellular respiration.

TERMS LIST

A. Flagellum	H. Nucleolus	O. Vacuoles
B. Endoplasmic reticulum	I. Chloroplasts	P. Golgi body
C. Exocytosis	J. Mitochondria	Q. Plasma membrane
D. Active transport	K. Prokaryotes	R. Chromosomes
E. Isotonic	L. Osmosis	S. Ribosomes
F. Lysosomes	M. Gradient	T. Endocytosis
G. Cytoplasm	N. Passive transport	

DIFFERENTIATING BETWEEN PROKARYOTES AND EUKARYOTES

You have been made aware of certain differences that exist between prokaryotic and eukaryotic cells. Fill in the table below by listing as many of those differences as you can.

PROKARYOTE CHARACTERISTICS	EUKARYOTE CHARACTERISTICS

MULTIPLE CHOICE QUESTIONS

Place the letter indicating your choice of answers in the space provided by each question. As you read through the selections, mentally note the reasons for your selection and rejection of the choices given.

_____ 1. How many microns are there in one millimeter?
   a. 10
   b. 100
   c. 1,000
   d. 10,000
   e. 100,000

_____ 2. As the size of a cell increases, which of the following decreases?
   a. Surface area
   b. Volume
   c. Surface area-to-volume ratio
   d. Both A and B
   e. A, B, and C

_____ 3. Which of the following is missing from a prokaryotic cell?
   a. Plasma membrane
   b. Nucleoid
   c. Ribosomes
   d. Membrane bound organelles
   e. Flagella or cilia

_____ 4. Endosymbiont organelles found in eukaryotic cells include which of the following?

   a. Mitochondria

   b. Ribosomes

   c. Chloroplasts

   d. A and B

   e. A and C

_____ 5. Which of the following statements would NOT be included as part of the cell theory?

   a. Cells are the smallest unit of living things.

   b. All cells contain a nucleus that controls every cellular function.

   c. Cells arise from other cells.

   d. All living things are made up of one or more cells.

   e. None of the above

_____ 6. If your doctor needed to administer saline intravenously, what would the concentration of salt in the saline solution be?

   a. Hypertonic to red blood cells to assure that the cells wouldn't lyse

   b. Hypotonic to red blood cells to assure that the cells wouldn't lyse

   c. Isotonic to red blood cells to assure that the cells wouldn't lyse

   d. The more salt in solution the better

   e. Salt concentration would not be a factor

_____ 7. Which of the following is NOT a passive transport process?

   a. Diffusion

   b. Endocytosis

   c. Osmosis

   d. Facilitated diffusion

   e. All the above are passive transport processes

_____ 8. Which method best describes how most fluid molecules enter the cell?

   a. Osmosis

   b. Endocytosis

   c. Through protein transport channels

   d. Through permanent pores in the plasma membrane

   e. Pinocytosis

_____ 9. Which cellular structure is depicted in the fluid mosaic model?
   a. Plasma membrane
   b. Ribosomes
   c. Chromosomes
   d. Endoplasmic reticulum
   e. Nucleus

_____ 10. Which of the following statements regarding the plasma membrane is true?
   a. The hydrophobic tails of the phospholipids face the aquatic environments on both sides of the membrane.
   b. The hydrophilic tails of the phospholipids face the aquatic environments on both sides of the membrane.
   c. Large globular proteins are embedded in the membrane.
   d. Sugars often attach to lipids and proteins on the outer surface of the membrane.
   e. B, C, and D

_____ 11. Which of the following organelles would NOT be found in all cells?
   a. Cytoplasm
   b. Cell wall
   c. DNA
   d. Ribosomes
   e. Plasma membrane

_____ 12. Which of the following would best describe the function of the nucleolus?
   a. The synthesis and storage of ribosomal components
   b. The formation of spindle fibers for cell division
   c. The condensation of chromosomes during cell division
   d. Directing the synthesis of DNA in a cell
   e. Packaging synthesized proteins for transport

_____ 13. What would the <u>volume</u> be in a cube-shaped cell measuring 1 cm on each side if this cell doubled in size?
   a. 2 cubic centimeters
   b. 4 cubic centimeters
   c. 6 cubic centimeters
   d. 8 cubic centimeters
   e. 24 cubic centimeters

_____ 14. What would the surface area be in a cube-shaped cell measuring 1 cm on each side if this cell doubled in size?
   a. 2 cubic centimeters
   b. 4 cubic centimeters
   c. 6 cubic centimeters
   d. 8 cubic centimeters
   e. 24 cubic centimeters

SHORT ANSWER QUESTIONS

1.   Why are chloroplasts found only in plant cells?

2.   What is the function of the Golgi bodies?

3.   State the cell theory.

4.   What prevents the cell from becoming very large?

CRITICAL QUESTIONS

1. In what way is information about cellular structure and function dependent on technology?

2.   Eukaryotic cells are compartmentalized. What advantage does this give these types of cells?

3. We have used the term "semipermeable" to describe the plasma membrane. How and why is the membrane selective?

4. If you were asked to provide evidence that active transport occurs in cells, what information would you offer?

5. Predict and support, with some rationale, the characteristics of the following cells:
    A. Bacterial cell
    B. Skeletal muscle cell
    C. Plant cell
    D. Sperm cell
        • Contain the largest number of mitochondria
        • Possess a flagella for motility
        • Have the greatest number of vacuoles
        • Reproduce the fastest

6. Explain the concept of a concentration gradient.

7. Cucumbers that are pickled sometimes become very flexible. Meat that has been cured, like beef jerky, becomes very hard. Offer an explanation as to why these results might occur.

TELLING THE STORY

To assure that you understand the concepts in this chapter, develop a story about the key concepts. Focus on linking the components of the concepts into meaningful construction that reflects confidence in your understanding.

CHAPTER REVIEW ANSWERS

IF.... THEN.... CHALLENGE QUESTIONS

1. The solution was hypertonic to the cellular solution.

2. The chromosomes, found in the nucleus, contain the DNA which has the codes for proteins, mRNA and tRNA, located in the cytoplasm. They also assist in the transport of the code and acquisition of the appropriate amino acids. The ribosomes, located in the cytoplasm and on the endoplasmic reticulum, provide a site for protein synthesis.

3. The passive process of facilitated diffusion occurs.

4. Hydrophobic portions repel water molecules.

5. This would be smooth endoplasmic reticulum that can synthesize carbohydrates and lipids.

6. Cell A would require the production of more ATP (cellular energy) because it expends more energy performing its functions.

7. The surface area-to-volume ratio would increase and the cell would have less opportunity to survive because there would be a greater demand for materials to enter and leave the cell than there would be membrane to facilitate such transport.

8. The cell is a plant cell which would be more protected and supported than that of an animal cell.

9. Fluid materials are being actively transported and the plasma membrane and vacuoles/vesicles would be used to assist this transport process.

10. The solute concentration is greater inside the cell. Osmotic pressure within the cell would continue to rise as water moves inward.

MATCHING

1. H	8. A	15. P
2. Q	9. I	16. E
3. D	10. R	17. G
4. S	11. M	18. N
5. L	12. F	19. C
6. O	13. B	20. J
7. T	14. K	

DIFFERENTIATING BETWEEN PROKARYOTES AND EUKARYOTES

PROKARYOTE CHARACTERISTICS	EUKARYOTE CHARACTERISTICS
• Nucleoid	• Nucleus
• No membrane-bound organelles	• Membrane-bound organelles
• Single chromosomes	• Chromosomes in pairs
• No cytoplasmic streaming	• Cytoplasmic streaming
• Cell division without mitosis	• Cell division by mitosis
• Simple flagella	• Complex flagella
• Simple cytoskeleton	• Complex cytoskeleton
• No cellulose in cell walls	• Cellulose in cell walls
• Smaller ribosomes	• Larger ribosomes

MULTIPLE CHOICE QUESTIONS

1. C	8. E
2. C	9. A
3. D	10. E
4. E	11. E
5. B	12. A
6. C	13. D
7. B	14. E

SHORT ANSWER QUESTIONS

1. Only organisms that are capable of photosynthesis would require an organelle that contains chlorophyll.

2. Molecules that are modified by enzymes in the Golgi complex are sorted and enclosed in vesicles. The vesicle is pinched off from the external membrane of the Golgi complex and moves through the endoplasmic reticulum towards its destination. Upon reaching the edge of this location, the vesicle will fuse with the membrane and empty the contents.

3. All living organisms are composed of cells. The smallest living units of structure and function of organisms are the cells. All cells are derived (come from) pre-existing cells.

4.   Generally, the given answer is the limit of a low surface area-to-volume ratio. As a cell increases in size, its volume increase to a greater extent than the surface area. This results in a cell unable to provide adequate surface area for membrane transport necessary to facilitate the metabolic and waste demands of the large volume. Additionally, the cell may lack to ability to control the activities associated with such a large volume.

CRITICAL QUESTIONS

1.   Scientists are limited by the technological capabilities they have to observe and experiment. Advancements allow them to explore in more detail or at a more acute level.

2.   Compartmentalization provides the cell with greater efficiency and utilization of resources. When the tasks are divided among specialized entities within a system, the operation is more effective and efficient.

3.   A plasma membrane is another control mechanism that protects the internal environment of the cell. It allows only certain sizes, charges, and types of materials to pass through more readily than others. This selectivity controls the rate of transport, concentration, volume, etc. for the cell.

4.   Active transport would result in an increase in the concentration of materials that would occur in the other direction if accomplished by osmosis or diffusion. Additionally, energy would be expended and some of this energy would be detectable in the form of heat.

5.   A. Bacterial cells reproduce the fastest and possess a flagella for motility.
     B. Skeletal muscle cells contain the largest number of mitochondria.
     C. Plant cells have the greatest number of vacuoles.
     D. Sperm cells possess a flagellum for motility.

6. A concentration gradient is the difference that exists in the concentration of a substance over a distance or across a semipermeable membrane. Molecules will tend to move from an area where the concentration is the highest to an area of lower concentration. This movement is said to occur with the gradient. When the distribution of the substance is equivalent, the net movement will be zero; however, there will still be equal amounts of the substance moving from one side of the membrane to the other.

7. In the pickling process, salt solutions are sometimes used. This concentration of salt is greater than that normally found in the cells of the cucumber. The cells will lose water to the salt solution due to osmosis and result in the "pickle" becoming more flexible. The meat will also loose water causing the cells to shrink. Because the salt is applied in a solid form, more water will be lost from the meat than from the cucumber. This will result in the meat becoming very dry and therefore very hard.

CHAPTER 4

THE FLOW OF ENERGY WITHIN ORGANISMS

KEY CONCEPTS OVERVIEW

1.   Energy Flow

Energy is the capacity to do work or to bring about some type of change. When the term "**work**" is used in this way, it means the ability to change or move matter against other forces acting on it. All activities associated with living things require energy and involve the conversion of energy. In living systems, the conversion process implies that energy is released from chemical bonds where it is stored. Energy exists in different forms, including mechanical, heat, sound, electrical, light, radioactive, and magnetic and occurs in two distinct states. Stored energy is known as **potential energy**, while energy that is in use (in motion) is known as **kinetic energy.**

Any change in energy that occurs is governed by two scientific laws known as the **laws of thermodynamics.** The first law states that energy can change from one form to another and from one state to another. Despite these changes, energy is never lost nor can new energy be produced. This implies that the energy in the universe is constant. The second law states that all objects in the universe tend to become more disordered and that such disorder is constantly increasing. For molecules this would mean that their random motion (disorder) would increase. When heat is added to a system, molecular motion increases. When energy is transferred, some energy is lost as heat and becomes useless for performing work.

In any chemical reaction the reactants, called **substrates**, require some kind of initial "jump start" or energy input to begin to react. This start up energy is called **free energy of activation.** Once the substrates are involved in the reaction, they undergo a chemical change that results in the breaking of existing chemical bonds and the making of new ones. The resulting molecules are called **products.**

The chains of reactions that store, move, and free energy are called **metabolic pathways.** When energy is released from the chemical bonds that were broken, the reaction is known as an **exergonic** reaction. The products of this spontaneous reaction contain less energy than the substrate. Such **catabolic** reactions occur in living organisms. Some reactions result in substrates being joined to form a product. Energy would be necessary to develop the required chemical bonds to hold the product together. This type of reaction is called an **endergonic** reaction. In such a reaction, the product contains more energy than the substrates. Such **anabolic** reactions do not occur spontaneously.

2.   Enzymatic Regulation of Chemical Reactions

To reduce the amount of free energy of activation necessary to cause a chemical reaction, reactions in biological systems are catalyzed by enzymes. **Enzymes,** biological catalysts, are proteins that speed up the rate of a reaction. These large molecules have very specialized sites (**active sites**) for interacting with a specific substrate. This active site is the location of catalysis. Since enzymes are highly specific, each enzyme typically catalyzed only one or a few chemical reactions (See Figure 4-9, P. 92 in your text)

Most enzymes operate within an optimum temperature and pH. Most human enzymes function best between temperatures of 36 and 38 degrees Celsius and between pH of 6 to 8. There are exceptions to this however (pepsin in the stomach). If these conditions are not met, the enzymes may become **denatured** and lose their ability to function. Certain chemicals may alter the shape of the enzyme and act as an **activator** or **inhibitor** of the enzyme. Enzyme action is often regulated by an inhibitor in a process called negative feedback. In this process, the product of the last reaction in a pathway of reactions binds to the active site of the enzyme catalyzing the first reaction in the pathway, preventing the enzyme from binding to the substrates for the first reaction. This type of negative feedback is called **end-product inhibition,** and turns off a metabolic pathway when enough product has been formed.

Enzymes also have additional components that assist enzymes when they are catalyzing a reaction. These components are known as cofactors (which are ions) and coenzymes (which are non-protein organic molecules, such as vitamins).

3.   Storing and Transferring Energy

The primary energy currency of all cells is **adenosine triphosphate (ATP)**. Each ATP molecule is composed of three subunits: (1) a ribose (5 carbon sugar), 2) an adenine (a double-ringed molecule), and (3) a chain of three phosphate groups called a triphosphate group. The ribose and the adenine combine to form adenosine.

Fig. 4-1 Adenosine triphosphate.

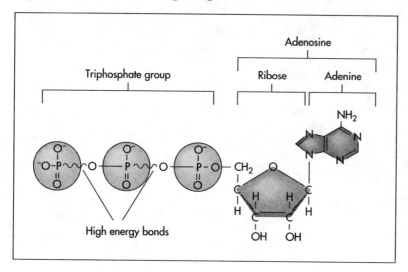

The last two phosphates have high energy bonds that, when broken, release energy. When ATP is used to drive a chemical reaction, the bond that links the last phosphate to the rest of the ATP is broken, releasing energy. The molecule that remains is called **adenosine diphosphate (ADP)**. Cells contain an abundance of ATP, ADP, and phosphate molecules. This supply provides the components for the resynthesis of ATP.

Fig. 4-2 Synthesis and Utilization of ATP.

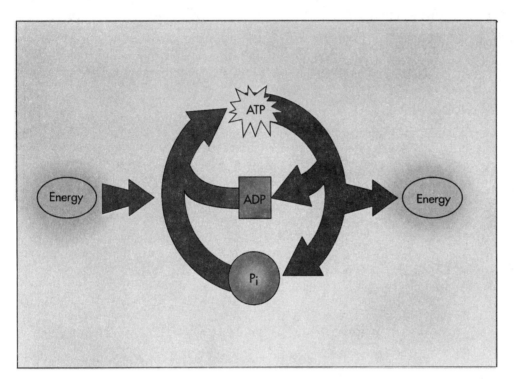

CHAPTER REVIEW ACTIVITIES

CONCEPT MAPPING

Construct a concept map that incorporates the main ideas in this chapter. Use as many of the terms in the Key Concept Review that will demonstrate your understanding of their relationship to one another and to the concepts of energy flow and metabolic processes in organisms.

MATCHING

Using the list of terms provided, select the items that best relates to each of the statements below. If you have difficulty answering any of the following, refer back to your text for review.

_____ 1. These are non-protein organic molecules, like vitamins, that assist enzymes in catalyzing reactions

_____ 2. This term refers to the state of energy when it is being stored.

_____ 3. This term refers to molecules that are entering into a chemical reaction.

_____ 4. This is the primary molecule used by cells to capture energy and later release it when needed.

_____ 5. This is the energy lost to disorder or a measure of the disorder of a system.

_____ 6. These biological catalysts reduce the free energy of activation needed to begin a chemical reaction and cause the rate of the reaction to increase.

_____ 7. These chemical reactions use energy to build complex molecules from simpler molecules.

_____ 8. This term refers to substrates that have undergone a chemical change.

_____ 9. This is a chemical reaction in which the products contain more energy than the substrates.

_____ 10. These are surface locations on an enzyme where reactions are catalyzed.

_____ 11. This term refers to energy actively involved in work.

_____ 12. This is a chemical reaction in which the products contain less energy than the substrate.

_____ 13. These non-protein enzymes assist other enzymes in catalyzing reactions.

_____ 14. These chemical reactions release energy by breaking down complex molecules into simpler molecules.

_____ 15. This occurs when the energy released in an exergonic reaction is used to drive an endergonic reaction.

TERM LIST

A. Coenzyme	F. Substrate	K. Products
B. Cofactor	G. Active sites	L. ATP
C. Coupled reaction	H. Enzymes	M. Potential energy
D. Entropy	I. Catabolic reactions	N. Kinetic energy
E. Exergonic	J. Anabolic reactions	O. Endergonic

IF.... THEN.... CHALLENGE QUESTIONS

Using the "IF" part of the statements below, complete the "THEN" part with guidance from the accompanying question.

1. If a chemical reaction occurs, then _____. What must happen to the substrate molecules?

2. If the energy required to start a reaction is minimized, then _____. What must be involved?

3. If stored energy in an ATP molecule is released, then _____. What alterations occur in the molecule?

4. If heat is added to a reaction, then _____. What happens to the rate of molecular motion?

5. If the temperature of humans was to rise above normal, then _____. What effect could be predicted on enzyme activity?

CELLULAR ENERGY

1.   Draw a diagram that represents the ATP-ADP cycle. Below it develop a statement that explains why this is a cyclic process.

MULTIPLE CHOICE QUESTIONS

_____ 1.   Which of the following pairs of terms are mismatched?
a. Exergonic-catabolic
b. ATP-biological energy
c. Enzyme-catalyst
d. Endergonic-spontaneous
e. Anabolic-energy storage

_____ 2.   The first law of thermodynamics addresses which of the following concepts?
a. The controlling of energy transfer
b. The conservation of energy
c. The control of entropy
d. The control of molecular motion
e. The idea that disorder increases with time

_____ 3. Which of the following statements is false?

   a. As heat in a substance increases, the speed of molecular motion increases.
   b. The amount of total energy is decreasing through time.
   c. As energy is converted from one form to another some energy is lost as heat.
   d. The second law of thermodynamics refers to what happens to energy when it is transferred.
   e. The amount of usable energy that comes out of a system is less than the amount of energy that goes into the system.

_____ 4. Which of the following statements about enzymes is true?

   a. Enzymes are biological catalysts.
   b. Enzymes reduce the amount of activation energy needed for a reaction to occur.
   c. Enzymes do not cause reactions to occur that wouldn't ordinarily occur without an enzyme.
   d. Enzymes are proteins that act to control the rate of biological reactions.
   e. All of the above

_____ 5. The juncture between an enzyme and its substrate is described by which of the following?

   a. An induced fit
   b. A bridge
   c. An adhesive strip
   d. Matching Velcro
   e. A crochet hook

_____ 6. The ATP molecule is composed of which of the following?

   a. A ribose sugar
   b. Three phosphates
   c. Adenine
   d. A, B, and C
   e. A and B only

_____ 7. What is free energy of activation?

   a. The energy freed up after you eat a meal
   b. The energy given off as heat
   c. The energy needed to initiate a reaction
   d. The energy required to raise one gram of water one degree Celsius
   e. None of the above

_____ 8. What is kinetic energy?

   a. Stored energy
   b. Energy in motion
   c. A heat producer
   d. A and C
   e. B and C

_____ 9. Which of the following refers to the total sum of all the chemical reactions within an organism ?

   a. Metabolism
   b. Entropy
   c. Negative feedback
   d. Homeostasis
   e. Cellular respiration

_____ 10. Which of the following best relates to anabolic reactions?

   a. Alter the effectiveness of enzymes
   b. Build complex molecules from simple ones
   c. Occur only between steroids
   d. Break complex molecules into simple ones
   e. These reactions do not normally occur in humans.

SHORT ANSWER QUESTIONS

1. Why is it necessary for enzymes to be specific?

2. What three factors affect the level of enzyme activity?

3. What are some of the functions of activation energy?

CRITICAL QUESTIONS

1. Energy can by measured in Calories. For example, one teaspoon of sugar has 16 Calories and a slice of apple pie contains approximately 365 Calories. Given what you now understand about the concept of energy, how will this energy be accounted for in your body? Additionally, how many teaspoons of sugar are there in that one slice of apple pie?

2. Explain why enzymes have such specificity to substrates.

3. Develop an analogy for ATP-ADP as a system that stores, transfers, and regains energy.

4. Discuss the factors that affect enzyme activities.

5. What are the laws of thermodynamics? Give a practice example of the application of these laws.

6. How can the amount of disorder in the universe continue to increase if the amount of energy is constant?

7. Explain the information presented in the following diagram.

Fig. 4-3

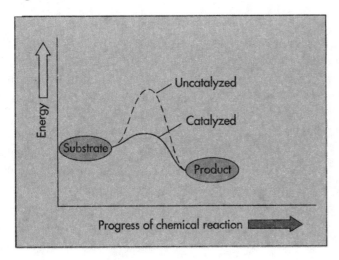

TELLING THE STORY

Using the information you have acquired in this chapter, write a description of your understanding of how energy flows within an organism and the importance of this process to maintain life. Try to focus on the concepts you have studied and link the components of these concepts in a meaningful way.

CHAPTER REVIEW ANSWERS

MATCHING

1. B	4. L	7. J	10. G	13. A
2. M	5. D	8. K	11. N	14. I
3. F	6. H	9. O	12. E	15. C

IF.... THEN.... CHALLENGE QUESTIONS

1. At least some of the chemical bonds that existed are broken and new ones are formed.

2. An enzyme that catalyzes the reaction must be involved causing a lesser amount of free energy of activation to be used.

3. The terminal phosphate molecule is broken off of the ATP molecule forming ADP and a free phophate.

4. The rate of molecular motion increases and there is greater randomness to their movement.

5. The effectiveness of some enzymes as catalysts may be diminished since enzymes have optimum temperature ranges in which to work. The reaction may still occur but at a much slower rate.

CELLULAR ENERGY

ATP is continually being broken down to form ADP, phosphate, and energy and also resynthesized from ADP, phosphate, and energy during coupled exergonic reactions.

MULTIPLE CHOICE QUESTIONS

1. D	3. B	5. A	7. C	9. A
2. B	4. E	6. D	8. E	10. B

SHORT ANSWER QUESTIONS

1. The action of enzymes on substrates involves a spatial relationship between the enzyme and the substrate. The shape of the enzyme dictates the type of substrate it will bind with to catalyze a reaction.

2. The three factors that influence the level of enzyme activity are temperature, pH, and the binding of specific chemicals. There are optimum temperatures and pH for different enzymes.

3. Activation energy can be used to break chemical bonds, to excite electrons so that they can achieve higher energy levels, and to help molecules overcome the mutual repulsion of their electrons so that they can get close enough to react.

CRITICAL QUESTIONS

1. This energy will be used to perform all the physiological functions required by the body to live. In the process, however, not all the available energy will be converted to usable energy. Some of the energy will be given off in the form of heat. The efficiency of converting food to usable cellular energy is not very high. There would be 22.8 teaspoons of sugar in that one slice of pie.

2. Enzymes are proteins that have one or more furrows on their surface. These furrows, called active sites, form a place for only certain substrates to fit. While the fit may not always be perfect, the enzyme will make the necessary adjustments to make the connection tight.

3. Answers may vary. An example would be a rechargeable flashlight battery. These batteries store energy in the form of electrical energy, transfer this energy to light energy, and can be recharged when the stored energy is used up.

4. The three factors that influence the activities of enzymes are pH, temperature, and the specificity of substrate/enzyme interaction. There are optimum levels of pH and temperature for different enzymes. The substrate/enzyme interaction is specific to assure the maintenance of physiological balance in the organism.

5. The first law of thermodynamics states that energy can neither be created nor destroyed, but may change from one form to another. The second law of thermodynamics states that as energy is transferred, some energy is lost in the form of heat. Examples of these occur in every metabolic reaction that takes place in living organisms.

6. While the total amount of energy in the universe is constant, the amount of useful energy declines because of the loss of energy in the form of heat .

7. Enzymes are able to catalyze particular reactions because they lower the amount of activation energy required to initiate the reaction and form products.

CHAPTER 5

CELLULAR RESPIRATION

KEY CONCEPTS OVERVIEW

1. Biological Activities Require Energy

All living organisms require energy to perform the functions necessary for life, such as the production of macromolecules, movement, reproduction, transport of materials through cellular membranes, etc. For most cells, the energy needed to fuel such activities originates from the sun. Through a process known as **photosynthesis**, glucose molecules are synthesized by plants, algae, and certain bacteria. These organisms use glucose for fuel and store what is not used in the form of starch. When these organisms are consumed by other organisms, chemical energy is transferred. Cells acquire chemical energy by systematically breaking down their food through complex series of reactions and releasing the stored energy.

Cellular respiration is the process that utilizes oxygen (**aerobic**) to catabolize glucose or other energy-rich organic compounds to obtain ATP. Following the complete breakdown of glucose, a free energy change of -686 kcal/mol is produced. The waste products of this process (water and carbon dioxide) are necessary chemical materials plants use in photosynthesis. In this way, the chemical elements are recycled in nature.

Fermentation, which occurs when there is not sufficient oxygen, is the partial degradation of glucose to release ATP. Since this process occurs without sufficient oxygen, the term **anaerobic** is used to describe the chemical reactions that take place.

All organisms use either cellular respiration or fermentative pathways to synthesize ATP. This suggests a strong evolutionary relationship among living things. Glycolysis is common to fermentation and the process of cellular respiration.

This means of glucose degradation probably evolved from ancient prokaryotes prior to oxygen being available in the atmosphere and used as a method of forming ATP.

2.   Cellular Respiration

The most common equation for cellular respiration is :

$$C_6H_{12}O_6 + 6O_2 \longrightarrow 6CO_2 + 6H_2O$$

This process occurs generally in three stages: (a) glycolysis, (b) the Krebs cycle, and (c) the electron transport chain and oxidative phosphorylation. Glycolysis (meaning to break apart glucose) occurs in the cytosol. This anaerobic process degrades glucose into two molecules called **pyruvate**. The Krebs cycle, which takes place in the mitochondria, converts a derivative of pyruvate (**acetyl CoA**) into carbon dioxide and water. Within this process, some of the electrons that are released are captured by electron acceptors (**NAD^+ and FAD**). These electrons are then passed through the electron transport chain. The energy released as these electrons are reorganized in chemical bonds is used to synthesize ATP by a process called **oxidative phosphorylation**. Each NADH will supply enough energy to resynthesize three ATPs and each $FADH_2$ supplies enough energy to resynthesize two ATPs. Review the process by closely examining the diagrams provided in your text.

3.   Fermentation

Fermentative processes synthesize ATP by substrate-level phosphorylation without the electron transport chain. The lack of oxygen limits the use of the electron transport chain because the final electron acceptor is missing. Additionally, the Krebs cycle reactions stop because the reduced form of NAD^+ (NADH) will not be oxidized back to NAD^+. Cells therefore rely on glycolysis to produce ATP. During anaerobic respiration, organisms produce an organic molecule from pyruvate that accepts the hydrogen from NADH. When no oxygen is available the final electron acceptor is pyruvate or one of its derivatives. Some organisms rely strictly on fermentation to produce ATP, others can use a combination of fermentation and respiration to accommodate their need for ATP. The alcohol fermentation of yeasts, a fungus, is used in brewing. The dairy industry makes use of lactic acid fermentation of certain fungi and bacteria to produce cheese and yogurt.

CHAPTER REVIEW ACTIVITIES

CONCEPT MAPPING

Create a concept map to organize your understanding of cellular respiration and fermentation. Since these are complex concepts, you may want to develop two different maps. Be as specific as you possibly can and choose the terms and linking words carefully.

MULTIPLE CHOICE QUESTIONS

_____ 1. Most organisms synthesize ATP through cellular respiration and fermentation. Which of the following is the exception to this?

a. Pneumococcus

b. Gonococcus

c. Chlamydia

d. Streptococcus

e. Brewers' yeast

_____ 2. Which of the following locations is the site for glycolysis?

a. Cytosol

b. Mitochondria

c. Golgi body

d. Plasma membrane

e. Outside the plasma membrane

_____ 3. Most of the ATP produced in cellular respiration is produced in which manner?

a. In anaerobic parts of the process

b. During the electron transport process

c. During glycolysis

d. During the Kreb's cycle

e. During fermentation

_____ 4. The efficiency of cellular respiration is approximately what percentage?

a. 80

b. 65

c. 50

d. 38

e. 18

_____ 5. What is another name for the Kreb's cycle?
   a. Chemiosmosis cycle
   b. Oxidative phosphorylation cycle
   c. Citric acid cycle
   d. Oxidation-reduction reaction cycle
   e. Fermentation cycle

_____ 6. How many molecules of CO_2 are generated for each acetyle CoA molecule that enters the Krebs cycle?
   a. 2
   b. 3
   c. 4
   d. 0
   e. 1

_____ 7. Why doesn't glycolysis yield as much energy as aerobic respiration?
   a. NAD^+ is regenerated by alcohol or lactic acid production, without the electrons passing through the electron transport chain.
   b. It is the pathway common only to yeast and they don't require as much energy.
   c. It does not take place in the mitochondria.
   d. Pyruvate is more reduced than CO_2 and it still contains much of the energy that was available from glucose.
   e. All of the above

_____ 8. Which of the following result from the conversion of pyruvate when skeletal muscle cells are oxygen deprived?
   a. NAD^+ and lactate
   b. ATP and lactate
   c. CO_2 and alcohol
   d. H_2O and lactate
   e. ATP and alcohol

_____ 9. Which of the following molecules is an electron acceptor?

    a. NAD^+

    b. CoA

    c. $C_6H_{12}O_6$

    d. ATP

    e. CO_2

_____ 10. What is the need for oxygen to break down lactic acid produced when muscle cells work in oxygen deprived conditions called?

    a. Anaerobic respiration

    b. Oxygen debt

    c. Oxygen deficit

    d. Lactic acid cycle

    e. Glycolysis

_____ 11. What is the organic molecule in yeast that is formed from pyruvate and then reduced to form alcohol called?

    a. Ethanol

    b. Acetyl Coenzyme A

    c. Serotonin

    d. Lactic acid

    e. Acetaldehyde

_____ 12. Natural wines contain approximately what percentage of alcohol?

    a. 5%

    b. 12%

    c. 3.2%

    d. 8%

    e. 15%

_____ 13. What happens when an atom or molecule gives up an electron?

    a. It is reduced.

    b. It is denatured.

    c. It is oxidized.

    d. It is destroyed.

    e. It is glycolyzed.

_____ 14. Which of the following statements about redox reactions is true?

    a.   The substance that is reduced loses energy.

    b.   The substance that is oxidized loses energy.

    c.   The charge of the oxidized atom is increased and the charge of the reduced atom is lowered.

    d.   A and C

    e.   B and C

_____ 15. Which of the following are part of the catabolic pathway of glycolysis?

    a.   Glucose mobilization

    b.   Oxidation

    c.   Cleavage

    d.   ATP generation

    e.   All of the above

_____ 16. Which of the following are the most significant changes that occur during glycolysis?

    a.   Glucose is converted to pyruvate.

    b. NAD^+ is converted to FAD.

    c.   ATP is converted to ADP and P_i.

    d.   Alcohol is produced.

    e.   All of the above

_____ 17. What are the two stages of the Krebs cycle?

    a.   Preparation reactions and fermentation

    b.   Fermentation and energy extraction

    c.   Electron transport and fermentation

    d.   Energy extraction and preparation reactions

    e.   None of the above

_____ 18. Each $FADH_2$ is responsible for providing enough energy to resynthesize how many ATP?

    a.   3

    b.   2

    c.   1

    d.   4

    e.   0

COMPLETION QUESTIONS

1. The compound that enters glycolysis is _____.
2. _____ pyruvate molecules are produced from one molecule of glucose.
3. Glucose contains _____ carbon atoms and pyruvate contains three carbon atoms.
4. _____ is the hydrogen by-product of the Krebs cycle.
5. In animal cells, the end product of fermentation is _____.
6. From one molecule of glucose, the Krebs cycle will produce _____ CO_2, _____ $NADH_2$, _____ $FADH_2$, and _____ ATP molecules.

OXIDATION OF GLUCOSE

Look at the diagram below and complete the information regarding the amount of ATP that can be produced as a result of the chemical activity occurring at each juncture. Fill the stars with the correct number of ATP, count up the total and record this value as well.

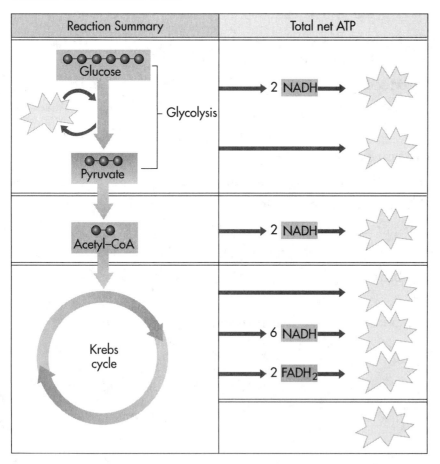

Figure 5-1

KREBS CYCLE DIAGRAM

Provide the appropriate information in the following diagram by filling in the numbered boxes. Carefully study the changes that are occurring in the process.

Figure 5-2

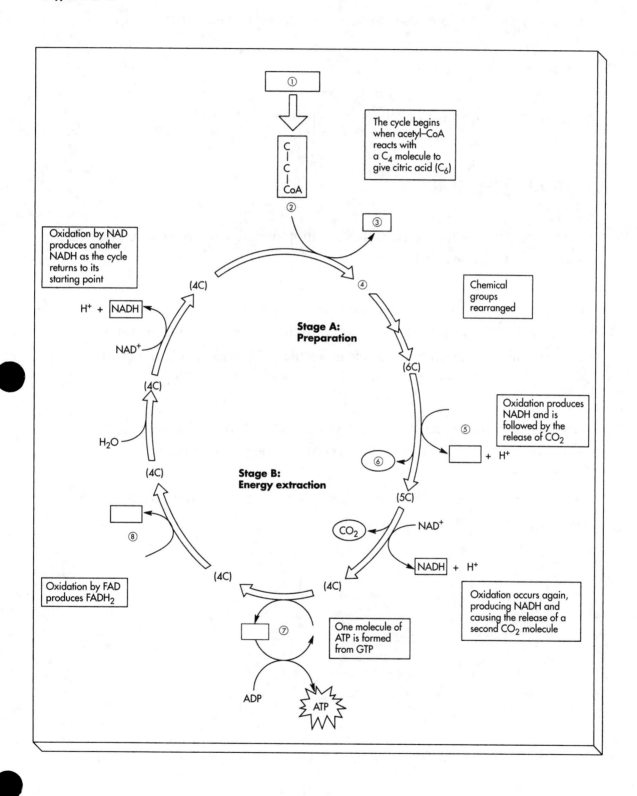

SHORT ANSWER QUESTIONS

1. What are coupled reactions?

2. What are some of the uses of the energy produced by cellular respiration?

3. What is a proton pump?

4. What are the three stages of cellular respiration?

CRITICAL QUESTIONS

1. What structural cellular differences might you expect between pure anaerobes and skeletal muscle cells?

2. One glucose molecule contains about 686 kcal of energy. A molecule of ATP contains about 7.3 kcal of energy. Given what you know about the number of ATP molecules produced per glucose molecule, calculate the efficiency of ATP production and explain what happens to the rest of the energy.

3. Cellular respiration serves other functions beyond the oxidation of glucose to produce usable cellular energy. Suggest other possible functions.

4. Why do the cells in complex organisms continue to carry out glycolysis when it yields such a small amount of ATP?

TELLING THE STORY

Write a description of your understanding of cellular respiration and fermentation within organisms and the importance of these processes to provide usable energy for all cellular functions. Try to focus on the way the components of these concepts link together to formulate a meaningful explanation of what they are and how they function.

CHAPTER REVIEW ANSWERS

MULTIPLE CHOICE QUESTIONS

1. C	7. D	13. C
2. A	8. A	14. E
3. B	9. A	15. E
4. D	10. B	16. A
5. C	11. E	17. D
6. A	12. B	18. B

COMPLETION QUESTIONS

1.   Glucose

2.   Two

3.   6

4.   Water

5.   Lactic acid

6.   2, 3, 1, 1

OXIDATION OF GLUCOSE

1.   2

2.   4

3.   6

4.   2

5.   18

6.   4

7.   Total = 36

KREBS CYCLE DIAGRAM

1. Pyruvate	5. $NAD^+ \longrightarrow NADH$
2. Acetyl CoA	6. CO_2
3. CoA	7. GTP $\longrightarrow$ GDP
4. Citric acid (6C)	8. FAD $\longrightarrow$ $FADH_2$

SHORT ANSWER QUESTIONS

1. A coupled reaction is an endergonic reaction matched with an exergonic reaction so that the exergonic reaction supplies the necessary energy to drive the endergonic reaction.

2. Some of the uses of this energy would be: active transport, maintenance of homeostasis, synthesis of macromolecules, cellular activities (muscle contraction, Na^+/K^+ pump, etc.)

3. A proton pump is an active transport mechanism that forces hydrogen ions out of cells.

CRITICAL QUESTIONS

1. The most obvious difference would be the presence of mitochondria. Since the mitochondria is used for the aerobic part of cellular metabolism, there would be no need for a purely anaerobic organism to have mitochondria. Anaerobic organisms produce energy through fermentation, and this process occurs in the cytosol.

2. The process is about 40% efficient. The remaining energy is lost as heat energy.

3. Not only does cellular respiration oxidize glucose, this process also can be used to produce other compounds. For example, amino acids can be converted to intermediates in the Krebs cycle and used as an energy source. They can also be used as an intermediate to form steroids, lipids, or other compounds. Unusual compounds such as alkaloids are produced by organisms that have the genes that code for the specific enzymes necessary to produce these compounds.

4.   Change in organisms occurs very slowly. During evolutionary changes, the process is systematically altered to improve on past successes. In catabolic metabolism, glycolysis was an improvement over what did occur. Cells that were unable to carry out catabolic reactions were at a disadvantage and had to compete for survival with cells that could catabolize. Later changes in metabolism improved on this success. Glycolysis was retained to serve as an initiating process for the extraction of chemical energy from glucose by producing the pyruvate and acetyl CoA.

CHAPTER 6

PHOTOSYNTHESIS: HOW PLANTS CAPTURE AND STORE ENERGY FROM THE SUN

KEY CONCEPTS OVERVIEW

1.   Producers and Consumers

The process of photosynthesis involves the conversion of light energy from the sun into chemical energy stored in organic molecules. These molecules are constructed from the inorganic molecules carbon dioxide and water. **Autotrophs** are capable of making these organic molecules and in a sense are able to produce their own food. All other organisms also feed on the food produced by autotrophs. This can occur directly by eating the autotrophs or indirectly by eating an organism that has eaten autotrophs. Because of their ability to produce their on food and that of others, autotrophs are termed **producers**.

Those organisms that are incapable of producing their own food are dependent on other organisms for food. These organisms are called **heterotrophs.** Because these organisms rely on the consumption of other organisms to acquire food, they are referred to as **consumers.** About 95% of the species of organisms on Earth are heterotrophs.

2.   Light Energy

We can think of the sun as a huge thermonuclear reactor. Fusion reactions produce radiation, called **electromagnetic energy**, that travels as wave disturbances. The distance from the **crest,** or top, of one electromagnetic wave to the crest of the next is the wavelength. There is an array of electromagnetic waves emitted from the sun and they vary in wavelength. The shortest wavelengths belong to gamma rays and the longest to radio waves. Waves that have lengths between 400 and 700 **nanometers** (one billionth of a meter) are visible to humans. These are referred to as **visible light.**

Despite the fact that electromagnetic energy travels in waves, it also behaves as if it consisted of discrete units called **photons (or quanta).** These "packets" have a fixed amount of energy. The amount of energy in each photon is inversely related to its wavelength. Therefore, those types of radiation that have shorter wavelengths have the greatest amount of available energy and vice versa. Within the visible spectrum of colors, this is also true.

Substances that absorb visible light are called **pigments.** We are able to see color because the pigments of an object absorb visible light energy and reflect the color you see back to special pigment containing cells in your eyes. The photons of this light are then absorbed. The brain "translates" the nerve signal produced by these cells as a specific color and you recognize that color. Therefore, you see a red flower as red, a blue flower as blue, etc. White objects reflect all the wavelengths of visible light and black objects absorb all wavelengths of visible light.

Carotenoids and chlorophylls are two important pigments in the process of photosynthesis. **Carotenoids** absorb photons of green, blue, and violet wavelengths and reflect photons of red, yellow, and orange. **Chlorophylls,** the primary light gatherers in all plants and algae, absorb photons of violet-blue and red wavelengths and reflect green and yellow photons. Since carotenoids absorb wavelengths of light that cholorphylls cannot, they pass this energy on to chlorophyll extending the spectrum of light that can be used to drive photosynthesis.

Chlorophyll is more visible than carotenoids until it begins to chemically break down. As the days shorten during the autumn months, less light is available. Carotenoids then become more visible and result in the yellows and oranges of fall foliage.

3.   Photosynthesis

The reaction of carbon dioxide, water, and light energy to produce glucose, oxygen, and water is called photosynthesis. It requires that light energy be captured, converted to chemical energy, and then stored in organic molecules that can be used by all living organisms for fuel.

The process of photosynthesis can be represented in the formula below. The equation show that water is both consumed and synthesized during this process.

$$6\,CO_2 + 6\,H_2O + \text{Sun's energy} \longrightarrow C_6H_{12}O_6 + 6\,O_2$$

The process of photosynthesis involves two sets of chemical reactions. Solar energy is converted into chemical energy in the chloroplasts. Chloroplasts have a system of internal membranes that are organized into flattened sacs called **thylakoids.** Stacks of thylakoids are known as **grana.** The fluid surrounding thylakoids is called the **stroma** and contains the enzymes necessary for the first set of photosynthetic reactions.

During this first set of reactions called **light-dependent reactions,** ATP and NADPH are formed. This occurs because the light energy (photons) absorbed by the photocenter in the chloroplasts drives the transfer of electrons and hydrogen from water to an electron acceptor ($NADP^+$). This then is reduced to NADPH. The energy is passed from one pigment molecule to another until it reaches a special molecule, known as chlorophyll a. There are two types of photocenters in the thylakoid membrane. **Photosystem I** uses a pair of chlorophyll a moelcules called P700 (after the wavelength of light they absorb the best) to transfer the energy of the photon. The subsequent events that follow are called **cyclic electron flow.** Cyclic electron flow occurs because electrons are excited to a higher energy level and are moved along a series of reactions through an electron transport chain until they return to P700. ATP is then synthesized from the energy released. NADPH and oxygen are not generated by this process.

Photosystem II uses chlorophyll a molecules called P680 (absorbs light energy best at a wavelength of 680 nm), and the chemical exchanges that follow are called **noncyclic electron flow.** In the noncyclic electron flow, electrons pass from water to $NADP^+$ which is then reduced to NADPH. Both photosystems are involved in this process. When P680 absorbs light, electrons are trapped by the primary acceptor of photosystem II, passed through the electron transport chain used in cyclic electron flow, and then to P700. Hydrogen ions are pumped across the thylakoid membrane, creating the force to resynthesize ATP. The net result of noncyclic electron flow is the movement of electrons from their low energy state in water to a higher energy state in NADPH. ATP is produced and oxygen is released.

Fig. 6-1 Photosystems I and II

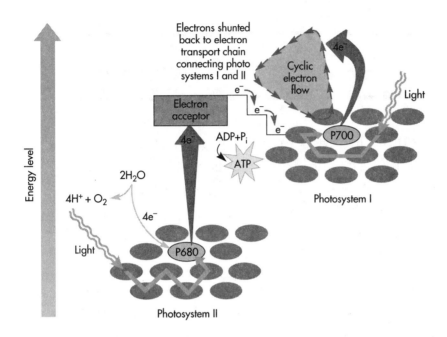

The second set of chemical reactions does not require light energy to drive it. This set of **light-independent reactions** is named the **Calvin cycle.** In this cycle, carbon from atmospheric carbon dioxide is incorporated into organic compounds by a process known as **carbon fixation.** These compounds are then reduced to form carbohydrates. The process occurs in three stages. Glyceraldehyde phosphate, a three-carbon sugar, is formed by three turns of the cycle. The cycle must turn three times to reduce three carbon dioxide molecules. This reduction process is powered by six NADPH molecules and nine ATP molecules that were produced from the light-dependent reactions. Some of the glyceraldehyde phosphate can be synthesized into many different carbohydrates. The rest is used to resynthesize the substrates in the Calvin cycle.

Fig. 6-2 The Calvin Cycle

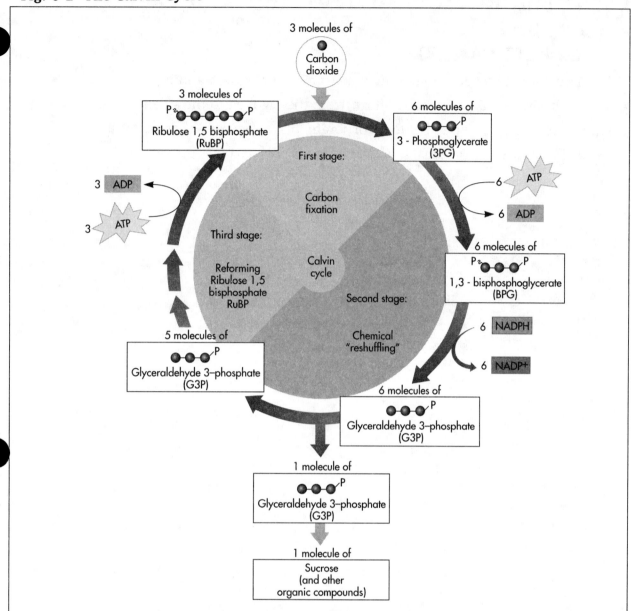

5. Relationship Between Photosynthesis and Cellular Respiration

It should be evident that the metabolic pathways of photosynthesis and cellular respiration are significantly linked. Water and carbon dioxide, the by-products of cellular respiration, are necessary substrates for photosynthesis. Oxygen and glucose, the by-products of photosynthesis, are the substrates of cellular respiration. The dependency of one on another is the basis for understanding the interrelationship between autotrophs and heterotrophs.

CHAPTER REVIEW ACTIVITIES

CONCEPT MAPPING

Develop a concept map that demonstrates your understanding of photosynthesis. Focus on creating "flow" from the big idea to the smaller components of the concept. Try to incorporate linking action words such as: involved in, occurs in, contains, is activated by, is powered by.

COMPLETION QUESTIONS

Fill in the spaces with a term that best fits. Read the part of the statement provided thoroughly before deciding on an answer. If you are unsure of the answer, look to the Key Concept Review or the text to review related materials BEFORE looking in the answer section.

1. All living things have energy needs. The origin of this energy is ultimately the _____.

2. Carbohydrates, one of the products of photosynthesis, are stored as _____ molecules in plant cells.

3. The flattened membranous sacs in chloroplasts are called the _____.

4. The shortest wavelength of radiant energy belongs to _____ rays.

5. The visible spectrum would have a wavelength range from _____ to _____ nanometers.

6. One of the electron acceptors in cellular respiration is NAD^+, and the primary electron acceptor in photosynthesis is _____.

7. A grana is a stack of _____ found in the _____ (organelle).

8. _____ are capable of manufacturing their own food, whereas _____ rely on these organisms, or organisms that have consumed them, for food.

9. _____ are discreet units or packages of fixed amounts of electromagnetic energy.

10. Blue, green, and violet light energy are absorbed by _____, a pigment found in chloroplasts. This pigment reflects _____ and _____ colors of the visible spectrum.

MULTIPLE CHOICE QUESTIONS

_____ 1. For every six turns of the Calvin cycle, how many glyceraldehyde phosphate molecules are gained?

     a.   1

     b.   2

     c.   3

     d.   6

     e.   8

_____ 2. Which of the following is NOT true of photosynthesis?

     a.   The process is more likely to occur during daylight hours.

     b.   Photosynthesis is an oxidation reaction.

     c.   Photosynthesis requires a supply of carbon dioxide.

     d.   B and C are both incorrect

     e.   None of the above are true.

_____ 3. Which of the following statements is false?

     a.   65% of all organisms on earth are autotrophs.

     b.   Autotrophs are able to convert solar energy into chemical energy.

     c.   The sun is the ultimate source of energy for all living things.

     d.   All organisms live on food produced by autotrophs.

     e.   Consumers cannot produce their own food.

_____ 4. Which of the following forms of electromagnetic energy has the least energy?

     a.   Visible light

     b.   Gamma rays

     c.   Radio waves

     d.   X-rays

     e.   Ultraviolet light

_____ 5. Which color of light would be the least desirable for growing plants?
   a.   Red
   b.   Yellow
   c.   Blue
   d.   Green
   e.   Violet

_____ 6. The oxygen released as a by-product of photosynthesis comes from which of the following?
   a.   Water
   b.   Glucose
   c.   Atmospheric gases
   d.   Ribulose bisphosphate
   e.   Carbon dioxide

_____ 7. Which of the following statements is false?
   a.   Chlorophyll *a* and chlorophyll *b* have the same absorption spectra.
   b.   Since carotenoids absorb photons of shorter wavelengths, they would reflect photons of longer wavelengths.
   c.   The carotenoids are known as accessory pigments and capture energy that they pass on to chlorophyll.
   d.   Boosting an electron to a higher energy level requires the right amount of energy - no more, no less.
   e.   Chlorophyll is used as the primary light gatherer in all plants.

_____ 8. Which of the following is not associated with chloroplasts?
   a.   Thylakoids
   b.   Lamella
   c.   Stroma
   d.   Grana
   e.   Cristae

_____ 9. Which of the following is not produced in the light-dependent reactions of photosynthesis?

    a.   Ribulose bisphosphate

    b.   Glucose

    c.   Glyceraldehyde phosphate

    d.   NADPH

    e.   An unstable six carbon compound

_____ 10. C_4 plants are most likely to be found in which type of environment?

    a.   Wet, mild

    b.   Hot, dry

    c.   Cold, moist

    d.   Salty

    e.   All of the above

_____ 11. Which of the following plants would exhibit CAM?

    a.   Pine trees

    b.   Corn

    c.   Members of the Crassulaceae

    d.   Aquatic weeds

    e.   Onions

_____ 12. Which of the following statements is true of photocenters?

    a.   They act as a receiver of energy and transfer it among chlorophyll molecules.

    b.   They are centers for accessory pigments.

    c.   They are the sites where photosystems I and II interact.

    d.   They store energy for use in the light independent reaction.

    e.   They fix carbon dioxide and thus convert solar energy into chemical energy.

_____ 13. Which of the following best describes P700?
  a. It refers to a molecule of chlorophyll *a*.
  b. It reacts with light of long wavelengths in the red portion of the visible spectrum.
  c. It is part of photosystem I.
  d. It accepts electrons after they have traveled through the electron transport chain of photosystem II.
  e. All of the above

_____ 14. Which of the following best describes ATP synthetase?
  a. It is an enzyme.
  b. It forms a channel for the escape of protons from spaces within the thylakoid membrane.
  c. It is a mechanism to produce energy in the form of ATP.
  d. It functions with hydrogen released from the splitting of water molecules.
  e. All of the above

_____ 15. Which of the following best describes the role of NADPH?
  a. It is used in the synthesis of glucose from glyceraldehyde phosphate.
  b. It is involved in the resynthesis of ribulose bisphosphate.
  c. It enables carbon dioxide to be fixed by ribulose bisphosphate.
  d. It reacts with the unstable six-carbon compound produced by the fixation of carbon.
  e. All of the above

_____ 16. Red light has less energy than blue light. Which statement below is true?
  a. Blue light has a longer wavelength than red light.
  b. Red light has a longer wavelength than blue light.
  c. Blue light contains more photons than red light.
  d. Blue and red light have the same range on the visible light spectrum.
  e. None of the above

_____ 17. Within photosystem I and II, electrons flow as chemical bonds are broken. In which of the following would the electrons have their lowest potential energy level?

a. Water

b. P680

c. P700

d. NADPH

e. Glucose

_____ 18. P680 is reduced by electrons from which of the following?

a. Photosystem I

b. Water

c. NADPH

d. Photosystem II

e. Glyceraldehyde phosphate

_____ 19. Which of the following best relates to the accessory pigments within chloroplasts?

a. They provide the energy for splitting water molecules.

b. They extend the absorption spectrum of chlorophyll _a_.

c. They absorb photons of different wavelengths than chlorophyll thereby increasing the energy to drive photosynthesis.

d. B and C

e. A and C

_____ 20. Noncyclic electron flow in the chloroplast forms which of the following?

a. ATP

b. Glucose

c. ATP and NAD^+

d. Glyceraldehyde phosphate

e. ATP, NADPH, and O_2

COMPARISION OF PHOTOSYNTHESIS AND CELLULAR RESPIRATION

Complete the table below by providing the missing information.

Photosynthesis	Cellular Respiration
Plants, algae	
	Mitochondria
Reduction reaction	
Carbon dioxide needed	
	ATP produced
	Carbon dioxide released
PGAL ------> glucose	
Overall Equation	Overall Equation

SHORT ANSWER QUESTIONS

1. Describe the benefits of having other pigments in addition to chlorophyll in the chloroplast.

2.   What are the main differences between light dependent reactions and light independent reactions?

3.   How does cyclic electron flow differ from noncyclic electron flow?

4.   Draw an overview of the photosynthesis process.

CRITICAL QUESTIONS

1.   Temperature becomes a limiting factor for photosynthesis as the intensity of light increases. Construct a graph that illustrates this fact.

2.   The concentration of carbon dioxide and oxygen influences the rate of photosynthesis inversely. An increase in carbon dioxide concentration increases the rate of photosynthesis to a point Draw a graph (two curves on one graph) to illustrate what happens to the rate of photosynthesis when carbon dioxide or oxygen concentrations increase.

3.   Following a nuclear bomb explosion, what would happen to the environmental temperature and why?

4.   You have been asked to provide evidence that white light is composed of a spectrum of colors. How might you demonstrate this?

5.   How does CAM photosynthesis differ from C4 photosynthesis?

TELLING THE STORY

Write a description of your understanding of the processes associated with photosynthesis. Include as many details as possible, but don't try to use terms that you don't really understand. Write the description as though you were preparing to explain the process to a classmate.

CHAPTER REVIEW ANSWERS

COMPLETION QUESTIONS

1. Sun	6. $NADP^+$
2. Starch	7. Thalakoid, chloroplasts
3. Thalakoid	8. Autotrophs/producers, Heterotrophs/consumers
4. Gamma rays	9. Photons or quanta
5. 400, 700	10. Carotenoid, yellow, orange

MULTIPLE CHOICE QUESTIONS

1. B	6. A	11. C	16. B
2. B	7. A	12. A	17. A
3. A	8. E	13. E	18. B
4. C	9. B	14. E	19. C
5. D	10. B	15. D	20. E

COMPARISON OF PHOTOSYNTHESIS AND CELLULAR RESPIRATION

Photosynthesis	Cellular Respiration
Plants, Algae	Animals and plants
Chloroplasts	Mitochondria
Reduction reaction	Oxidation reaction
Carbon dioxide needed	Oxygen needed
ATP produced and used	ATP produced
Oxygen released	Carbon dioxide released
Pgal ------> Glucose	Glucose ------> Pgal
$6CO_2 + 6H_2O ----> C_6H_{12}O_6 + 6O_2$	$C_6H_{12}O_6 + 6O_2 ----> 6CO_2 + 6H_2O + ATP$

SHORT ANSWER QUESTIONS

1. Other pigments provide additional photon gatherers to increase the amount of solar energy that can be captured by the chloroplasts. Since pigments such as carotenoids can absorb wavelengths of light that chlorophyll cannot, they are able to collect this energy and transfer it to chlorophyll.

2.   Besides the obvious light requirement, the light-dependent reactions produce ATP and NADPH, split water molecules in noncyclic photophosphorylation, and produce oxygen as a by-product. The light-independent reactions involve the reduction of carbon dioxide in the Calvin cycle and utilize ATP and NADPH produced in the light-dependent reactions.

3.   In cyclic electron flow P700, (photosystem I) energy is collected independently of P680 (photosystem II). In cyclic electron flow, water is not involved nor is NADPH produced as they are in noncyclic electron flow. Additionally, in cyclic electron flow the electrons ejected from P700 are involved in a long electron transport chain that produces ATP.

4.   Your diagram should resemble the one below.

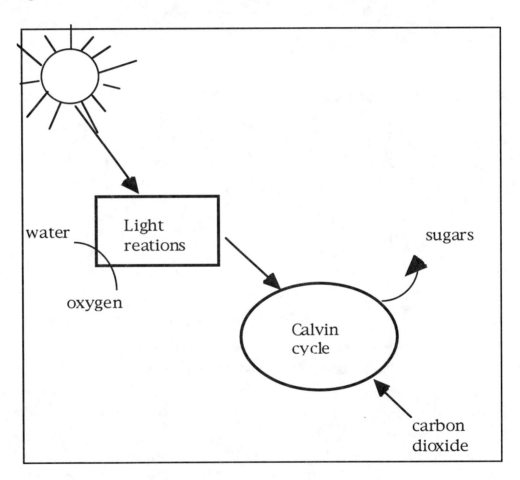

CRITICAL QUESTIONS

1. The graph should look similar to the one below.

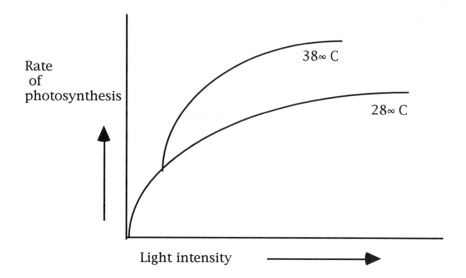

2. Your graph should resemble the one below.

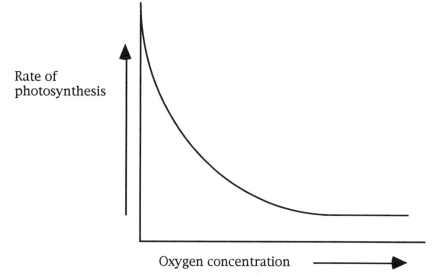

3.   When the bomb exploded the resulting smoke and particulates in the atmosphere would be dense enough to prevent sunlight from reaching the earth. The lack of sunlight would cause the environmental temperature to decrease.

4.   There are a number of ways this could be done, but the simplest might be to pass a beam of white light through a prism. The light will refract as it passes through the prism and break the white light into the different energy levels ranging from the shortest violet light to the longest red light.

5.   The stomata of CAM plants are only open at night when carbon dioxide is absorbed and converted to a four carbon compound. Instead of immediately being converted to a three-carbon compound and entering the Calvin cycle, the four-carbon compound is delayed until the light is available the next day. It then is converted to a three-carbon compound and enters the Calvin cycle.

CHAPTER 7

LEVELS OF ORGANIZATION IN THE HUMAN BODY

KEY CONCEPTS OVERVIEW

1. Levels Of Organization

    The human body is composed of highly varied cells. In fact, there are over 100 different cell types that form the structural and functional integrity of humans. These cells rarely act independently and therefore can be categorized according to the functions they serve. Groups of similar cells that work together to perform a function are called **tissues**. The broad array of cell types within the human organism can be placed into four tissue categories: **epithelial**, **connective**, **muscle**, and **nervous.**

    Tissues are not functionally independent either. Several tissue types will work together as a functional unit. This unit is referred to as an **organ**. Each organ performs highly specialized and complex tasks.

    Because the anatomical and physiological needs of human organisms are numerous and complex, you might expect that organs do not always function alone to facilitate certain tasks. An **organ system** is a group of organs that functions together to carry out the predominant functions of the body. It would be expected then that these functions could be carried out in highly ordered, systematic, and synergistic fashion within the total human organism.

2. Tissues

    Tissue groups were developed to classify the functions of cells that work collaboratively to accomplish tasks. Since there are over 100 different cell types and only four tissue categories, the many cells must contribute similarly to these functional groups.

Epithelial tissue is made up of varying types of epithelial cells. All epithelia function in a general sense to guard and protect the body. More specifically, they protect, absorb, provide sensation, secrete, excrete, and provide surface transport. Epithelial tissue is described in two ways: by the shape of the cells that make up the tissue, and by the arrangement of the cells within the tissue. There are three main epithelial cell shapes. **Squamous** cells are very thin, flat cells, **cuboidal** cells are cube shaped, and **columnar** cells are taller than they are wide. There are four epithelial cell arrangements used for classification. **Simple epithelium** is only one cell thick, while **stratified epithelium** is composed of two or more layers of cells. **Pseudostratified epithelium** is composed of one layer of cells, but the cells vary in height, giving the impression of more than one layer. Finally, the **transitional epithelium** is composed of cube-like cells organized into a formation suitable for stretching.

Connective tissue contains cells that are spaced apart from each other and have a supporting, nonliving matrix around them. One way to group the cell types within this tissue category is by function. There are three distinct functional groups: defensive, structural, and isolating connective tissues.

Defensive connective tissue contains three cell types. **Lymphocytes** are a type of white blood cell that function by attacking foreign cells or viruses that enter the body. They also produce antibodies that can work against these invaders. **Macrophages** act in the defense against foreign substances by phagocytizing (engulfing) and digesting invading material. **Mast cells** produce substances that produce an inflammatory response.

Structural connective tissue also contains three different cell types. **Fibroblasts** are the most numerous type. These flat, irregular, branching cells secrete different types of protein fibers (collagen, reticulin, and elastin) into a matrix between them. Because of these fibers, there are some structural differences among fibroblasts. **Dense fibrous** connective tissue is made up of collagen fibers which are strong and wavy. **Reticulin** fibrous connective tissue is made up of reticulin fibers which are also made of collagen. However, it has very fine branching known as networks. **Elastic** connective tissue is composed of elastic fibers that are more flexible and stretchy. Fibroblasts are also found in **loose** connective tissue, which can be found throughout the body. The second type of cell found in structural connective tissue is the **chondrocytes**. These cells produce **cartilage** (hyaline, elastic, and fibrocartilage). The third type of cell found in structural connective tissue is the **osteocyte** from which bone is produced. There are two types of internal structure to bone tissue. The ends and interior of long bones, as well as the interiors of flat and irregular bones, are composed of **spongy bone.** The exterior surface of most bones is **compact bone,** a dense, hard structure that gives bone strength.

Within the **isolating** connective tissue types there are two major cell types. **Blood cells** (red blood cells, white blood cells, and platelets) and **adipose cells,** which are storage cells for fat.

Muscle tissue is specialized for different types of movement. There are three different types of muscle tissue: **cardiac,** which is found in the heart, **smooth,** which is found in the tube-like structures of the body, and **skeletal,** which provides the necessary accessory to the skeleton for locomotion.

Nervous tissue is composed of two types of cells. Those called **neurons** conduct nerve impulses. These neurons have supporting cells that provide them with protection and support called **glial cells.**

3. Organ Systems

There are eleven organ systems in humans: **digestive, respiratory, circulatory, endocrine, immune, urinary, nervous, skeletal, integumentary, muscular,** and **reproductive.** While these systems can be identified and discussed separately, remember that the physiology of humans is an orchestration between these systems to maintain a healthy organism.

4. Homeostasis

The rate of exchange between the internal and external environments of cells changes constantly. Therefore, cells, and ultimately the organism they make up, must have mechanisms for keeping the internal environment in a range of acceptable variation. The tendency to regulate the internal environment within this acceptable level is called **homeostasis.** As we move from the simplest organisms to the most complex, these mechanisms become much more important. Homeostatic mechanisms are controlled and maintained through feedback loops. **Negative feedback** mechanisms interrupt processes by signaling the process to slow or stop. **Positive feedback** mechanisms tend to enhance or continue a particular process.

CHAPTER REVIEW ACTIVITIES

CONCEPT MAPPING

Using the terms that were highlighted in the overview, construct a concept map that demonstrates your understanding of the levels of organization in the human body.

MATCHING

Using the list of terms provided, select the term that most appropriately relates to the following statements. If you have difficulty answering the questions, refer back to the text for assistance.

_____ 1.   These are the mechanisms by which information regarding the status of a physiological situation are fed back to the system so necessary adjustments can be made.

_____ 2.   This epithelial tissue is made up of two or more layers.

_____ 3.   This function of epithelial cells removes waste products (such as during the formation of urine or the removal of carbon dioxide from the lungs).

_____ 4.   This group of similar cells work collaboratively to perform a function.

_____ 5.   This somewhat strong and very flexible tissue is formed by various connective tissue cells and fibers within a semifluid matrix that attaches the skin to the layers beneath it.

_____ 6.   This shape of epithelial cells has each side approximately the same length and the nucleus centrally located.

_____ 7.   These are the cells that produce cartilage.

_____ 8.   This is the term for two or more tissues grouped together to form a structural and functional unit.

_____ 9.   These phagocytic cells are abundant in the bloodstream and in the fibrous mesh of tissues such as the lungs, spleen, and lymph nodes.

_____ 10.   This is another term for red blood cells.

_____ 11.   This issue allows for stretching and is formed by branching elastic fibers with fibroblasts interspersed throughout.

_____ 12. These highly organized strands of specialized thick and thin myofilaments are capable of shortening a muscle fiber.

_____ 13. These are cells that produce bone tissue.

_____ 14. This term refers to the maintenance of a stable, internal environment despite what may be a very different external environment.

_____ 15. These are supporting nerve cells of the brain and spinal cord.

_____ 16. This type of cartilage, which generally functions as a shock absorber, has elastic fibers embedded in its matrix and is therefore a very tough structure.

_____ 17. This type of feedback loop results when the regulating mechanism moves away from the stable level.

_____ 18. This is the general term for any of several types of white blood cells.

_____ 19. These cells produce substances involved in the body's inflammatory response to physical injury or trauma.

_____ 20. This type of connective tissue cells secrete the protein fibers (collagen, reticulin, or elastin) into their surrounding matrix.

_____ 21. This type of white blood cell, circulating in the blood or residing in the organs, vessels, and lymph nodes, produces specific antibodies against foreign substances.

_____ 22. This type of bone tissue is loosely constructed into an open lattice.

_____ 23. This is the nonliving substance in which connective tissue cells are embedded.

_____ 24. This group of organs functions together to carry out the principal activities of the body.

_____ 25.   This is the feedback loop in which the body's regulatory mechanisms move the body toward a stable level.

_____ 26.   This structure separates the thoracic and abdominal cavities.

TERMS LIST

a. Matrix	j. Glial cells	s. Leukocytes
b. Tissues	k. Spongy	t. Positive feedback
c. Lymphocytes	l. Mast cells	u. Homeostasis
d. Myofibrils	m. Fibroblasts	v. Stratified
e. Fibrocartilage	n. Osteocytes	w. Macrophages
f. Feedback loops	o. Erythrocytes	x. Excretes
g. Organ systems	p. Organs	y. Cuboidal
h. Chondrocytes	q. Loose connective	z. Diaphragm
i. Elastic connective	r. Negative feedback	

IDENTIFICATION

Carefully observe the following pictures. Look for identifying characteristics that would support your identification of the type of tissue represented. Write the name of the tissue type in the space provided and the specific characteristic that made you decide on the type.

1.   Fig. 7.1

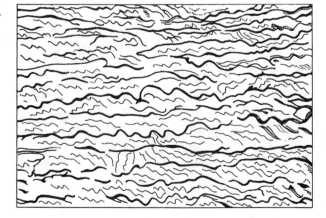

Type: _____

Specific characteristic:

2.　　　　　Fig. 7.2

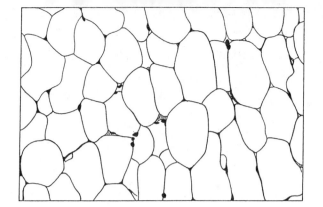

Type: _____

Specific characteristic:

3.　　　　　Fig. 7.3

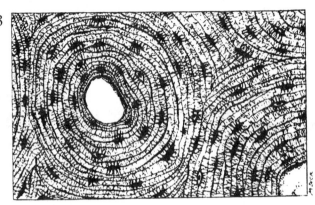

Type: _____

Specific characteristic:

4.　　　　　Fig. 7.4

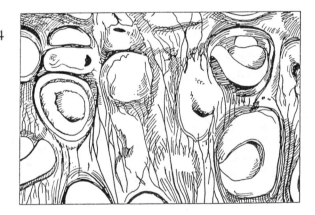

Type: _____

Specific characteristic:

5.

Fig. 7.5

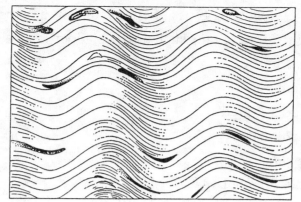

Type: _____

Specific characteristic:

6.   Fig. 7.6

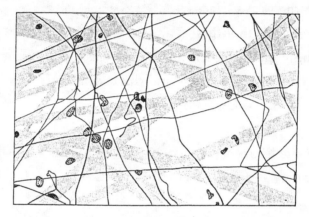

Type: _____

Specific characteristic:

MULTIPLE CHOICE QUESTIONS

_____   1.   Which of the following is NOT one of the major tissue groups?

   a.   Blood

   b.   Muscle

   c.   Nervous

   d.   Epithelial

   c.   Connective

_____ 2. Which of the following types of epithelial cells are specialized for stretching to allow organs such as the bladder to stretch to accommodate collecting urine?

   a.   Columnar
   b.   Pseudostratified
   c.   Transitional
   d.   Cuboidal
   e.   Simple squamous

_____ 3. Cuboidal epithelial cells would be found in which of the following areas?

   a.   On the skin surface
   b.   Lining the mouth and anus
   c.   Lining blood vessels and capillaries
   d.   Lining kidney tubules and ducts of glands
   e.   Lining the digestive tract

_____ 4. Epithelial cells are most closely related to which of the following?

   a.   Linings and surfaces of structures
   b.   Bone tissue
   c.   Brain and spinal cord
   d.   Muscle tissue
   e.   The immune system

_____ 5. Which type of tissue exhibits the greatest amount of diversity in structure?

   a.   Muscle
   b.   Epithelial
   c.   Connective
   d.   Nervous
   e.   Blood

_____ 6. Cells that are responsible for cleaning up the debris of dead cells or fragments are known as which of the following?
   a. Mast cells
   b. Collagen
   c. Schwann cells
   d. Osteocytes
   e. Macrophages

_____ 7. In which of the following structures would you most likely find fibrocartilage?
   a. Nose
   b. Ear
   c. Vertebral disks
   d. All of the above
   e. None of the above

_____ 8. Which of the following would LEAST characterize bone tissue?
   a. Provides a source of blood calcium
   b. Contains both compact and spongy architecture
   c. Lacks living cells
   d. Has a very rich blood supply
   e. Has nerve pathways

_____ 9. Which of the following types of muscle tissue is characterized by being striated and involuntary?
   a. Skeletal
   b. Smooth
   c. Cardiac
   d. Both B and C
   e. Both A and C

_____ 10. Which of the following would NOT be considered a function of epithelial layers of the body?
   a.   Absorption
   b.   Locomotion
   c.   Excretion
   d.   Surface transport
   e.   Protection

HUMAN BODY ARCHITECTURE

Label the structure in the diagram below.

Fig. 7-7 Human Body Architecture

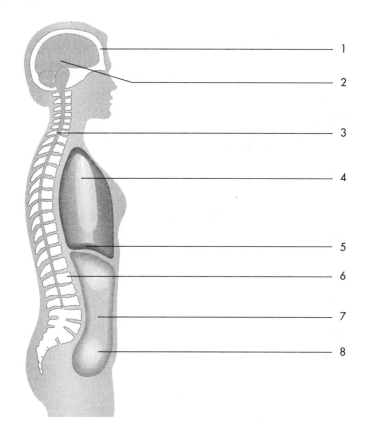

CRITICAL QUESTIONS

1. The following diagrams represent collagen fibers that make up tendons having a parallel orientation (on the left) and collagen fibers in skin having a multidirectional orientation (on the right). What advantages are there for these variations in fiber arrangements?

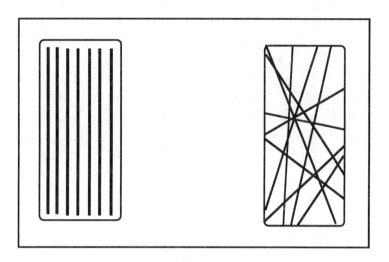

2. During childbirth hormones are released that increase the contraction rate of the uterus. Explain how this is considered a positive feedback mechanism.

3. The skeleton seems to serve a rather limited function (support) in the human body. However, this would be a serious misconception. Describe the various functions of the bone tissue.

4. Describe how physiological negative feedback loops work to maintain a stable body temperature.

TELLING THE STORY

Write a descriptive text about the levels of organization within the human body. Write as if you were explaining the key concepts to a classmate. Try to focus on the flow of the text, making sure your thoughts link together and flow appropriately.

CHAPTER REVIEW ANSWERS

MATCHING

1. F	10. O	19. L
2. V	11. I	20. M
3. X	12. D	21. C
4. B	13. N	22. K
5. Q	14. U	23. A
6. Y	15. J	24. G
7. H	16. E	25. R
8. P	17. T	26. Z
9. W	18. S	

IDENTIFICATION

* Reasons for selecting the tissue type will vary depending on how you learned to recognize them.

1. Elastic connective tissue	6. Loose connective tissue
2. Adipose connective tissue	
3. Bone tissue	
4. Elastic cartilage tissue	
5. Dense fibrous connective tissue	

MULTIPLE CHOICE QUESTIONS

1. A	3. D	5. C	7. C	9. C
2. C	4. A	6. E	8. C	10. B

HUMAN BODY ARCHITECTURE

1. Cranial cavity	5. Diaphragm
2. Brain	6. Vertebral column
3. Spinal cord	7. Abdominal cavity
4. Thoracic cavity	8. Pelvic cavity

CRITICAL QUESTIONS

1. The tension applied to the collagen fibers in tendons tends to be pulled along the length of the tendon. Since the force applied to these structures requires significant strength to withstand the tension, this parallel arrangement is

best suited for these structures. Skin does not receive forces from the same direction therefore the arrangement of collagen fibers reflects the need to withstand tension from many different directions. The trade-off is strength. The tendons are much stronger due to the density and arrangement of fibers.

2. Oxytocin is the hormone that controls the rate of contractions during labor. As it is released, contractions occur. The resulting contractions signal the release of more oxytocin to cause more contractions to occur. The contractions occur at closer intervals and with greater force until the cycle is interrupted by the birth of the baby. This mechanism demonstrates how a feedback loop can move the system away from steady state conditions.

3. Bones are multifunctional structures within vertebrate organisms. They provide movement, points of attachment, support, protection, storage for minerals, blood cell production, shock absorption, and articular surfaces to allow the organism to bend, rotate, etc.

4. Organisms maintain internal conditions through systems of homeostatic control, which work to preserve the physical and chemical aspects of the body's interior within a specific range of tolerance. Since the human body cannot withstand significant changes in temperature, one of the negative feedback control mechanisms within the body is that of temperature regulation. There are three basic components to feedback mechanisms: receptors, integrators, and effectors. Thermal receptors detect changes in external temperature, send neural signals to the brain through the integrators. Then response signals are sent to effectors (such as muscles) causing the body to shiver or sweat and the blood vessels to constriction or dilation, depending on if the temperature is too cold or too hot. These signals are also received and monitored to determine if these effects should continue or be stopped.

• CHAPTER 8

DIGESTION

KEY CONCEPTS OVERVIEW

1. Nutrition-Digestion Connection

Since heterotrophs require food from other sources, they must be able to acquire, metabolize, and use or store nutrients. The resulting effect is sufficient energy to fuel all the necessary cellular activities to maintain life. These nutrients, or raw materials from which the food is composed, can be classified into six distinct groups: **carbohydrates, fats, proteins, vitamins, minerals,** and **water.** This list can be subdivided into organic compounds that supply the necessary energy, and inorganic compounds that assist in the numerous chemical reactions required to extract the energy stored in the organic molecules. The energy currency that will be used to quantify the amount of energy available or the energy spent is the **Calorie or kilocalorie.**

2. Digestion

The conversion of complex organic molecules into simple molecules that can be absorbed by cells occurs during the process of **digestion.** This process is carried out mechanically, chemically, and physically. Mechanically, the teeth crush and grind food into smaller pieces, the stomach churns and mixes the food, and the smooth muscles in the walls of the digestive tube propel material through the system. Chemically, digestion occurs in three phases: (1) the denaturing of proteins by the action of **hydrochloric acid** in the stomach, (2) by **bile salts** which emulsify large lipid droplets, and (3) by a number of very specific enzymes that catalyze chemical reactions. Each group of macromolecular nutrients has a basic type of enzyme. **Proteases** break down proteins into smaller polypeptides and amino acids. **Amylases** break down starches and glycogen to sugars. **Lipases** break down the triglycerides in lipids to fatty acids and glycerol.

You can think of the digestive process as a journey for nutrients through a pathway that narrows and widens and that is straight and curved. This process begins in the mouth where food is chewed and mixed with saliva. The saliva contains **salivary amylase** which is released by the salivary glands. This enzyme initiates the break down of starches. As food is chewed, the tongue acts as a mixing tool to form the food/saliva mixture into a spherical mass called **bolus**. Once this material is ready, the tongue assists in the swallowing process. A small flap of tissue called the **epiglottis** covers the **glottis**, which is the opening to the airway, and assures that the bolus proceeds through to the **esophagus**. Bolus material travels through this and the remaining digestive tract by a smooth muscle contraction known as **peristalsis**.

The esophagus is attached to the upper portion of the stomach at a juncture regulated by a dense, oval muscle called the **cardiac sphincter**. When food reaches the stomach, specialized cells in the lining respond by secreting the hormone **gastrin**. Gastrin stimulates the release of hydrochloric acid and the inactive form (**pepsinogen**) of a protein called **pepsin**. Hydrochloric acid causes this conversion of pepsinogen to pepsin so that protein nutrients can begin to be broken down. Additionally, hydrochloric acid assists in protein digestion by denaturing proteins and by killing any bacteria that may have been ingested with the food. The bolus material remains in the stomach between 2 to 6 hours. While there, the bolus is mixed with these gastric juices and mucus and prepared to be released into the small intestine in a liquid form called **chyme**. The release of chyme is regulated by another oval muscle called the **pyloric sphincter**.

Chyme has a very low pH because of the hydrochloric acid in the gastric juices. Therefore, to avoid tissue damage, it must be neutralized. The first segment of the small intestine where the chyme is initially released from the stomach is called the **duodenum**. When chyme enters this section, it stimulates the release of the hormone secretin. **Secretin** signals the release of sodium bicarbonate from the pancreas to neutralize the chyme. Additionally, secretin stimulates the release of three different proteases (**trypsin, chymotrypsin,** and **carboxypeptidase**), lipases to break down lipids, and **pancreatic amylase** that continues to effect the break down of carbohydrates. The small intestine contributes another type of enzyme called **peptidase** that digests short peptide chains, and three different **disaccharidases** that work to finalize the digestion of sugars.

Cholecystokinin (CCK) is another hormone released when chyme enters the duodenum. CCK stimulates the gallbladder to release bile. **Bile** is made by the liver and assists with the digestion of lipid materials.

There would be little reason for all the digestive processes if the tract was not capable of absorbing these nutrients. The lower two sections of the small intestine (jejunum and ileum) function to capture these nutrients through small projections on the internal surface of the tract. These projections are called **villi** and greatly enhance the surface area for absorption. The surfaces of the villi are also covered with microscopic projections called **microvilli**. Within each of these structures is a network of capillaries and a lymphatic vessel called a **lacteal**, which provide the mechanism for absorption.

Undigested materials are moved through the remaining portions of the digestive tract called the **large intestine**. Here any remaining water and vitamins are reclaimed and the rest of the material (feces) moves through the **rectum** and is then eliminated through the **anus**. The removal process is termed **defecation**.

3.   Diet And Nutrition

While eating has become a social event in our culture, the physiological effects of diet have become a notable health issue. Diets vary significantly, but should be composed of approximately 12% protein, 30% fats, and 58% carbohydrates. Additionally, complete proteins must be ingested because some of the amino acids necessary for protein synthesis are not manufactured in humans. There are eight such **essential amino acids**.

Food that is not metabolized for the synthesis of ATP will be stored in the form of fat. Therefore, paying attention to the caloric intake and expenditure of energy helps to regulate body composition and avoid unwanted weight gain.

CHAPTER REVIEW ACTIVITIES

CONCEPT MAPPING

Develop a concept map that demonstrates your understanding of the process of digestion. Challenge yourself to be as thorough as possible, selecting the terms and linking phrases carefully.

MATCHING

Select the term from the list provided that best relates to the following statements. Reflect on your confidence in selecting the term and review those you hesitate on when answering.

_____ 1. This hormone signals the release of bile from the gallbladder.

_____ 2. These structures are found on the internal surface of the small intestine and increase the absorptive surface area.

_____ 3. Excess glucose is converted to this molecule by the liver.

_____ 4. This oval muscle controls the entrance to the stomach at the attachment site of the esophagus.

_____ 5. The majority of nutrient absorption occurs in this portion of the digestive tract.

_____ 6. This is another term for the large intestine.

_____ 7. Saliva contains this enzyme which initiates the digestion of carbohydrates.

_____ 8. This material is required to convert pepsinogen to pepsin.

_____ 9. This is the nutrient that should constitute about 12% of the adult diet.

_____ 10. The term used to refer to separating large fat droplets into smaller ones.

_____ 11. Lipases break triglycerides into these molecules and glycerols.

_____ 12. A collection of molecules secreted by the liver that helps in the digestion of lipids.

_____ 13. This is the initial segment of the small intestine that is actively involved in digestion.

_____ 14. This enzyme is produced by the pancreas, which assists with the digestion of proteins.

_____ 15. This flap of tissue folds back over the opening to the larynx, preventing food from entering the airway.

_____ 16. This digestive hormone of the stomach controls the production of gastric juices.

_____ 17. This is the last section of the small intestine.

_____ 18. This is the process in which food particles are broken down into small molecules that can be absorbed by cells.

_____ 19. These successive waves of smooth muscle contractions are needed to move food from the esophagus to the stomach.

_____ 20. These are the raw materials of food consisting of carbohydrates, lipids, proteins, vitamins, minerals, and water.

TERMS LIST

a. Proteins	f. Villi	k. Bile	p. Nutrients
b. Ileum	g. Colon	l. CCK	q. Fatty acids
c. HCl	h. Emulsify	m. Gastrin	r. Trypsin
d. Epiglottis	i. Digestion	n. Duodenum	s. Peristalsis
e. Cardiac sphincter	j. Salivary amylase	o. Small intestine	t. Glycogen

ENZYMATIC ACTIVITY

If each of the following substances were mixed together, would digestion occur? Explain the reason for your answer.

1.   Pancreatic amylase, bile, egg whites
2.   Salivary juices, starch, boiling water
3.   Trypsin, sodium bicarbonate, water, fish, (solution slightly warmed)

VITAMINS AND MINERALS

Complete the missing information in the following chart . Try to answer as many as you can, but if you need to assistance refer to information provided in your text.

Vitamin or Mineral	Function	Deficiency
1. Potassium		
2.	Increased absorption of calcium, promotes bone formation	
3. Vitamin K		Severe bleeding
4.		Scurvy
5.	Visual pigment production, maintenance of epithelial tissue	
6.		Deficient carbon dioxide transport and protein metabolism
7.	Hemoglobin synthesis, oxygen transport	Anemia, energy loss

MULTIPLE CHOICE QUESTIONS

_____ 1. Which substrate does amylase digest?

   a. Proteins

   b. Starches

   c. Fats

   d. Vitamins

   e. Amino acids

_____ 2. Which of the following compounds is digested in the stomach?

   a. Fats

   b. Disaccharides

   c. Polysaccharides

   d. Proteins

   e. Triglycerides

_____ 3. Which of the following is the sphincter muscle that regulates chyme release into the small intestine?

   a. Pyloric sphincter

   b. Cardiac sphincter

   c. Anal sphincter

   d. Fundal sphincter

   e. Pancreatic sphincter

_____ 4. Which of the following enzymes does NOT act as a catalyst for protein catabolism?

   a. Chymotrypsin

   b. Amylase

   c. Lipase

   d. Pepsin

   e. B and C

5. Which of the following would NOT be found in bolus?
   a. Pancreatic lipase
   b. Chymotrypsin
   c. Carboxypeptidases
   d. Cholecystokinin
   e. All of the above

6. The lymphatic vessel found networked with capillaries in microvilli is known as which of the following?
   a. Epiglottis
   b. Glottis
   c. Lacteal
   d. Villi
   e. Duodenum

7. In which of the following areas are vitamins absorbed?
   a. Stomach
   b. Liver
   c. Jejunum
   d. Colon
   e. Esophagus

8. Which of the following is the hormone that signals the release of proteases, lipases, and pancreatic amylases?
   a. Secretin
   b. Gastrin
   c. Bile
   d. Hydrochloric acid
   e. Cholecystokinin

9. Which of the following nutrients leaves the intestinal tract by osmosis?
   a. Fatty acids
   b. Triglycerides
   c. Water
   d. Monosaccharides
   e. Amino acids

_____ 10. Bile salts function as which of the following?

    a. Lubricants to eliminate waste

    b. Enzymes to break down proteins

    c. Buffers to increase pH

    d. Detergents to emulsify fats

    e. Chemical signals for hormone release

CRITICAL QUESTIONS

1. Food products sold in the United States must have labels that provide the nutritional information for the consumer. The ingredients are listed from the most predominant to the least predominant. Nutritional information for three different products has been provided for you. Read and study the lists carefully, then answer the questions.

LIGHTEN UP PEANUT BUTTER

Nutritional Facts

Serving Size 2 Tbsp (36g)
Servings Per Container about 14
Per Serving
Calories 190   Calories from Fat 100

Total Fat 12 g	18%
Saturated Fat 2g	10%
Cholesterol 0 mg	0%
Sodium 150 mg	6%
Total Carbohydrate 15 g	5%
Dietary Fiber 0 g	
Sugars 5 g	
Protein 8 g	

Vitamin A < 2%   Vitamin C < 2%

Calcium < 2%   Iron 2%

```
┌─────────────────────────────────────────────┐
│         POWER UP PORK                         │
│         AND BEANS                             │
│                                               │
│  Nutritional Facts Per Serving                │
│    Serving size  1/2 cup (130g)               │
│    Calories 110                               │
│    Calories from Fat 10                       │
│  ───────────────────────────────────────────  │
│    Total Fat  1.5 g            2%             │
│       Saturated Fat  0.5 g       2%           │
│         Polyunsaturated Fat 0.5 g             │
│         Monounsaturated Fat 0.5 g             │
│       Cholesterol 0 mg                 0%     │
│       Sodium 490 mg                    21%    │
│       Total Carbohydrate 24 g          8%     │
│          Fiber 6 g            24%             │
│          Sugars 7 g                           │
│       Protein 6 g                             │
│  ───────────────────────────────────────────  │
│    Vitamin A  0%        Vitamin C  2%         │
│    Calcium    4%          Iron     10%        │
└─────────────────────────────────────────────┘
```

a.   Which product provides the best source of carbohydrates for nutrients?
Support your answer.

b.   How do the two products compare with regard to saturated fats?

c.   Both products contain nutrients from legumes. Which gives you the best
source of protein?

d.   What percentage of each product (per serving) comes from fat in both
products? From the information given on the labels, determine the caloric
value for 1 gram of fat.

2.   Why do you think there are different opinions regarding the recommended
          proportions of proteins, fats, and carbohydrates in a dietary plan?

TELLING THE STORY

Write a description of your understanding of the concept of digestion. Complete the writing by summarizing the relationship between the process of digestion and the nutritional requirement of the human organism. Remember to organize your thoughts before you begin to write to assure that there is sequencing and good transition in your text.

CHAPTER REVIEW ANSWERS

MATCHING

1. L	6. G	11. Q	16. M
2. F	7. J	12. K	17. B
3. T	8. C	13. N	18. I
4. E	9. A	14. R	19. S
5. O	10. H	15. D	20. P

ENZYMATIC ACTIVITY

1.   No digestion. The list lacks an enzyme that breaks down proteins which are the components of egg whites.

2.   No digestion. The excessive heat will destroy the salivary amylase which works most effectively at body temperature.

3.   Digestion. The sodium bicarbonate will neutralize the pH and allow the trypsin to break down the proteins in fish.

VITAMINS AND MINERALS

1. Potassium	Muscle and nerve function	Muscle weakness, abnormal EKG
2. Vitamin D	Increased absorption of calcium, promotes bone formation	Rickets, bone deformation
3. Vitamin K	Essential for blood clotting	Severe bleeding
4. Vitamin C	Important for collagen formation	Scurvy
5. Vitamin A	Visual pigment production, maintenance of epithelial tissue	Night blindness, flaky skin
6. Zinc	Component of several enzymes, carbon dioxide transport, protein metabolism	Deficient carbon dioxide transport and protein metabolism
7. Iron	Hemoglobin synthesis, oxygen transport	Anemia, energy loss

MULTIPLE CHOICE QUESTIONS

1. B	3. A	5. E	7. D	9. C
2. D	4. E	6. C	8. A	10. D

CRITICAL QUESTIONS

1. a. The beans. Notice that the dietary fiber constitutes 24% of the total carbohydrate nutrients available. Humans cannot digest this material. The amount of sugar from carbohydrates is greater in the bean product by 2 grams per serving.

b. The beans have only 1/5 of the saturated fat that the peanut butter has. Even though the peanut butter is "light," it still contains significantly more saturated fat than the beans.

c. The misconception here might be that the bean product would contain more protein than the peanut butter. The peanuts actually contain more protein per serving than the beans.

d. The total amount of calories per serving from fats in the peanut butter is approximately 52%. In the beans, the calories from fats are about 9%. According to the information provided on the label, the bean product suggests that each gram of fat provides about 7 calories, whereas the peanut butter suggests that each gram of fat provides about 8 calories. Nutritional guides suggest that the average value per gram of fat is approximately 9 calories of energy.

2. Answers may vary. There are very different opinions about the nutritional make up of a "good" diet. Many nutritionists believe that the diet should be proportioned 65% carbohydrates, 20% fats, and 15% proteins. The differences relate to the way nutritionistS believe each fuel source contributes to the metabolic demands of the body and the physiological mechanisms for uptaking, converting, and utilizing these fuels. The food pyramid should be used as a guide for preparing and consuming foods that will help develop and maintain a balance of nutrients in an individual's diet.

Fig. 8.1 Food pyramid

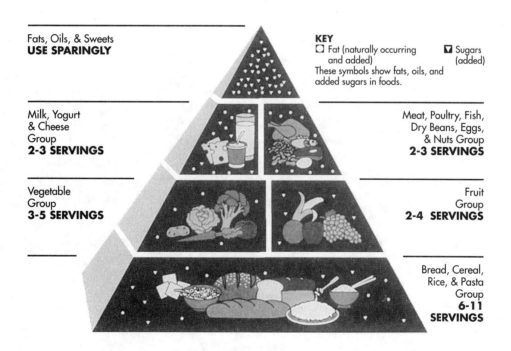

CHAPTER 9

RESPIRATION

KEY CONCEPTS OVERVIEW

1. Respiration

One of the necessary requirements for life is oxygen. In humans, every cell utilizes oxygen to oxidize fuel and produce energy for all cellular activities. One of the waste products of this process is carbon dioxide. Since both of these substances are gases, the human must have internal mechanisms that will extract, transport, and release these gases. The term **respiration** can be used to refer to the processes of ventilation, gas exchange, and gas transport. These processes occur systematically during breathing, external respiration, and internal respiration. Anatomically, the human has a system of structures that provides the passageways for these activities to occur. This system is referred to as the **respiratory system**. This system consists of the **nasal cavities, pharynx, larynx, trachea, bronchi** and their branches, and **alveoli**.

2. Breathing

Breathing, also called ventilation, is the movement of air into and out of the lungs. This process is a muscular activity that includes the processes of inspiration and expiration. During **inspiration**, the diaphragm and intercostal muscles contract, causing the thoracic cavity to enlarge and thereby reduce the pressure within the cavity. The reduced pressure is now less than the atmospheric pressure and therefore air is forced inward. When the stretch receptors in the lungs recognize the expansion caused by the inspiration, a signal is sent to "turn off" the contraction of the diaphragm and intercostal muscles. As these muscles relax, the throacic cavity space is reduced and air is pushed out into the atmosphere. This process is called **expiration**.

Fig. 9-1 Inspiration and Expiration.

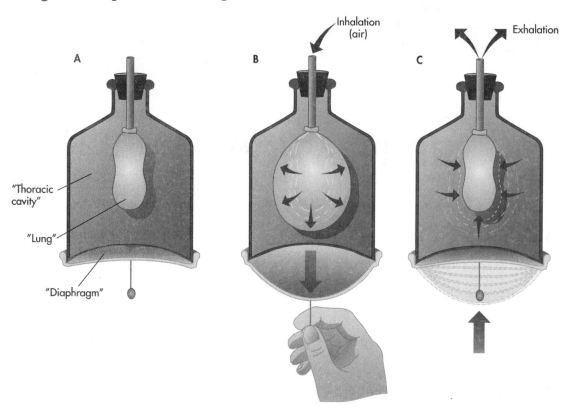

The volume of air that can be exchanged with the environment is dependent on the size of the individual, muscle control, and health of the lungs, just to name a few. A healthy adult male breathes out/in approximately 500 ml of air during a normal expiration/inspiration. This value is a measure of **tidal volume.** Only about 350 ml of this air is actually involved in respiration, the remaining occupies the spaces within the respiratory tract.

3. Gas Exchange And Transport

External respiration is a process of diffusion. Oxygen moves from areas of high concentration in the alveoli across the epithelial membranes and into a capillary. Once the oxygen enters the capillary, hemoglobin molecules on red blood cells pick up the oxygen. The oxygenated blood is now ready to leave the lungs, with the help of the heart, and move to other parts of the body.

Internal respiration occurs after the oxygenated blood leaves the lungs. In internal respiration, oxygen is delivered to the tissues of the body and is exchanged for carbon dioxide, which will be transported back to the lungs for expiration. Carbon dioxide is transported to the lungs is three ways. Bicarbonate ions carry about 70%, hemoglobin carries about 23%, and the remaining 7% is dissolved in the blood. Once carbon dioxide reaches the lungs, it diffuses into the alveoli and is released into the atmosphere during expiration. The amount of carbon dioxide in the respiratory system is constantly monitored by the autonomic nervous system. When carbon dioxide enters the blood, it reacts with water to form carbonic acid. This acid dissociates and releases hydrogen ions, which cause a decrease in blood pH as it becomes more acidic. Chemoreceptors sense this change and signal the medulla and pons (respiratory control center) of the brainstem to speed up the ventilation process.

4.   Respiratory Disease

Asthma is a chronic lung disease that causes the airways to narrow making it difficult for a person to inspire and expire. Besides asthma, there are other **chronic obstructive pulmonary diseases** that affect millions of people in the United States. An example of these includes **chronic bronchitis** and **emphysema.**

CHAPTER REVIEW ACTIVITIES

CONCEPT MAPPING

Construct a concept map that demonstrates your understanding of the process of respiration. Be sure your map clearly differentiates among the processes of pulmonary ventilation, external respiration, and internal respiration.

MATCHING

Select the most appropriate answer from the list of terms provided. Reflect on your choice and note whether you selected it with hesitation or confidence. If you have difficulty selecting an answer, refer to Chapter 9 in your text.

_____ 1. This is the movement of gases from an area of high concentration to an area of low concentration.

_____ 2. These hair-like projections in air passageways assist in moving materials in the respiratory system.

_____ 3. This is the quantity of air moved during a normal inspiration or expiration.

_____ 4. This is the air that remains in the lungs after expiration.

_____ 5. This material is produced by epithelia to help trap foreign particles.

_____ 6. This term means "full of air" and refers to a chronic pulmonary disease that causes individuals to work harder to exhale.

_____ 7. This is caused when a foreign object becomes lodged in the airway.

_____ 8. During this process, the diaphragm contracts causing the thoracic cavity to expand.

_____ 9. This thin delicate membrane lines the interior walls of the thoracic cavity.

_____ 10. This structure forms a partition between the thoracic and abdominal cavities.

_____ 11. This is the oxygen and carbon dioxide carrying molecule on red blood cells.

_____ 12. This occurs when the volume of the thoracic cavity is decreased and pressure begins to increase.

TERMS LIST

a. Residual air	e. Diffusion	i. Tidal volume
b. Hemoglobin	f. Inspiration	j. Diaphragm
c. Pleura	g. Cilia	k. Choking
d. Emphysema	h. Expiration	l. Mucus

IF .... THEN .... CHALLENGE QUESTIONS

Using the "IF" part of the following statements, complete the "THEN" part with guidance from the accompanying question.

1. If a person is choking, cannot talk, breathe, or cough, then _____. What intervention should be applied?

2. If an individual's rate of respiration increases following exercise, then _____. What can you say about the carbon dioxide levels and the effects on that individual?

3. If red blood cells carry oxygen and carbon dioxide, then _____. What effects would anemia have on energy levels?

4. If an individual exhibits periodic episodes of wheezing and difficulty breathing, then _____. What chronic disorder of the respiratory system would this individual most likely have?

5. If oxygen is needed in the mitochondria to generate ATP, _____. What would happen if oxygen were not available in sufficient amounts?

MULTIPLE CHOICE QUESTIONS

_____ 1. Which of the following statements about smoking cigarettes is true?
   a. Smoking has numerous negative effects on the respiratory system.
   b. Smoking is an addictive habit due to the chemical nicotene?
   c. Smoking has been linked directly to emphysema and cancer.
   d. Smoking has been linked to low birth weights in babies.
   e. All of the above are true.

_____ 2. The internal respiratory process involves which of the following?
   a. The chemical breakdown of fuel molecules to produce energy
   b. The movement of air in and out of the lungs
   c. The exchange of carbon dioxide and oxygen between the blood and tissue fluid
   d. The exchange of oxygen and carbon dioxide between the blood and alveoli
   e. None of the above

_____ 3. What percentage of carbon dioxide is transported by water?
   a. 50%
   b. 70%
   c. 23%
   d. 7%
   e. 0%

_____ 4. Which of the following is characteristic of the respiratory control center?
   a. Located in the medulla and pons
   b. Part of a feedback loop
   c. Can be stimulated by carbon dioxide
   d. Controls the contraction of the diaphragm
   e. All of the above

_____ 5. Which of the following is the MOST inclusive term?
   a. External respiration
   b. Internal respiration
   c. Breathing
   d. Cellular respiration
   e. Respiration

_____ 6. Blood going to the lungs leaves from which area of the heart?
   a. Left atrium
   b. Left ventricle
   c. Right atrium
   d. Right ventricle
   e. Blood does not leave the heart to go to the lungs.

7. How is disease that occurs over an extended period of time described?
   a. Acute
   b. Chronic
   c. Epidemic
   d. Epistatic
   e. Pandemic

8. Which of the following is the smallest structure in the respiratory system?
   a. Alveoli
   b. Trachea
   c. Pharynx
   d. Bronchi
   e. Larnyx

9. Which of the following structures prevents food from entering the trachea?
   a. Pharynx
   b. Bronchus
   c. Diaphragm
   d. Epiglottis
   e. Larynx

10. Which of the following is often referred to as the "Adam"s apple?"
    a. Pharynx
    b. Bronchus
    c. Diaphragm
    d. Epiglottis
    e. Larynx

LABELING

Label the following diagram with the appropriate terms. Use arrows to indicate the direction of blood flow.

Fig. 9.2 Route of blood during gas transport.

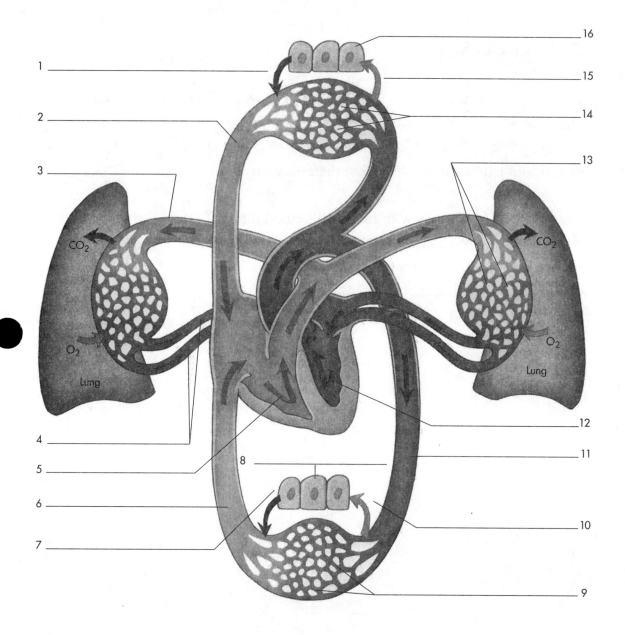

SHORT ANSWER QUESTIONS

1. What is respiration?

2. How does the brain monitor the rate of breathing?

CRITICAL QUESTION

1. What is the advantage of having very small, but very numerous, alveoli in the lungs?

2. What would happen if the lungs lost their elasticity?

3. Bobby became very angry with his mother and told her that he was going to hold his breath until he died. His mother was upset at Bobby's behavior, but was not really concerned that he would in fact die if he held his breath. Why was she not concerned about this?

TELLING THE STORY

Write about your understanding of respiration. Describe the differences in the respiratory processes and note where each process occurs. At the end of your writing, reflect on the causes and effects of pulmonary diseases.

CHAPTER REVIEW ANSWERS

MATCHING

1. E	5. L	9. C
2. G	6. D	10. J
3. I	7. K	11. B
4. A	8. F	12. H

IF .... THEN .... CHALLENGE QUESTIONS

1. The Heimlich maneuver should be adminstered to attempt to dislodge the obstruction and allow the individual to breathe.

2. The levels of carbon dioxide are higher than normal and are signals to the respiratory centers in the medulla and pons to speed up ventilation.

3. The energy levels would be decreased because the transport of oxygen to cells and the removal of carbon dioxide from cells would be diminished. The effect would be less synthesis of ATP than normal.

4. The person would more than likely suffer from asthma.

5. The person would exhibit signs of fatigue. ATP is necessary to fuel energy required cellular activities. Aerobic respiration yields ATP through a more efficient oxidation of glucose than that produced by anaerobic respiration.

MATCHING

1. E	3. B	5. E	7. B	9. D
2. C	4. E	6. D	8. A	10. E

LABELING

1. Tissue cells	10. Tissue cells
2. Carbon dioxide	11. Oxygen
3. Inferior vena cava	12. Pulmonary capillaries
4. Right ventricle	13. Left atrium
5. Pulmonary veins	14. Left ventricle
6. Pulmonary artery	15. Aorta
7. Superior vena cava	16. Oxygen
8. Systemic capillaries	17. Systemic capillaries
9. Carbon dioxide	

SHORT ANSWER QUESTIONS

1. Respiration is the uptake and exchange of oxygen for carbon dioxide by the body. It includes the process of cellular respiration, external respiration, internal respiration, and breathing.

2. Stretch receptors are located in the lungs and monitor the degree of inflation in each the lungs. This information is then relayed to the respiratory centers located in the medulla and pons of the brain.

CRITICAL QUESTIONS

1. The number of alveoli increase the available surface area for gas exchange to occur.

2. Loss of elasticity in the lungs would prevent the expanded lung from recoiling and forcing the air out. Carbon dioxide is therefore not released in the appropriate quantity and oxygen intake is reduced. This is the classic symptom of individuals suffering from emphysema.

3. The primary trigger for breathing is the signal received from the chemoreceptors. These receptors monitor the pH of the blood. As the level of carbon dioxide in the blood increases, the more bicarbonate and hydrogen ions are formed. As hydrogen ions enter the blood, it causes the pH to lower. This information is received by the chemoreceptors and signals the exhalation process. If the level of oxygen to the brain is too low, unconsciousness will result and the involuntary reaction to breathe would take over.

CHAPTER 10

CIRCULATION

KEY CONCEPTS OVERVIEW

1.   Circulatory System Functions

The circulatory system has four basic functions: (1) to transport nutrients and waste materials, (2) to transport oxygen and carbon dioxide, (3) to maintain body temperature, and (4) to facilitate hormone transport. The movement of water between the blood and the tissues is also regulated by proteins and ions in blood plasma. Additionally, specialized cells within the circulatory system provide immunity to the body. While these functions may be more closely associated with blood tissue, the circulatory structures are necessary to move blood through the body so that all cells can reap the benefits.

2.   The Heart and Blood Vessels

The structures associated with the circulatory system consist of three parts: the **heart**, the **blood vessels**, and the **blood**. The heart and blood vessels together make up what is known as the **cardiovascular system.**

The heart is a hollow organ with surrounding walls composed of cardiac muscle. The human heart is basically a double pump with a **pulmonary circuit** that provides a means for gas exchange in the lungs and a **systemic circuit** that moves blood throughout the rest of the body. As blood is pumped from the heart, it is distributed throughout the body via an intricate series of different sized blood vessels. Oxygen-depleted blood is returned through the **venous system** of vessels to the right side to the heart. Upon its arrival through the **superior** and **inferior vena cavae**, blood enters the **right atrium**. It then passes through the tricuspid valve and empties into the **right ventricle**. When the thick muscular walls of the right ventricle contract, they provide sufficient force to move the blood through the **pulmonary semilunar valve** into the **pulmonary artery** and then to both

lungs. Here the deoxygenated blood releases carbon dioxide and picks up oxygen. This oxygenated blood returns to the heart through the **pulmonary vein** to the **left atrium.** From here the **bicuspid valve** opens to allow blood to pass into the **left ventricle.** When the left ventricle contracts, blood is ejected from the left ventricle through the **aortic semilunar valve** to the **aorta.**
(See Figure 10-9, P. 206 in your text)

As you can imagine, the walls of the arteries receiving oxygenated blood must be elastic enough to withstand the surges of blood coming from the left ventricle. The arterial walls have tissue layers that are **elastic** and **smooth muscle.** The elastic layer allows the arteries to expand and recoil in response to the pulsation of the blood. The smooth muscle layer provides contractile forces to assist with the transport process. The walls of the veins have much thinner layers of elastic and smooth muscle tissue, because they are not subjected to the same forces. The veins also contain small one-way valves that prevent any blood from back-flowing on its return to the right side of the heart.

Fig. 10-1 Artery, Vein, and Capillary Structures

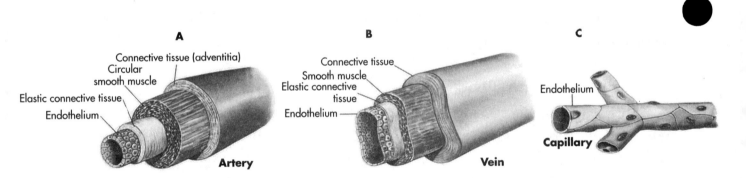

The contractile process of both the atria and ventricles must be orchestrated to assure that they work in synchrony. Contraction is initiated by autonomic nerve impulses in a cluster of specialized cells in the right atrium call the **sinoatrial node** (SA node or pacemaker). The impulse spreads throughout the atria and causes contraction. Then the impulse is transmitted to a second cluster of specialized cells called the **atrioventricular node** or AV node. These cells transmit the impulse through a region of the heart called the bundle of His, located inferiorly to the apex of the heart. At the apex, the impulse is conducted through the Purkinje fibers and spread throughout the ventricles, causing them to contract.

The conduction of these impulses can be traced by an instrument called an electrocardiogram. The resulting graphic display provides descriptive information about the condition of the heart as the impulses are received and transmitted. (See Figure 10-11. p. 208 in your test)

3.   Blood

The cardiovascular system is responsible for many things. However, without blood there would be no need to have such a system. Blood is a connective tissue that consists of formed elements, proteins, electrolytes (ions), and a straw-colored fluid called plasma. The average adult body contains about 5 liters of blood, which constitutes 8% of the total body make-up.

The formed elements consist of **erythrocytes** (red blood cells), **leukocytes** (white blood cells), and **platelets** (cell fragments). Each of these components have very specific functions. The erythrocytes have oxygen carrying molecules called hemoglobin. The mature erythrocytes lack nuclei or the ability to synthesize proteins. There are five types of leukocytes classified according to their staining characteristics. The **granulocytes** (having visible granules) are the **neutrophils**, the **basophils**, and the **eosinophils**. The **agranulocytes** (lacking visible granules) are the **monocytes** and the **lymphocytes**. As a group the leukocytes provide a substantial defense mechanism against foreign substances. The platelets assist in maintaining homeostasis and function during blood clotting.

4.   Lymphatic System

Plasma does not remain entirely in the blood vessels. It can diffuse through the capillary walls into the surrounding tissues. This plasma becomes the main component of intercellular fluid, bathing cells to prevent their drying out. The **lymphatic system** collects this intercellular fluid not collected by blood, and returns it to the general circulation. This fluid is referred to as **lymph** (tissue fluid). The lymphatic system contains a network of vessels, organs, and nodes. The vessels carry lymph to veins of the circulatory system near the base of the neck. The lymph nodes are located throughout the body and provide a filtering system. Additionally, the nodes house some cells that are important in the immune response. The organs of the lymphatic system- the spleen and the thymus- also have specific tasks. The spleen stores an emergency blood supply and contains leukocytes. The thymus is the site of T cell maturation. These cells are critical players in the immune response.

5. Cardiovascular Diseases

Cardiovascular diseases are the leading cause of deaths in the United States. These diseases collectively cause the death of more people every year than AIDS and cancer combined. Statistically, one out of every five people in the United States has or will have some form of cardiovascular disease. Of the many types of cardiovascular diseases, atherosclerosis and heart attacks are significant contributors to the high mortality rate. **Atherosclerosis** is a form of arteriosclerosis in which masses of cholesterol and other lipids build up in the walls of large and medium-sized arteries. These masses are called **plaque**. **Heart attacks** may be caused by a blood clot forming in one of the supplying vessels of the heart or as a result of blockage caused by fatty deposits.

Fig. 10.2 Atherosclerotic plaque.

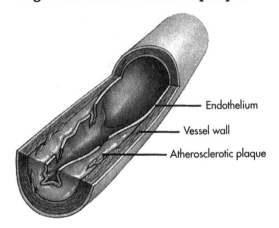

Endothelium

Vessel wall

Atherosclerotic plaque

CHAPTER REVIEW ACTIVITIES

CONCEPT MAPPING

Construct two separate concept maps that demonstrate your understanding of (a) the cardiovascular system and (b) cardiac, pulmonary, and systemic blood flow.

MATCHING

Select from the term list the answers that best correlate to the following statements. Reflect on your ability to confidently match the terms with the statements and review those you have difficulty matching appropriately.

_____ 1. This artery extends from the right ventricle of the heart and then branches into two smaller arteries that carry deoxygenated blood to the lungs.

_____ 2. This is the one-way valve within the heart through which blood passes from the right atrium to the right ventricle.

_____ 3. This is the collective term for the heart and blood vessels.

_____ 4. This is the disease in which the inner walls of the arteries accumulate fat deposits which narrow the passageways and lead to elevated systolic blood pressure.

_____ 5. This solid portion of blood is suspended in blood plasma.

_____ 6. These masses of cholesterol and other lipids can collect on the walls of large and medium-sized arteries.

_____ 7. This organ tstores an emergency blood supply and contains white blood cells.

_____ 8. These cell fragments play a role in blood clotting.

_____ 9. This strand of impulse-conducting muscle is located in the septum of the heart and conducts nerve impulses to the cells of the right and left ventricles.

_____ 10. This upper left chamber of the heart receives oxygenated blood from the lungs.

_____ 11. This is the heart valve through which blood passes from the contracting right ventricle into the pulmonary artery.

_____ 12. This is the collective term for all the white blood cells.

_____ 13. This pathway of blood vessels delivers blood to the body regions and organs other than the lungs or heart.

_____ 14. This is the one-way valve in the heart through which blood flows from the left atrium to the left ventricle.

_____ 15. This is a group of specialized cells that receives the impulses initiated by the sinoatrial node and conducts them by way of the bundle of His.

_____ 16. These disk-shaped cells tfunction as oxygen-carrying cells in the blood.

_____ 17. This largest artery in the body receives blood ejected from the left ventricle.

_____ 18. These blood vessels collect blood from the capillary beds and carry it to the heart.

_____ 19. This part of the circulatory system circulates blood to and from the lungs.

_____ 20. This is the network of very narrow tubes in which gases are exchanged between blood and tissues.

_____ 21. This term refers to a blood clot that has traveled to the brain from another location.

_____ 22. This tissue fluid diffuses out of the blood through capillaries and bathes the surrounding cells.

_____ 23. These small spongy structures are located throughout the body and provide a filter for lymph.

_____ 24. This small cluster of specialized cells in the right atrium receives an autonomic nerve impulse that initiates each cardiac cycle.

_____ 25. This small gland is located near the neck and plays a role in the maturation of T cells.

TERM LIST

a. Veins	j. Tricuspid valve	s. Lymph nodes
b. Thymus	k. Embolus	t. Plaque
c. Spleen	l. Pulmonary artery	u. Left atrium
d. Capillaries	m. Leukocytes	v. Pulmonary semilunar
e. Cardiovascular system	n. Sinoatrial node	w. Erythrocytes
f. Atrioventricular node	o. Aorta	x. Systemic circulation
g, Lymph	p. Atherosclerosis	y. Bicuspid valve
h. Formed elements	q. Platelets	
i. Bundle of His	r. Pulmonary circulation	

TYPES OF BLOOD CELLS

FORMED ELEMENTS	LIFE SPAN IN BLOOD	FUNCTION
	120 Days	
Basophil		
		Blood clotting
	7 Hours	
Monocytes		
		Defense against parasites
		Antibody production
	Memory cells may survive for years	

ELECTROCARDIOGRAM

Identify the events that occur at each marked site.

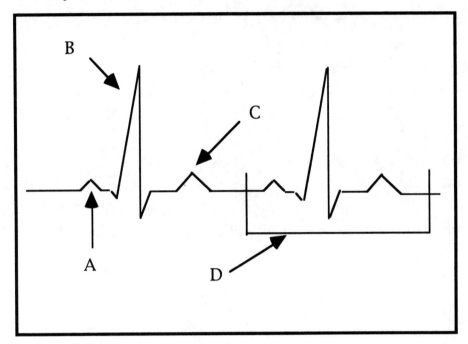

MULTIPLE CHOICE QUESTIONS

_____ 1. Where is the control center that regulates body heat located?
   a. Cerebellum
   b. Medulla
   c. Hypothalamus
   d. Cerebrum
   e. Pons

_____ 2. Which blood vessels are generally the largest in diameter?
   a. Arteries
   b. Veins
   c. Venules
   d. Capillaries
   e. Arterioles

3. Which of the following is mismatched?
   a. The left ventricle leads to the aorta.
   b. The tricuspid valve is between the right atrium and right ventricle.
   c. The bicuspid valve is between the left atrium and left ventricle.
   d. The semilunar valve is between the left atrium and left ventricle.
   e. The the vena cavae drain into the right atrium.

4. Which of the following are chief ions in plasma?
   a. Chloride
   b. Sodium
   c. Bicarbonate
   d. Only A and B
   e. All of the above

5. Which of the following are cholesterol carrying molecules?
   a. CCK
   b. LDL
   c. HDL
   d. B and C
   e. All of the above

6. What does angina pectoris mean?
   a. Angel wings
   b. Chest pains
   c. Heart beats
   d. Brain embolism
   e. None of the above

7. Which of the following constitutes 50% to 70% of all leukocytes in the body?
   a. Neutrophils
   b. Eosinophils
   c. Basophils
   d. Monocytes
   e. Lymphocytes

_____ 8. Megakaryocytes within the bone marrow pinch off bits of their cytoplasm to form which of the following?
   a. Eosinophils
   b. Platelets
   c. Basophils
   d. Erythrocytes
   e. Lymphocytes

_____ 9. Which of the following substances would be involved in blood clotting?
   a. Vitamin K
   b. Calcium ions
   c. Platelets
   d. A, B, and C
   e. A and C

_____ 10. Which one of the following percentages of the cells in human blood are leukocytes?
   a. Less than 10%
   b. Less than 20%
   c. Less than 1%
   d. Less than 44%
   e. Less than 35%

_____ 11. Which of the following are large amoeba-like cells that phagocytize foreign particles?
   a. Monocytes
   b. Lymphocytes
   c. Platelets
   d. Eosinophils
   e. Basophils

_____ 12. Fats constitute what percentage of the typical American diet?
   a. 15%
   b. 37%
   c. 20%
   d. 10%
   e. 55%

_____ 13. What amount of plasma is made up of water?

    a. 25%

    b. 37%

    c. 90%

    d. 44%

    e. 70%

_____ 14. Which of the following values would be considered normal for a systolic blood pressure?

    a. 80

    b. 40

    c. 110

    d. 60

    e. 150

_____ 15. Which of the following values would be considered normal for a diastolic blood pressure?

    a. 78

    b. 100

    c. 45

    d. 30

    e. 120

SHORT ANSWER QUESTIONS

1. What are the three components of the circulatory system?

2. Why is the circulatory system called a closed loop system?

3. Why is the human heart called a double pump?

4. What are the functions of valves in the heart?

CRITICAL QUESTIONS

1. The lymphatic system is subject to cancer. During some radical cancer surgery involving the lymphatic system, the nodes are removed and the attached vessels are tied off. What effects might be expected as a result of this procedure?

2. The heart can sometimes develop what is commonly called a "murmur" because of the sounds that are heard through a stethoscope. What might cause heart sounds to produce a murmuring sound?

TELLING THE STORY

Write about your understanding of the circulatory system. Focus on the key concepts that were highlighted in the chapter. Use as much of the appropriate terminology as you can.

CHAPTER REVIEW ANSWERS

MATCHING

1. L	6. T	11. V	16. W	21. K
2. J	7. C	12. M	17. O	22. G
3. E	8. Q	13. X	18. A	23. S
4. P	9. I	14. Y	19. R	24. N
5. H	10. U	15. F	20. D	25. B

TYPES OF BLOOD CELLS

FORMED ELEMENTS	LIFE SPAN IN BLOOD	FUNCTION
Erythrocytes	120 DAYS	Oxygen and carbon dioxide transport
BASOPHIL	A few hours to a few days	Inflammatory response
Platelets	7 - 8 days	BLOOD CLOTTING
Neutrophils	7 HOURS	immune defenses
MONOCYTES	3 days	Immune surveillance
Eosinophil	8 - 12 days	DEFENSE AGAINST PARASITES
B-lymphocyte	Memory cells may survive for years. Length of time depends on sub-type.	ANTIBODY PRODUCTION
T-lymphocyte	MEMORY CELLS MAY SURVIVE FOR YEARS. Length of time depends on sub-type.	Cellular immune response

ELECTROCARDIOGRAM

1. Atrial excitation or P wave
2. Ventricular excitation or QRS complex
3. Ventricular relaxation or T wave
4. One complete cardiac cycle or heart beat

MULTIPLE CHOICE QUESTIONS

1. C	4. E	7. A	10. C	13. C
2. A	5. D	8. B	11. A	14. C
3. D	6. B	9. D	12. B	15. A

SHORT ANSWER QUESTIONS

1.   The three parts of the circulatory system are the heart, blood vessels, and the blood.

2.   In humans, the erythrocytes are retained inside blood vessels. Plasma and white blood cells are capable of moving across the epithelial lining and traveling through the intercellular spaces. Although the vascular structures are branched networks, they ultimately attach to the heart. There is no "open" end.

3.   The heart is an organ with four chambers. The left side accommodates the oxygenated blood and the right side accommodates the deoxygenated blood. For this reason the heart is described as a double pump.

4.   The valves allow the heart to be efficient by preventing backflow or backup. Without functioning valves, contractions of the heart would result in blood flow in two directions rather than one.

CRITICAL QUESTIONS

1.   The patient will begin to develop swelling or edema caused by the collection of lymph. When the node is removed and the vessel tied off, it diminishes the return mechanism and the collecting fluids cause the tissue to expand.

2.   A murmur is generally caused by a valve that does not close completely or sufficiently. The result is the leaking of blood through the valve. Since the heart sounds originate from the closing of the valves, blood being squeezed through the valve will cause an additional and detectable sound.

Chapter 11

DEFENSE AGAINST DISEASE

KEY CONCEPTS OVERVIEW

1. Defense Mechanisms

    The term "immune" means protection. The human body has ways to defend itself against foreign material that can cause infection. **Infection** is an invasion of the body by foreign organisms, such as viruses, bacteria, and fungi, all of which have the potential to cause disease. To protect against infection, there are two different types of defense mechanisms. The **nonspecific defense mechanism** consists of mechanical and chemical barriers that are directed against any foreign invaders, and the **specific defense mechanism**, which protects against particular invaders. The specific defense is referred to as the immune system, and it consists of cellular and molecular responses to these foreign invaders.

    Nonspecific resistance can be accomplished in several ways. The **first line of defense** against potential infection is the skin and mucus membranes. The lining of the respiratory tract contains cells with cilia that sweep away debris and microorganisms that enter the system. The stomach secretes acid that presents a very harsh environment for microorganisms. Antibacterial enzymes are found in saliva, tears, and sweat. There are even bacteria that are normal inhabitants of the body that act as protectors. Additionally, the **inflammatory response** occurs when cells are damaged by microbes, chemicals, or physical injury. This response is characterized by redness, pain, heat, and swelling.

    Specific resistance is available in the event foreign invaders manage to get through the nonspecific defense barriers. Invaders that enter the blood, lymph, or tissues are met by the **second line of defense**: the immune system. This defense mechanism was discovered by Edward Jenner in 1796, when he experimented with the body's ability to produce a resistance to small pox. Jenner was the first to use a **vaccine** containing similar microbes to initiate the body's natural ability to produce resistance against that same organism should the person ever come in contact with the microbe again. Some years after Jenner, Louis Pasteur developed a vaccine against the fowl cholera using weakened (attenuated) cholera microbes. The basis of the immune response is that the cells of a person's immune system recognize foreign invaders as different from themselves. The protein surfaces of foreign cells are distinctly different and are recognized by the immune system as foreign or **antigens.**

2.   Cells Of The Immune System

Throughout the evolution of multicellular organisms, subgroups of some cells developed specialized functions. One subgroup became the cells of the immune system. These are known as the leukocytes. This system has no organs, but instead is made of cells that are scattered throughout the body. These cells all have a common function, that is, to react to specific foreign antigens.

There are two types of white blood cells involved in the immune system: the **phagocytes** (macrophages and neutrophils) and the **lymphocytes.** There are two classes of lymphocytes that offer specific resistance: **T cells** and **B cells.** These cells originate in the bone marrow, move through the blood and lymph, and reside in the lymph nodes, spleen, liver, and thymus.

3.   Immune Response

The immune system responds to the presence of an antigen in several ways. The response can be immediate or can be established for future needs.

The phagocytic cells are constantly monitoring the presence of any unwanted invaders. Of these phagocytes, the **macrophage**, a matured monocyte, has the key role in the immune response. Macrophages act as scavengers, attacking anything not normally found. They engulf and break down these antigens and then display parts of them on their surfaces. Macrophages secrete a number of different proteins. One is a protein called **interleukin-1** that causes the body temperature to rise. Increasing the body's core temperature aids the immune response and inhibits microbial growth. Another protein triggers the maturation of monocytes into macrophages.

Once the cell surface changes occur on any infected cells, natural killer cells (NK) attack the microbes by secreting proteins (**complement**) that degrade the membrane and cause the cells to burst open and die. This initial response gives the rest of the immune system time to respond.

Interleukin-1 also triggers the activation of **helper T cells** tht have been bound to the antigen on the antigen-displaying macrophage. These helper T cells can stimulate both branches of the immune system: the **cell-mediated immune response** and the **humoral immune response**.

During the cell-mediated immune response, activated helper T cells produce chemicals called lymphokines. **Lymphokines** stimulate the T cells that are bound to antigens causing them to divide rapidly, and form clones of T cells that can also recognize the antigen. Lymphokines also attract macrophages to the infection site. **Cytotoxic T cells** kill the body cells that have been invaded by the antigen. As the infection subsides, **suppressor T cells** shut down the production of new T cells. A few **memory T cells** remain to initiate the cell-mediated response, should the same invader appear again.

Activated helper T cells also stimulate the humoral (antibody) immune response. This response is more long-range. The key players in the humoral response are the B cells. **B cells** have protein receptors on their surfaces. Each one has a different protein receptor, therefore each B cell recognizes a different antigen. B cells bind to antigens on invading microbes or to antigens displayed by macrophages. Helper T cells then bind to the B cell to divide. Clones of the B cells are produced. Following the formation of these clones, the B cells stop dividing and begin to produce antibodies, which are copies of the receptor protein that responded to the antigen. The secreting B cells are referred to as **plasma cells**. Antibodies mark the invading microbe for destruction by the macrophages or complement. After the infection has subsided, some clones that did not become plasma cells live on as **memory B cells**. These cells respond quickly should another invasion of the same microbe occur.

The molecular structure of antibodies, or immunoglobulins, consists of two short chains (**light**) and two long chains (**heavy**). Antibodies are Y-shaped with binding sites on the "arms" of the Y. These binding sites are able to recognize only one specific antigen. The "stem" of the Y may be one of five different types of immunoglobulins, each of which has a different function.

Immunity is **active** if the body of the individual makes appropriate antibodies. The body enhances its ability to manufacture appropriate B and T cells by having the disease or by receiving a vaccination. **Passive immunity** occurs when individuals do not produce their own antibodies, but acquire them from another source. For example, antibodies are passed to a fetus from the mother across the placental wall. Also infants acquire antibodies through breast-feeding.

Acquired immune deficiency syndrome (**AIDS**) is caused by the human immunodeficiency virus (**HIV**). It was first recognized because of increasing numbers of Kaposi's sarcoma and *Pneumocystis* pneumonia cases during the early 1980s. In 1983, researchers in the United States and in France identified the causative agent as HIV. This virus is transmitted through body fluids and incapacitates the immune system by destroying the helper T cells that are important in both cell-mediated and humoral immune responses. Glycoproteins on the surface of HIV bind to specific receptor sites on these helper T cells. The virus invades the cell and begins to replicate, forming new viruses from the host cells. These budding viruses can remain latent for extended periods of time.

Because HIV can undergo very rapid antigenic changes during the replication process, the immune system has difficulty responding to any infections that may be caused by these pathogens. Helper T cells are required to activate other T cells and B cells. As they become infected and die, the decreasing numbers of cells involved in cell-mediated immunity severely impair the immune system. This results in the development of opportunistic infections. The time course for process to occur is about ten years.

This virus is transmitted by the transfer of body fluids (semen, blood, and in rare cases, saliva). Individuals who have been exposed to HIV develop antibodies to the virus. The presence of these antibodies can be used to screen blood, tissue donors, or to determine if an individual is HIV positive. Early in the awareness stage of this disease, people associated AIDS with homosexuals and IV drug users. Today, we know that anyone can be at risk for the disease. Increasing numbers of cases are being diagnosed in college-age women who engage in unprotected sex and babies born to infected mothers. While research continues to develop drug therapies to slow or suppress the progression of this virus, no cure has been determined at this time.

An **allergic reaction** is an excessive immune response to an antigen. Allergic reactions include, sneezing, fever, runny nose, rashes, and even anaphylactic shock. Anaphylactic shock is a very severe response that causes a disruption in the normal size of blood vessels. These vessels abruptly dialate due to the release of histamines from cells in the connective tissues called **mast cells**. When the blood vessels dialate to this extent, it results in a rapid decrease in blood pressure which can be life-threatening.

CHAPTER REVIEW ACTIVITIES

CONCEPT MAPPING

Construct a concept map that demonstrates your understanding of immunity. Make a conscious effort to include each of the relevant topics associated with the concept of immunity and link them to the specific components that clarify their importance to this overarching concept.

MATCHING

From the list provided, select the terms that best correlates with the each of the following statements. Reflect on your choices and the confidence you have in selecting the answer. If you are unsure of an answer or hesitate because you don't know the answer, review the information in your text and attempt this activity again.

_____ 1. This is an immune response that results from the immune system doing its job too well, and then mounts a major defense against an antigen.

_____ 2. This type of white blood cell is the precursor to the macrophage.

_____ 3. This is a set of initial defenses that the body uses to keep out any foreign invaders.

_____ 4. This type of immune response results in the production of antibodies.

_____ 5. These receptor protein molecules are produced after the B cells stop dividing.

_____ 6. This is a description of the characteristic of some organisms that are good at establishing an infection or damaging the body.

_____ 7. This response occurs when cells are damaged by microbes, chemicals, or physical injuries and can cause redness, swelling, heat, and/or pain.

_____ 8. An English physician whose experiments marked the origin of immunology.

_____ 9. These enzymes found in tears are deadly to most bacteria.

_____ 10. This is one of the two types of white blood cells involved in the immune system. It destroys other cells by engulfing and ingesting them.

_____ 11. These foreign molecules invade organisms and induce the production of antibodies.

_____ 12. This is the second line of defense or collective term for the body's specific defenses.

_____ 13. This injection of harmless microbes confers resistance against a specific invader.

_____ 14. These are the key cellular players in the body's antibody-mediated immune response.

_____ 15. This type of immunity in a vaccination causes the body to build up antibodies against a particular disease without getting the disease.

_____ 16. These white blood cells develop in the bone marrow, migrate to the thymus for maturation, and develop the ability to identify invading bacteria and vireses by the foreign molecules displayed on their surfaces.

_____ 17. This type of cell responds quickly and vigorously if the same foreign antigen reappears.

_____ 18. These proteins signal the brain to raise the body's temperature, producing a fever to combat invaders.

_____ 19. These cells limit the immune response.

_____ 20. This term describes a variety of chemical substances produced when a helper T cell has been activated.

_____ 21. This system of proteins kill foreign cells by creating a hole in their membranes.

_____ 22. This type of cell breaks apart cells that have been infected by viruses. It also breaks apart foreign cells such as those found in incompatible organ transplants.

_____ 23. This kind of immunity can be used when protection is needed more quickly than vaccination can provide. Antitoxins used against snakebites are examples.

TERMS LIST

a. Passive immunity	i. Vaccination	q. Immune system
b. B cells	j. Monocyte	r. Virulent
c. Lymphokines	k. Antibodies	s. Complement
d. Jenner	l. Phagocytes	t. Supressor T cells
e. Antigens	m. T cells	u. Cytoxic T cells
f. Allergic reaction	n. Inflammation	v. Humoral
g. Interleukin-1	o. Nonspecific defenses	w. Active immunity
h. Lysozymes	p. Memory T cells	

ANTIBODIES

Label the parts of the following structure. Beneath the diagram, write a brief explanation of the function of antibodies.

Fig. 11.1 Antibody structure.

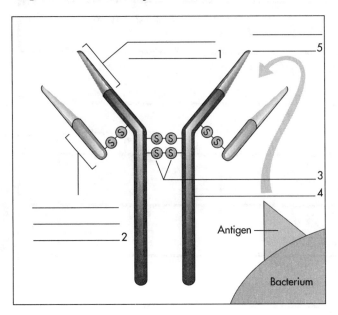

Explanation Statement:

CELLS OF THE IMMUNE SYSTEM

Complete the following table by supplying the missing information.

Cell Types	Function	Type of Immunity
Lymphocytes	Multiple	Specific resistance
•		
•inducer		
•suppressor		
•memory		
•cytotoxic		
•helper		
•		Antibody-mediated
Macrophages	Phagocytosis	Activation of immune response
•neutrophils		
•monocytes		

MULTIPLE CHOICE QUESTIONS

_____ 1. The area surrounding an infection is NOT characterized by all which of the following?
   a. Redness and swelling
   b. Lack of fluids and dryness
   c. Increased temperature
   d. Increased numbers of white blood cells
   e. Pain

_____ 2. The first vaccination was developed against which of the following diseases?
   a. Flu
   b. Bubonic plague
   c. Smallpox
   d. Measles
   e. Rabies

_____ 3. Pasteur developed a vaccine against which of the following?
   a. Smallpox
   b. Measles
   c. Flu
   d. Fowl cholera
   e. None of the above

_____ 4. Why did Pasteur's attenuated bacteria generated a resistance to disease?
   a. They still had the same surface molecules.
   b. They had undergone a mutation.
   c. They were virulent forms of bacteria.
   d. They had produced non-virulent offspring.
   e. They combined with other resistant forms.

_____ 5. Which of the following is NOT considered to be part of the immune system?

a. Thymus gland
b. Spleen
c. Bone marrow
d. Lymph nodes
e. Liver

_____ 6. Which of the following is NOT one of the T cells?

a. Mast
b. Cytotoxic
c. Suppressor
d. Inducer
e. Helper

_____ 7. Which of the following events occurs before the others?

a. Populations of memory T cells become established.
b. Cytotoxic T cells recognize and destroy infected body cells.
c. Lymphokines attract macrophages to infection sites.
d. Activated helper T cells produce lymphokines.
e. Suppressor T cells multiply rapidly and shut down the immune response.

_____ 8. Which of the following is NOT one of the conditions that is used to diagnose AIDS when an individual has HIV?

a. Karposi's sarcoma
b. Helper T cell count falls below 200/mm^3
c. Increased red blood cell count
d. Wasting syndrome
e. Dementia

_____ 9. Which of the following is a mismatched pair?

a. Macrophages - memory cells
b. B cells - antibodies
c. Interleukin-1 - fever
d. Thymus - T cells
e. Complement - membrane degradation

10. Which of the following antibodies are produced in abundance during the first week of an infection?
   a. IgE
   b. IgM
   c. IgA
   d. IgD
   e. IgG

SHORT ANSWER QUESTIONS

1. Describe the inflammatory response.

2. What are the different types of T cells?

3. What types of leukocytes are involved with the immune system?

4. What organs in the body are involved with the immune system?

CRITICAL QUESTIONS

1. Women who undergo mastectomies often have swelling in the upper and lower arm on the same side as the breast surgery. Explain what might be the cause of this swelling.

2. Why are tissue transplant patients prescribed immunosuppressant drugs?

3. Some individuals don't see the importance of vaccinations since few cases of the diseases that the vaccination prevents actually occur in the United States. What problems might result if vaccinations against certain diseases were not required?

TELLING THE STORY

Write about your understanding of the immune system. Focus your writing on the key concepts that were presented in Chapter 11. Use as many of the terms from the chapter as possible. Note at which points you have difficulty making connections as you write. Review these areas in the text.

CHAPTER REVIEW ANSWERS

MATCHING

1. F	9. H	17. P
2. J	10. L	18. G
3. O	11. E	19. T
4. V	12. Q	20. C
5. K	13. I	21. S
6. R	14. B	22. U
7. N	15. W	23. A
8. D	16. M	

ANTIBODIES

1. Heavy chain

2. Light chain

3. Disulfide bonds

4. "Stem"

5. Antigen binding site

Explanation Statement: Answers may vary. Antibodies are the immune response to specific types of antigens. Their role is to mark antigens for destruction by either complement or macrophages.

CELLS OF THE IMMUNE SYSTEM

Cell Types	Function	Type of Immunity
Lymphocytes	Multiple	Specific resistance
•T cells	Recognize antigens	Cell mediated
•Inducer	Oversees T cell development	Cell mediated
•Suppressor	Limits immune response	Cell mediated
•Memory	Responds to same antigen reappearance	Cell mediated
•Cytotoxic	Breaks apart infected cells	Cell mediated
•helper	Initiates immune response	Cell mediated
•B cells	Humoral response	Antibody-mediated
Macrophages	Phagocytosis	activation of immune response
•Neutrophils	Phagocytosis	Antibody mediated
•Monocytes	Phagocytosis	Antibody mediated

MULTIPLE CHOICE QUESTIONS

1. B	3. D	5. E	7. D	9. A
2. C	4. A	6. A	8. C	10. B

SHORT ANSWER QUESTIONS

1. Inflammation is a response of the injured area to physical, chemical, or biological injury. The affected area swells and turns red. An increase in temperature in the affected area occurs. Inflammation is a nonspecific response that helps to eliminate the source of the irritations. During the response, the dilation of the blood vessels will enhance the blood flow to the affected area and transport white blood cells to increase the defense

mechanism. When these cells die, they accumulate in the area and form pus.

2. The T cells include helper T cells, cytotoxic T cells, inducer T cells, and suppressor T cells.

3. The white blood cells involved in the immune response are: phagocytes (neutrophils and monocytes) that engulf foreign materials and lymphocytes (including T cells, B cells, and natural killer cells).

4. The organs associated with the immune system include the lymph nodes and vessels, the bone marrow, spleen, and the thymus.

CRITICAL QUESTIONS

1. This surgery often results in some of the lymphatic vessels being cut and tied off. The results include the collection of tissue fluids in the interstitial spaces and swelling. Massage, diet, hydration, compression, and exercise are some ways to minimize the swelling.

2. Transplanted tissue is often rejected because the immune system recognizes it as a foreign material. All cells have surface proteins that are used for recognition and these proteins are unique to individuals. Tissues can be typed on the basis of these proteins in a process similar to the way blood is type matched. Following transplantation, the immune system will monitor the new tissue for foreign proteins. An immune response will be initiated if the proteins do not appear familiar. To reduce the chance of an immune response, immune suppressant drugs are given to turn off the mechanisms in the thymus and bone marrow that generate leukocyte production. While these drugs are being administered, the individual may be very susceptible to minor infections because the normal defense system is suppressed.

3. Increase cases of these diseases will begin to occur. Because of the increased mobilization of people in the world, more opportunities exist to reinstate a pathogen into a population. It is therefore important to keep a population vaccinated and protected even though the disease "seems" to be eradicated.

CHAPTER 12

EXCRETION

KEY CONCEPTS OVERVIEW

1.   Excretory Organs

Excretion regulates the concentrations of ions and molecules in the body and eliminates metabolic wastes. The kidneys and lungs are the primary organs of excretion; however the skin, liver, and large intestine play a minor role in the excretion of metabolic wastes from the body. The lungs excrete carbon dioxide and water vapor and the kidneys excrete the ions of salts, nitrogenous wastes, assorted other materials based on general health and diet, and water. In the body, most nitrogenous wastes are produced from the breakdown of nucleic acids and proteins. The liver breaks down amino acids by removing their amino groups (a process known deamination) and converting the amino groups into ammonia. Because ammonia is highly toxic, it is immediately combined with carbon dioxide to form urea. Uric acid is a waste product formed from the breakdown of nucleic acids. Creatinine is a waste product derived from creatine, a molecule found in the muscle cells.

2.   Kidneys: Structure, Functions, and Problems

Humans have two bean-shaped kidneys about the size of a small fist. These organs are found in the lower back region. They are both **excretory** and **regulatory** organs. They excrete waste products in the form of **urine**. They are also responsible for maintaining the concentrations of necessary ions within a stable, narrow limit.

A vertical sectional view of a kidney displays the arrangement of tissue organization. The outer portion of the kidney is called the **cortex** and the inner portion is the **medulla**. Blood enters each kidney through the **renal** arteries. Within the medulla, there are small triangular regions of tissue known as the **renal pyramids.** The cortex and pyramids contain the functional filtering units called the **nephrons.** The urine produced by the nephrons is carried by collecting ducts of the nephrons to a space between the cortex and the medulla called the **renal pelvis.** From there, urine leaves the kidney.

The kidneys' actions are controlled by two hormones. **Antidiuretic hormone (ADH)** functions to reduce the loss of water by controlling the permeability of water at the loop of Henle. **Aldosterone** increases the reabsorption of sodium ions and decreases the reabsorption of potassium.

Urine formation can result in some of the salts crystallizing into solid formations. **Kidney stones** are crystals formed from calcium salts or uric acid. They are now treated by a shock-wave therapy that breaks the stones into smaller pieces so that they can be naturally eliminated in the urine.

Renal failure occurs when the filtration of the blood at the glomerulus either slows or stops. Renal failure can be **acute** (sudden) or **chronic** (gradual) and can be treated with dialysis, a process that uses membranes and osmosis to mimic the action of the kidneys. The only alternative to dialysis in severe cases is transplantation.

3. Urine Formation

Within the kidney, each nephron works independently to facilitate filtration, reabsorption, and secretion. A nephron is a long tubule that descends into the medulla and emerges again. The "entrance" structure of the kidney is a filtration apparatus called the **Bowman's capsule,** which surrounds a network of capillaries known as the **glomerulus.** Materials filtered through vascular capillaries enter at the Bowman's capsule and move through a continuous tubular structure with specialized sections. These sections are, from the Bowman's capsule, **the proximal convoluted tubule, the descending arm of the loop of Henle, the loop of Henle, the ascending arm of the loop of Henle, and the distal convoluted tubule.** These sections have different permeabilities to different substances and thus play distinct roles in the filtration and reabsorption process.

Nephron function can be divided into three stages.

- **Filtration:** During this process, blood moves through capillaries. These capillaries are permeable and blood pressure forces approximately 1/6 of the water and smaller molecules from the lumen of these vessels into the Bowman's capsules which then passes through the glomerulus. Larger molecules, 5/6 of the water, and formed elements in the blood, such as red blood cells, remain in the capillaries as blood. The water and small substances that were forced into the Bowman's capsule and into the glomerulus are called the **filtrate**.

- **Selective reabsorption:** The process of reabsorption is driven by two factors: (1) the development of a high osmotic gradient surrounding the loop of Henle and (2) the varying permeability of the membranes of the cells lining the kidney tubules. As the filtrate moves through the tubular structure of the nephron, selected substances move in and out of the filtrate. Each section of the tubule reabsorbs different substances. The proximal convoluted tubule facilitates the reabsorption of substances such as small proteins and glucose. The amount is dependent on the physiological needs of the body. In the descending loop of Henle, the tubule is impermeable to salt and urea and permeable to water. Water moves out of the descending arm and the filtrate in the tubule becomes more concentrated. At the ascending loop of Henle and beyond, the tubule becomes more permeable to salts and less permeable to water. As a result, salts move out of the filtrate and into the surrounding tissues. High in the ascending arm, active transport channels pump out even more salt. Water diffuses out and the remaining filtrate contains a very high concentration of urea. The collecting duct is permeable to urea and as the filtrate moves through the it, some of the urea diffuses out into the surrounding tissue. This higher concentration of urea in the surrounding tissue is what causes water to leave the filtrate at the descending arm.

- **Tubular secretion:** The distal convoluted tubule collects wastes and other substances, such as drugs and toxins, from the blood and prepares them for excretion in the urine. This is one reason that drug testing can be performed on urine samples.

There are approximately a million nephrons per kidney in humans. They collectively receive a flow of approximately 2000 liters of blood per day. Remember that an adult human has about 5 liters of blood.

4.   Urinary System
      From the collecting duct, the urine is delivered to the renal pelvis. Then, urine is collected from here by the **ureters,** which connect each of the kidneys to the **urinary bladder.** Urine is excreted from the body, emptying the bladder through the **urethra.**

Fig. 12-1   Male urinary system.

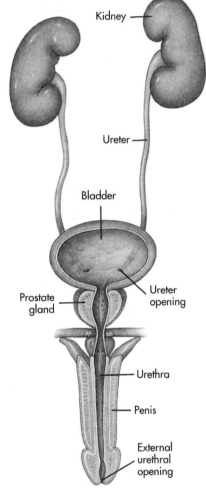

CHAPTER REVIEW ACTIVITIES

CONCEPT MAPPING
      Construct a concept map of your understanding of excretion in the human body. Remember that while the chapter focused mainly on excretion from the urinary system, the concept is more inclusive. Include as many of the new terms in the key concept review as possible in your map.

MATCHING

Select a response from the list provided that best correlates with the following statements. Reflect on the level of confidence you have to select the appropriate answer. If you are unsure of the answer, go back to the chapter and review the information associated with that statement.

_____ 1. This is one of the primary organs of excretion that removes carbon dioxide and water vapor.

_____ 2. A muscular tube that brings urine from the bladder to be released outside the body.

_____ 3. This hormone regulates the level of sodium and potassium ions in the blood, while retaining sodium and excreting potassium.

_____ 4. This nitrogenous waste, found in small amounts in the urine, is formed from the breakdown of proteins and nucleic acids.

_____ 5. This is a process occurring within the kidney nephrons in which materials such as potassium and hydrogen ions, ammonia, and potentially harmful chemicals are added to the filtrate from the blood.

_____ 6. This is a process whereby unabsorbed digestive wastes leave the body during defecation.

_____ 7. This disease occurs when the filtration of the blood at the glomerulus either slows or stops. It can be either acute or chronic.

_____ 8. This method of treating renal failure filteres blood through a machine called an artificial kidney.

_____ 9. This is one of the three tasks accomplished within each nephron by which their membranes act to separate blood cells and proteins from much of the water and molecules dissolved in the blood.

_____ 10. These substances produced by the liver from the breakdown of old red blood cells are responsible for the color in the feces

_____ 11. This hollow muscular organ acts as a storage pouch for urine.

_____ 12. A hormone, secreted by the pituitary gland, regulates the rate at which water is lost or retained by the body.

_____ 13. This describes the water or dissolved substances that are removed from the blood during the formation of urine.

_____ 14. This tube extends from each kidney and carries urine to the bladder.

_____ 15. Urine is formed in these microscopic filtering systems, which number in the millions within each kidney.

_____ 16. These molecules, containing nitrogen, are produced as waste products from the body's breakdown of proteins and nucleic acids.

_____ 17. This process removes metabolic wastes and excess water and salt from the blood and passes them out of the body.

_____ 18. This is an excretion product formed by water and substances excreted by the kidneys.

_____ 19. Crystals of certain salts can develop in the kidney and block urine flow.

_____ 20. This set of interconnected organs not only removes wastes, excess water, and excess ions from the blood, but stores this fluid until it can be expelled from the body.

_____ 21. A process occurring within kidney nephrons during which desirable ions and metabolites and most of the water from blood plasma are recaptured from the filtrate, leaving nitrogenous wastes and excess water and salts behind for later elimination.

_____ 22. A nitrogenous waste, found in small amounts in the urine, is formed from a nitrogen-containing molecule in muscle cells.

_____ 23. This is the primary excretion product from the deamination of amino acids.

_____ 24. This is one of the primary organs of excretion that removes salt ions, nitrogenous wastes, water, and small amounts of other waste products.

_____ 25. This is the cup-shaped space which collects filtrate and serves as the entrance to the each nephron.

a.	Kidney stones	j.	Urea	s.	Lungs
b.	Aldosterone	k.	Tubular secretion	t.	Urinary bladder
c.	Kidneys	l.	Bile pigments	u.	Creatinine
d.	Urethra	m.	Nephron	v.	Excretion
e.	Nitrogenous waste	n.	Urine	w.	Uric acid
f.	Elimination	o.	Filtrate	x.	Dialysis
g.	Ureter	p.	Renal failure	y.	Filtration
h.	Bowman's capsule	q.	Urinary system		
i.	Selective reabsorption	r.	Antidiuretic hormone		

LABELING

Label the figures below.

Fig. 12-2 Human kidney.

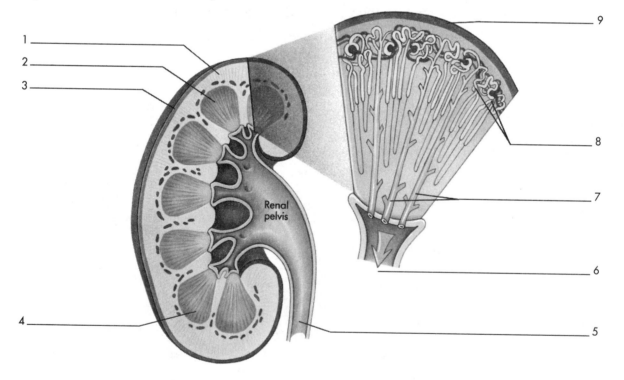

Fig. 12-3   Nephron.

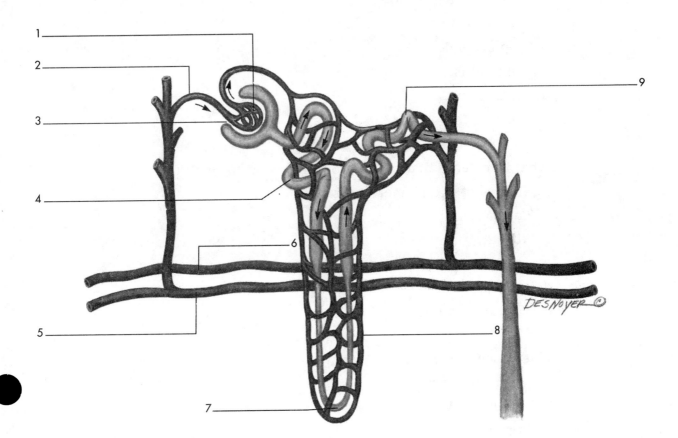

MULTIPLE CHOICE QUESTIONS

_____   1.   Which of the following is NOT an excretory organ?
a. Liver
b. Heart
c. Lungs
d. Large intestine
e. Kidney

_____ 2.	Which of the following structures is most closely associated with the process of filtration?

a. Collecting duct

b. Loop of Henle

c. glomerulus

d. Renal pelvis

e. Ureter

_____ 3.	What is material that passes from the blood through the capillaries of the glomerulus called?

a. Plasma

b. Filtrate

c. Urine

d. Urea

e. Formed elements

_____ 4.	Where does reabsorption begin?

a. Bowman's capsule

b. Descending loop of Henle

c. Glomerulus

d. Ascending loop of Henle

e. Proximal convoluted tubule

_____ 5.	Which of the following is found in the medullary region of the kidney?

a. Glomerulus

b. Proximal convoluted tubule

c. Bowman's capsule

d. Loop of Henle

e. Distal convoluted tubule

_____ 6. The pH of the blood is maintained by which of the following?
a. Process of filtration
b. Movement of water and urea in the loop of Henle
c. Tubular secretion of hydrogen ions into the filtrate
d. Tubular secretion of sodium ions into the filtrate
e. Concentration of mineral ions

_____ 7. What does aldosterone promote?
a. Retention of sodium, chlorine, and potassium
b. Retention of sodium and the excretion of chlorine
c. Retention of sodium and the excretion of potassium
d. Excretion of potassium and sodium
e. None of the above

_____ 8. Which of the following is the last structure that urine passes through as it leaves the excretory system and exits the body?
a. Loop of Henle
b. Urinary bladder
c. Urethra
d. Ureter
e. Renal pelvis

_____ 9. Which of the following statements is FALSE?
a. Any glucose detected in urine samples means kidney disease.
b. Blood pH imbalance can be associated with kidney dysfunction.
c. Excessive protein in urine can denote kidney disease.
d. High blood pressure and kidney disease are often related.
e. None of the above

_____ 10. Sodium ions are removed from the nephrons by which of the following?
a. Tubular excretion
b. Filtration
c. Active reabsorption
d. Passive reabsorption
e. Urination

_____ 11. Kidneys are considered organs of homeostasis. This is true because they perform which of the following functions?
a. Excrete nitrogen-containing waste materials
b. Assist in maintaining correct concentration of ions in blood
c. Help maintain plasma volume
d. Help regulate blood pH
e. All of the above

_____ 12. Where are the collecting ducts of the kidneys primarily found?
a. Medulla
b. Cortex
c. Ureters
d. Glomerulus
e. Loop of Henle

_____ 13. What is the force that drives filtration to occur?
a. Active transport
b. Blood pressure
c. Osmotic pressure
d. Atmospheric pressure
e. Diffusion

_____ 14. A person who is unable to produce enough ADH is subject to which of the following?
a. Decreased urine production
b. Increased blood volume
c. Too much glucose in urine
d. Production of too much urine
e. Ooss of control of protein excretion

SHORT ANSWER QUESTIONS

1. Describe how ADH regulates water balance in humans.

2. Explain why females tend to have more urinary tract infections than males.

3. The kidneys are organs of homeostasis. Explain why.

4. What prevents excessive loss of cells, proteins, glucose, etc., from being excreted in the urine?

CRITICAL QUESTIONS

1. The regulation of blood pressure is related to the effectiveness of the kidneys to control ionic concentration and water volume. Explain this statement.

2. Excessive exercise can deplete the body of necessary substances. What are these and what is the most effective way to replace them?

TELLING THE STORY

Develop a textual description of your understanding of the concept of excretion. Organize your information so that there is appropriate sequencing in the text. Write as though you were explaining your understanding to a classmate with the same general knowledge you have.

CHAPTER REVIEW ANSWERS

MATCHING

1. S	6. F	11. T	16. E	21. I
2. D	7. F	12. R	17. V	22. U
3. B	8. X	13. O	18. N	23. J
4. W	9. Y	14. G	19. A	24. C
5. K	10. L	15. M	20. Q	25. H

LABELING

Fig. 12-2

1. Cortex	4. Pyramid	7. Collecting ducts
2. Medulla	5. Ureter	8. Nephrons
3. Capsule	6. Renal pelvis	9. Capsule

Fig. 12-3

1. Glomerulus	4. Proximal convoluted tubule	7. Loop of Henle
2. Afferent arteriole	5. Artery	8. Capillary network
3. Bowman's capsule	6. Vein	9. Distal convoluted tubule

MULTIPLE CHOICE QUESTIONS

1. B	3. B	5. D	7. C	9. A	11. E	13. B
2. C	4. C	6. C	8. C	10. C	12. A	14. D

SHORT ANSWER QUESTIONS

1. Antidiuretic hormone (ADH) increases the collecting duct cells' permeability to water. By doing so, this increases the amount of water that is returned to the circulatory system, thereby preventing severe water loss.

2.	Due to the length of the urethra in females, the relationship between the bladder and the external environment is very close. Bacteria have a much better chance of gaining entrance and access to the urinary bladder and establishing infection.

3.	The functions of the kidneys go far beyond the production of urine. Because each kidney is responsible for maintaining the balance of water and ions (particularly sodium) in the body, they impact every other system. In this respect, the kidneys are very important homeostatic organs.

4.	The activities associated with the excretory process (filtration, reabsorption, and secretion) are mechanisms that occur through the membrane of the kidney tubules. Since membranes are essential components of the transport process, one of their roles is that of selectivity. Being selective about what crosses the membrane barrier allows each system within the body to have control over the maintenance, gain, or loss of materials. In this way, cells (red and white blood cells for example), necessary proteins and ions, and glucose are spared from being eliminated.

CRITICAL QUESTIONS

1.	The process of reabsorption and secretion, which both take place in the proximal convoluted tubule, provides a stable means of controlling necessary ion concentration and water volumes. The cells within these tubules actively transport sodium across the membrane into the spaces surrounding cells outside the tubules; chlorine ions are transported passively out of the cell to balance this increasing positive charge occurring and water ultimately follows by osmosis. This mechanism assures the appropriate plasma volume is maintained in the blood and therefore sufficient water is maintained in cells. Excessive water acquired would increase the plasma volume and increase blood pressure. The converse of this is also true.

2.   Increased sweat rates that occur for adequate evaporation during cooling can lead to excessive loss of water and electrolytes. The most serious consequences of excessive sweating is water loss. Excessive water loss can decrease plasma volume and can decrease the sweat rate causing the body temperature to be elevated. The best replacement fluid is one that contains as much ions and water as is lost through sweating. This is usually about 1 to 2 grams of salt per liter of water. Many commercial ergogenic aids are available for fluid replacement. Additionally, good hydration habits during exercise will prevent excessive water loss and potential distress to the body. Water should never be restricted during any type of physical exercise.

CHAPTER 13

NERVE CELLS AND HOW THEY TRANSMIT INFORMATION

KEY CONCEPTS OVERVIEW

1.   Communication Systems

Since the human body is composed of trillions of cells and these cells form structures that collaboratively work toward the survival of the individual, it makes sense that some form of communication must exist among them. Because of the importance of cell-to-cell communication, there are several mechanisms used. One means is **direct contact**. Open channels between adjacent cells allow certain ions and other small molecules to freely move from one cell to another. While this method would not be very effective for cells lacking proximity to one another, it is a necessary process for those that do.

Another form of communication occurs by means of chemical communicators or **hormones**. These chemical signals are produced by one of several endocrine glands, secreted into the blood, and then transported to specific target tissues in the body. While this is another important means of communication, it could not serve all the communication needs because it occurs too slowly for certain types of information.

The nervous system is designed to facilitate all "quick" information needs of your body. This system is composed of nerve cells, called **neurons**, that transmit signals throughout the body. Just like phone cables, these cells are bundled in groups, called **nerves**, that allow certain areas to be serviced without having to alert all areas. The information transmitted through these bundles of cells are called **nerve impulses** and vary in their frequencies, points of origins, and destinations. Each nerve is controlled from the central command post within the nervous system - the **brain**. This mass of interconnected neurons is connected to both the hormone-producing glands and the individual muscles and other tissues by connecting pathways of neurons.

2. Nerve Cells

The fundamental unit of the nervous system responsible for transmitting nerve impulses is the neuron. Structurally, a neuron consists of a **cell body** and cytoplasmic extensions called **dendrites** and **axon**. The cell body contains a nucleus and other organelles. The dendrites bring messages to the nerve cell body and the axons carry information away from the cell body. While the structural arrangement of neurons are not all the same, these differences can be accounted for by the functions each type of neuron has and its location in the body.

Fig. 13-1 Types of neurons.

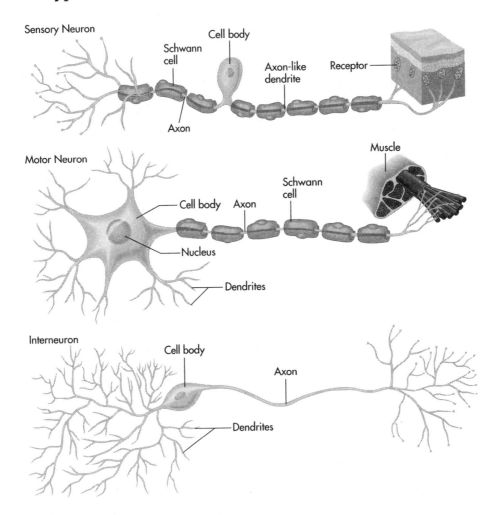

There are three types of neurons. The **sensory neurons** carry sensory information (gathered by sensory organs) from the receptor sites to the central nervous system. The **interneurons**, which are located in the central nervous system and spinal cord, organize and facilitate the information that has been transmitted by the sensory neurons. This information will then be sent to the spinal cord and brain. The **motor neurons** are pathways for information being carried from the brain and spinal cord to the **effectors** (glands and muscles).

Companion cells, called neuroglia, provide support and protection for neurons. These types of cells can be found within and outside the central nervous system. For example, the **Schwann cells** are types of neuroglia that wrap around many of the axons of sensory and motor neurons. Since these neurons are outside or peripheral to the spinal cord and brain, this part of the nervous system is called the **peripheral nervous system**. The configuration of the Schwann cell wrapped around the axon can be seen in the following diagram. Schwann cells contain a fatty material called **myelin** that provides an insulation for the wrapped neuron, hence the name **myelin sheath**. The spaces between the Schwann cells are bare and referred to as the **nodes of Ranvier**. These nodes and the myelin sheaths provide conditions that increase the rate of nerve impulse conduction by allowing the impulse to jump from one node to the next. This type of neural impulse is called **saltatory conduction**.

Fig. 13-2 Myelinated axon.

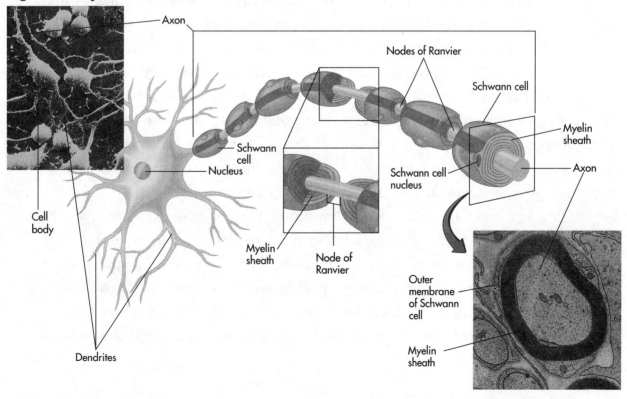

3. Action Potential

The neuron has certain characteristics that make it easier to understand how impulses are transmitted. At rest, the internal side of the membrane is negatively charged and the external side is positively charged as a result of the action of **sodium-potassium pumps.** These pumps are **transmembrane proteins** that assist in moving potassium ions (K^+) into the neuron and sodium ions (Na^+) out of the neuron. Potassium ions diffuse back into the neuron through special "open gates" in the membrane; however, the sodium gates are not open and therefore prevent sodium from reentering. Additionally, there are negatively charged chlorine ions (Cl^-) within the neuron that add to the internal negative charge. Given these conditions, there is a net charge, called the **resting potential,** on the inside of the neuron that measures -70 millivolts. The resting neuron is said to be **polarized** because there is a charge difference between the external and internal aspects of the neuron.

When a stimulus is provided that is of sufficient nature, a neuron will transmit an impulse. To be sufficient enough to cause an impulse to be transmitted the stimulus must meet or exceed the **threshold potential.** Threshold potential assures that the sodium channels in the membrane will open. This process is called **depolarization** because as sodium ions are allowed back into the neuron, the internal charge changes from negative to positive and the potential can be measured at 40 millivolts. The rapid change in the membrane's potential is referred to as the **action potential** and lasts for only a few milliseconds because sodium gates close very quickly. They remain closed until the sodium-potassium pump reestablishes the resting potential. This period of inactivity is known as the **refractory period** and no new impulse can be transmitted during this time. Nerve impulse transmission occurs in an **all-or-nothing response,** meaning that the stimulus must be sufficient enough to open enough sodium channels (threshold) and generate an action potential. All action potentials are the same; there is no gradation in intensity. However, they may be transmitted at different speeds.

Propagation of an action potential means that the impulse must travel along the dendrites, through the cell body, and down the axon of the neuron until it reaches another neuron or an effector. On neurons with Schwann cells surrounding their axons, an action potential can only be generated at the nodes of Ranvier. The ion pumps are concentrated in these areas and enhance ion movement across the membrane at these locations. When the nerve impulse reaches the end of an axon,

Fig. 13-3 Propagation of an action potential.

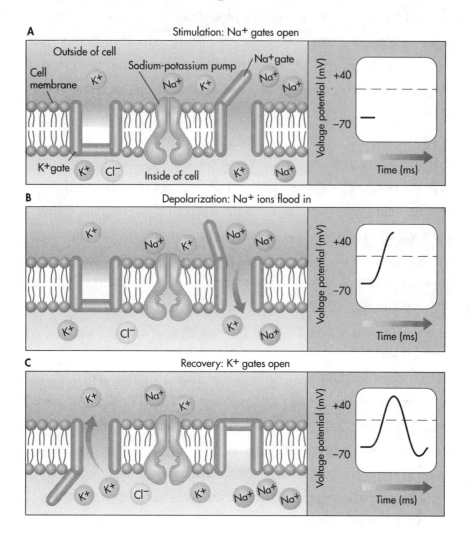

there is a gap between the transmitting neuron and the receiving neuron or effector. This gap is called a **synapse** (see section 4 of Key Concept Review). In humans specialized chemicals called **neurotransmitters** are used to bridge the synapse. When the impulse reaches the synapse, it triggers the release of neurotransmitters from vesicles near the membrane. This side of the membrane is called the **presynaptic membrane**. The neurotransmitter binds to receptors across the synapse on the **postsynaptic membrane** causing the sodium gates on the receptor neuron or effector to open. This allows the continuation of the action potential or results in the depolarization of an effector, such as a muscle cell.

Neurotransmission can be altered by some chemicals. Some drugs work by decreasing the amount of neurotransmitter that is released from a presynaptic membrane. Others cause neurotransmitter molecules to leak out of their container vesicles and be degraded by enzymes. There are also drugs that increase the amount of neurotransmitter or its effects at the synapse. And there are other drugs that chemically resemble specific neurotransmitters, bind to their normal receptor sites and prevent the appropriate neurotransmitter from attaching to the site.

4.   Synapses

The impulse from one neuron to another can be accomplished by several different neurotransmitters. Some of these neurotransmitters depolarize the postsynaptic membrane. This type of synapse is an **excitatory synapse.** Other neurotransmitters reduce the ability of the postsynaptic membrane to depolarize. This type of synapse is known as an **inhibitory synapse.** The postsynaptic neuron is an integrator that keeps track of these various signals. The summed effect of the inhibitory and excitatory signals either facilitate depolarization or inhibit it.

5.   Psychoactive Drugs

Drugs within this class are chemical substances that affect neurotransmitter transmission in specific parts of the brain. Chronic use of these drugs can lead to addiction. These chemicals are classified into four broad areas: depressants, opiates, stimulants, and hallucinogens. **Depressants** (sedative-hypnotics, alcohol) work by slowing down the central nervous system activities. **Opiates** (opium, morphine) are derived from the secretions of the poppy plant or can be synthetic derivatives. These drugs stop or reduce pain without causing a person to lose consciousness by chemically mimicking endorphins. **Stimulants** (amphetamines, cocaine) enhance the activity of two specific neurotransmitters: norepinephrine and dopamine, both function in brain pathways that regulate emotions, sleep, attention, and learning. **Hallucinogens** (LSD, marijuana), or psychedelic drugs, produce sensory perceptions that have no external stimuli. It is thought that these types of drugs disrupt the sorting process and allow inappropriate sensory data to overload the brain.

CHAPTER REVIEW ACTIVITIES

CONCEPT MAPPING

Construct a concept map of your understanding of nerve impulse transmission in the human body. Include as many of the new terms in the key concept review as possible in your map. Try to incorporate both causes and effects.

MATCHING

Using the list provided, select the term that best relates to the following statements. Focus your attention on your level of confidence as you select your answer. If you are unable to answer a selection appropriately, review the related concept in your text.

_____ 1. This is the fatty wrapping created by multiple layers of many Schwann cell membranes.

_____ 2. These are neurons that are specialized to transmit information to the central nervous system.

_____ 3. This is an electrical potential difference along the membrane of a neuron that is not transmitting an impulse.

_____ 4. These drugs enhance the activity of two neurotransmitters: norepinephrine and dopamine, resulting in an alerting effect.

_____ 5. No myelin is found at this site on a myelinate axon.

_____ 6. These projections extend from a neuron that conducts impulses away from the cell body.

_____ 7. These nerve cells, located within the brain or spinal cord, integrate incoming messages with outgoing messages.

_____ 8. These compounds are derived from the milky juice of the poppy plant and act as narcotic analgesics, reducing or stopping pain without causing the loss of consciousness.

_____ 9. This chemical signal is released from cells within endocrine glands and targeted for specific tissues elsewhere in the body.

_____ 10. This describes the decrease in the effects of the same dosage of a drug in a person who takes the drug over time.

_____ 11. Neuroglial cells wrap around the axon of some neurons in the peripheral nervous system, providing insulation and increasing the rate of neural impulse transmission.

_____ 12. This is the rapid change in a membrane's electrical potential caused by the depolarization of a neuron to a certain threshold.

_____ 13. These drugs slow down the activity of the central nervous system.

_____ 14. This space or gap between two adjacent neurons, acts as a bridge for nerve impulses to cross.

_____ 15. These projections extend from neurons and act as antennae for the reception of nerve impulses. Then they conduct these impulses to the cell body.

_____ 16. This is the level of nerve membrane depolarization in which sodium-specific channels in the membrane open, allowing sodium ions to diffuse into the cell.

_____ 17. Neurons are specialized to transmit commands away from the central nervous system toward an effector.

_____ 18. These chemicals are released when nerve impulses reach the axon tip of a neuron and act as chemical bridges to facilitate the continuation of the neural impulse.

_____ 19. These enzymes are embedded within cell membranes of neurons that transport sodium and potassium ions across the membrane.

_____ 20. Psychedelic drugs cause sensory perceptions that have no external stimuli.

TERMS LIST

a. Neurotransmitters	h. Sensory neurons	o. Resting potential
b. Depressants	i. Schwann cells	p. Hormones
c. Node of Ranvier	j. Hallucinogens	q. Dendrites
d. Threshold potential	k. Drug tolerance	r. Action potential
e. Myelin sheath	l. Stimulants	s. Interneurons
f. Opiates	m. Synaptic cleft	t. Axons
g. Na^+-K^+ pumps	n. Motor neurons	

MULTIPLE CHOICE QUESTIONS

_____ 1. Which of the following nerve cells is found only within the central nervous system?

a. Motor neuron

b. Interneuron

c. Sensory neuron

d. Schwann cell

e. All the above are found only in the CNS

_____ 2. Which of the following cell types would NOT conduct a neural impulse?

a. Schwann cell

b. Motor neuron

c. Sensory neuron

d. Interneuron

e. None of the above

_____ 3. A disease associated with the destruction of the myelin sheath is known as which of the following?

a. Graves disease

b. Myastenia gravis

c. Parkinson's disease

d. Polio

e. Multiple sclerosis

_____ 4. Which of the following is true about a saltatory conduction in a neuron?

a. It involves rapid impulse transmission by jumping.

b. It allows depolarization at the nodes of Ranvier.

c. It occurs on myelinated axons.

d. It can occur in both sensory and motor neurons.

e. All of the above.

_____ 5. An action potential is the result of which of the following?

a. Influx of sodium and potassium

b. Efflux of sodium and potassium

c. Efflux of potassium and influx of sodium

d. Efflux of sodium and influx of potassium

e. Efflux of sodium and influx of chlorine

_____ 6. What does the electrical potential of a nerve cell at rest measure?

a. -70 mV

b. 40 mV

c. 125 mV

d. -85 mV

e. 70 mV

_____ 7. The movement of sodium and potassium ions across the nerve cell membrane involves which of the following?

a. Active transport

b. Osmosis

c. Simple diffusion

d. Facilitated diffusion

e. Phagocytosis

_____ 8. Which of the following statements are true of neurotransmitters?

a. They are transported in vesicles to the end of an "active" neuron.

b. They are released into the synaptic cleft.

c. They bind to specific receptors on the postsynaptic membrane.

d. A, B, and C

e. B and C only

_____ 9. Which of the following ions contribute negative charges to a neuron?

    a. Sodium

    b. Magnesium

    c. Chlorine

    d. Potassium

    e. Hydrogen

_____ 10. Which of the following describes acetylcholine?

    a. Hallucinogen

    b. Neurotransmitter

    c. Depressant

    d. Opiate

    e. Stimulant

LABELING

Label the following diagrams by supplying the correct terms.

Fig. 13. 4 Presynaptic neuron.

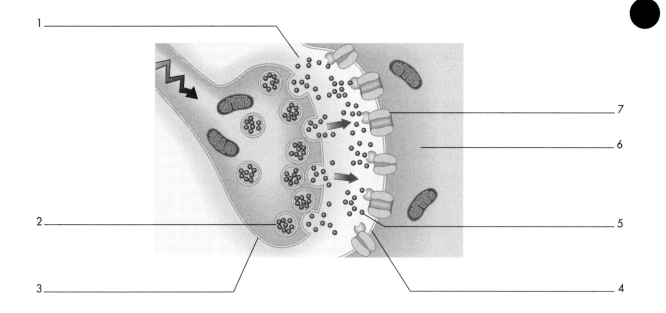

Fig. 13-5 Cell to cell impulse transmission.

Using the events below, identify what happens at each numbered site.

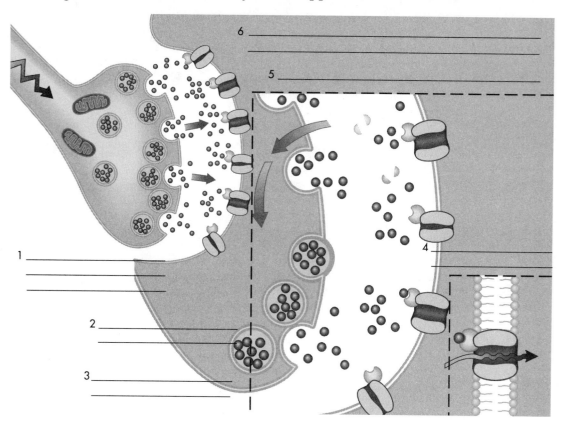

a. _____ Acetylcholine is released from vesicles.

b. _____ Influx of sodium causes postsynaptic action potential.

c. _____ Acetylcholine is resynthesized.

d. _____ Acetylcholine binds to receptors, causing sodium channels to open.

e. _____ Acetylcholine is broken down by enzyme (acetylcholinesterase) into acetic acid and choline, then choline is transported back to presynaptic cell.

f. _____ Presynaptic action potential arrives at synapse.

Fig. 13.6A Sensory neuron.

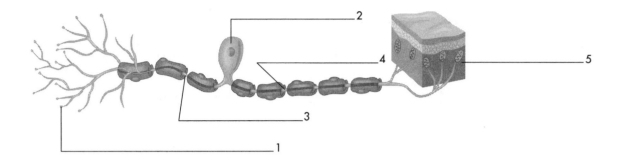

Fig. 13.6B Interneuron.

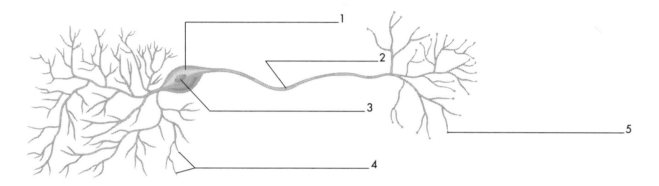

Fig. 13.6C Motor neuron.

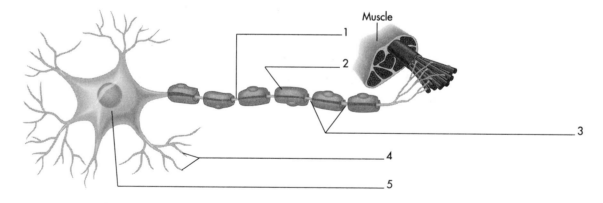

SHORT ANSWER QUESTIONS

1.   What are some distinguishing differences between dendrites and axons?

2. What determines if a neuron will transmit a neural impulse or depolarize?

3. Define the all-or-none law. Why doesn't this law apply to nerves?

4. Why is the conduction speed greater in myelinated axons?

5. Distinguish between a neuron and a nerve.

6. Distinguish between efferent pathways and the afferent pathways with regard to impulse transmission.

CRITICAL QUESTIONS

1. Neurons that release acetylcholine are called cholinergic neurons. One of the cholinergic systems in the brain has a major role in memory and learning. Individuals with Alzheimer's disease exhibit deterioration of the neurons in this system. How would decreased acetylcholine result in the symptoms associated with this disease?

2. How does nerve gas work? Give an example of the effect.

TELLING THE STORY

Write a description of your understanding of nerve impulse transmission. Be sure to organize your thoughts before you begin constructing your text and incorporate as many of the related terms from Chapter 13 as possible. Focus on the areas where you have difficulty constructing meaning in your writing and review that material again.

CHAPTER REVIEW ANSWERS

MATCHING

1. E	5. C	9. P	13. B	17. N
2. H	6. T	10. K	14. M	18. A
3. O	7. S	11. I	15. Q	19. G
4. L	8. F	12. R	16. D	20. J

LABELING

Fig. 13.4

1. Synaptic cleft	5. Neurotransmitter molecules
2. Presynaptic vesicles	6. Effector
3. Presynaptic membrane	7. Receptor molecule
4. Postsynaptic membrane	

Fig. 13.5

1. F	3. D	5. E
2. A	4. B	6. C

Fig. 13.6

SENSORY NEURON	INTERNEURON	MOTOR NEURON
1. Presynaptic membrane	1. Cell body	1. Axon
2. Cell body	2. Axon	2. Schwann cell
3. Axon	3. Nucleus	3. Nodes of Ranvier
4. Dendrites	4. Dendrites	4. Dendrites
5. Receptor cells	5. Presynaptic membrane	5. Cell body

MULTIPLE CHOICE QUESTIONS

1. B	3. E	5. C	7. A	9. C
2. A	4. E	6. A	8. D	10. B

SHORT ANSWER QUESTIONS

1.   Dendrites receive impulses from another neuron and transmit it toward the cell body. Axons receive impulses from the cell body and carry them away toward another neuron or an effector. Dendrites form branched cytoplasmic

extensions from the cell body and increase the cell's receptive surface area. The axon is a single extension from the cell body and may have branched collaterals along its length.

2. In order for a neuron to transmit an impulse, several requirements must be met. First, the neuron must receive a stimulus when it is not being depolarized or in what is called absolute refractory. Secondly, the action potential must be of sufficient strength, called threshold potential, to initiate depolarization through the entire neuron. Thirdly, sufficient sodium and potassium ions must be present and able to be transported across the membrane as the appropriate channels are opened.

3. For an action potential to be initiated, the stimulus must exceed the threshold potential for that neuron. The all-or-none law states that a stimulus must reach a critical level to cause the impulse to begin being conducted down the length of the neuron. An impulse stronger than the threshold potential will not cause a stronger impulse.

4. The conduction speed is increased in a myelinated axon because the amount of surface membrane that must depolarize is significantly reduced due to the presence of the neuroglial cells (Schwann cell). Since these cells insulate the membrane and leave only small bare spaces, known as the nodes of Ranvier, depolarization occurs only at these intervals and the impulse travels much faster.

5. A neuron is a single nerve cell. Structurally, it can be a unipolar, bipolar, or multipolar type. A nerve is a group or bundle of neurons.

6. Afferent pathways transmit impulses toward the central nervous system. Efferent pathways transmit impulses from the central nervous system to effectors, such as muscles or glands.

CRITICAL QUESTIONS

1. Alzheimer's disease is a brain disease that is usually age-related and is the most common cause of diminishing intellectual function in older people. It is reported to affect 10% to 15% of humans over the age of 65 and 50% of humans over the age of 85. Due to the degeneration of cholinergic neurons, the

symptoms associated with the disease are thought to be caused by a decrease in the amount of acetylcholine in certain areas of the brain. If no acetylcholine is released because the neuron is damaged, then impulses are not able to be transmitted to associated neurons and the impulse pathway is disrupted. Postsynaptic neurons are then lost because they receive no stimulus. These defects are presumably related to the decline in language, perceptual abilities, clear thinking, and memory.

2. Nerve gas blocks the action of acetylcholinesterase. This enzyme destroys acetylcholine and without it, continual neuromuscular depolarization occurs. The results are continuous contraction of muscles and no opportunity for relaxation. If, for example, the diaphragm could not relax, it would be impossible to exhale.

CHAPTER 14

THE NERVOUS SYSTEM

KEY CONCEPTS OVERVIEW

1.   Nervous System Organization

The nervous system regulates many internal functions and organizes and controls activities that we collectively refer to as human behavior. The activities are very diverse, some more obvious than others. The various parts of the nervous system are networked and interconnected in a very complex way. To simplify this web of nerve cells, the system is divided into two branches: the **central nervous system (CNS)** and the **peripheral nervous system (PNS)**. The CNS is composed of the brain and spinal cord and the PNS consists of the nerves that extend from the brain and spinal cord out to all areas of the body. The brain consists of four basic parts. The cerebrum, the cerebellum, the diencephalon, and the brainstem. Because of the critical and sensitive nature of the CNS, it is protected by three layers of membranes known as the meninges. There are no nerves in the central nervous system. A group of nerve fibers in the CNS is called a pathway or a **tract**. Cell bodies that have similar functions are generally clustered together and, within the CNS, are usually called nuclei.

The nerves of the peripheral nervous system are made up of neurons of two types: **sensory** or **afferent neurons** that carry messages toward the CNS, and the **motor** or **efferent neurons** that carry messages away from the CNS to an effector. These **effectors** are either muscles or glands within the body.

Within the PNS, further subdivisions help in understanding the level of organization. Motor neurons that control voluntary responses make up what is known as the **somatic nervous system.** The term somatic means "body." Motor neurons that control involuntary responses make up the **autonomic nervous system.** Since there are very complex control mechanisms that must work together to regulate involuntary functions, the **autonomic nervous system** can be divided into two distinct branches: the **sympathetic nervous system,** which generally is associated with excitatory responses, and the **parasympathetic nervous system,** which is associated with inhibitory responses.

2.   Central Nervous System

The human brain contains billions of neurons that collaboratively form very intricate structures. One of these is the cerebrum. For organizational purposes, the cerebrum is divided into right and left halves or **hemispheres** connected by a thick bundle of nerve fibers called the **corpus callosum.** This "bridge" allows information to pass from one side of the brain to the other. Although each of these hemispheres have motor, sensory, and association areas, each is responsible for different activities. The **right side** is associated with spatial relationships, artistic skills, and expression of emotions. The **left side** is generally associated with verbal and analytical skills. The right hemisphere controls functions on the left side of the body and the left hemisphere controls functions on the ride side. Each hemisphere can be further divided into specific regions: the **frontal,** the **parietal,** the **temporal,** and **occipital lobes.** By identifying these regions, it allows us to "map" the functional control areas of the cerebrum. For example, the **motor area**, which sends signals to effectors, is located in the most posterior portion of the frontal lobe. Sensory areas are dispersed. For example, speech centers are located in the parietal and frontal lobes, hearing and smell are located in the temporal lobes and vision centers are in the occipital lobes. The remaining areas of the brain are association areas, that is, areas of the brain that perform higher cognitive activities.

Most of these activities take place in the **cerebral cortex,** which is a thin layer of tissue that covers the surface of the cerebrum. Within the cerebral cortex, there are tightly packed neurons that appear dark in color. Hence the term "**gray matter.**" The outer surface of the cerebral cortex is highly convoluted, greatly increasing the available surface area. Deep within the cerebrum are masses of gray matter called basal ganglia that control the subconscious movements of the skeletal muscles. Additionally, there is a layer of tissue beneath the cortex made up of a "highway" of myelinated tracts. These pathways, referred to as "**white matter**" (because of the white, fatty covering), run in three directions: (1) from a place in one hemisphere to another place in the same hemisphere, (2) from a place in one hemisphere to a corresponding place in the other hemisphere, and (3) to and from the cerebrum to other parts of the brain and spinal cord.

The **diencephalon** is composed of two important structures, the thalamus and the hypothalamus. The **thalamus** routes sensory information to the appropriate areas of the brain and monitors the body's internal environment. The **hypothalamus** is the control center for homeostatic mechanisms. It controls hormonal secretions from the pituitary gland. There is an obvious need for the hypothalamus to communicate with the cerebral cortex. This is accomplished through a network of tracts called the **limbic system**. This system is the center for drives and emotions associated with sex, hunger, thirst, anger, pain, and pleasure.

The **cerebellum**, which is located just beneath the occipital lobes of the cerebrum, is the center for coordinating subconscious activities associated with skeletal muscles. Damage to this area would result in reduced motor function.

The **brainstem** is the oldest portion of the brain and certainly a very vital part. It consists of the **midbrain**, the **pons**, and the **medulla**. Twelve pairs of cranial nerves provide pathways for messages from the brainstem to the spinal cord and vice versa. Nuclei located through the midbrain are responsible for reflexive activities such as breathing and heart rate. The **reticular formation**, a web of neurons located throughout the midbrain, monitors incoming information and routes important messages to the correct part of the brain. The reticular formation also plays a role in maintaining consciousness.

Tracts of neural fibers (white matter on the surface, gray matter on the interior), collectively called the **spinal cord**, are located within the vertebral column. The cord is hollow in the center and contains cerebrospinal fluid that bathes the brain and offers some protection. This fluid is also found on the outside of the CNS between the brain and spinal cord and the meninges. The myelinated tracts receive information from the peripheral nervous system and pass it up to the brain through **spinal nerves**. Likewise, information coming from the brain is transmitted through the spinal cord to the peripheral nervous system. Additionally, the gray matter integrates responses to certain types of stimuli. These pathways are called **reflex arcs**.

Fig. 14-1 Spinal cord cross-section.

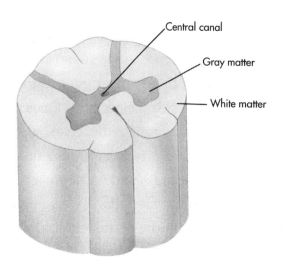

3.   Peripheral Nervous System

In order to assure that every part of the body can communicate with the CNS, the PNS provides a highly branched network of nerves that relay messages. A combination of neurons are needed to facilitate this relay system. Sensory neurons provide receptor sites that detect stimuli and pathways for this information to be sent to the CNS. Except for the simplest type of reflex arcs, information that comes from the brain is carried by interneurons to motor neurons. Motor neurons then transmit the information to effectors that respond to the initial stimulus. Most spinal nerves are mixed nerves that contain both sensory and motor neurons. These neurons however, separate before entering the spinal cord.

The somatic branch of the PNS contains motor pathways that relay messages to and from the brain and skeletal muscles. Reflex arcs are also part of this branch. Reflexes that involve only two neurons with a single synapse are called **monosynaptic reflex arcs.** These types of arcs do not require an interneuron and are not controlled by the CNS. Monosynaptic reflex arcs are relayed through the spinal cord. There are more complex types of reflex arcs , which would be controlled by the CNS.

The autonomic branch of the PNS controls all the involuntary responses of glands and smooth muscles. The sympathetic neurons generate the "fight or flight" response to stimuli, causing various notable signs, such as pupil dilation, increased blood pressure, pulse, and respiration. The parasympathetic system is antagonistic to the sympathetic system, providing a balancing mechanism to assist in maintaining homeostasis.

CHAPTER REVIEW ACTIVITIES

CONCEPT MAPPING

Develop a concept map that demonstrates your understanding of the nervous system. Constructing the map is an excellent way to visualize your understanding of the organization and significant components of the nervous system. Try to incorporate as many of the key terms as possible and select linking phrases that are not repetitive.

MATCHING

From the list provided, select the term that best relates to the following statements. Decide on a term BEFORE looking through the list, then search the list for the one you selected. If you are unsure of the correct answer or chose an incorrect response, review the related concept in Chapter 14.

_____ 1. This subdivision of the autonomic nervous system generally mobilizes the body for greater activity or enhances response.

_____ 2. These cranial nerves are made up of both sensory and motor neurons.

_____ 3. This part of the central nervous system, having white matter on the outer surface and gray matter on the interior aspect, consists of tracts that relay messages to and from the PNS and CNS.

_____ 4. This part of the brain consists of the thalamus and hypothalamus.

_____ 5. This thin layer of tissue consists of gray matter that forms the outer surface of the cerebrum and in which higher order cognitive processing occurs.

_____ 6. These are nerves by which the spinal cord receives information from the body.

_____ 7. This network of neurons is responsible for the drives and emotions associated with sex, hunger, thirst, pain, anger, and pleasure.

_____ 8. This material found within and around the CNS and serves as a protective mechanism that bathes the tissues with nutrients.

_____ 9. This is the largest and most dominant portion of the human brain.

_____ 10. These sections of the cerebral cortex house the center for hearing.

_____ 11. This branch of the peripheral nervous system consists of motor neurons that send messages to the skeletal muscles and control voluntary responses.

_____ 12. This single, thick bundle of nerve fibers connects the two hemispheres of the cerebrum.

_____ 13. These neurons are specialized to transmit information to the CNS.

_____ 14. This part of the brain is located beneath the occipital lobes and coordinates subconscious movements of the skeletal muscles.

_____ 15. These areas of the brain appear to be the sites of higher cognitive activities, such as planning and contemplation.

_____ 16. This is the lowest portion of the brainstem and houses the center for respiration and heart rate.

_____ 17. This branch of the PNS consists of motor neurons that control the involuntary and automatic responses of the glands and the nonskeletal muscles.

_____ 18. Any of three membranes covering the CNS are known by this term.

_____ 19. Any of the twelve pair of nerves that enter the brain through holes in the skull are known by this term.

_____ 20. These nerve cells found in the brain and spinal cord are situated between other neurons and can receive and transmit messages.

TERMS LIST

a. Temporal lobes	h. Cerebral cortex	o. Corpus callosum
b. Mixed nerves	i. Spinal nerves	p. Sensory nerves
c. Cerebellum	j. Cerebrum	q. Cranial nerves
d. Diencephalon	k. Medulla	r. Limbic system
e. Interneurons	l. Spinal cord	s. Cerebrospinal fluid
f. Meninges	m. Somatic nervous system	t. Sympathetic nervous system
g. Association areas	n. Autonomic nervous system	

IF.... THEN.... CHALLENGE QUESTIONS

Using the "IF" part of the statements below, complete the "THEN" part with guidance from the accompanying questions.

1. If the threshold potential is 15 mV and Cell A receives a stimulus equal to that, and Cell B receives one of greater potential, then _____. What effect will the strength of the stimulus have on the force of the action potential?

2. If an individual is described as being predominantly "right brain," then _____. What predominant characteristics will be demonstrated?

3. If damage occurs to the inferior (lower) aspect of the frontal lobe, then _____. What sensory information might be compromised?

4. If injury occurs to the left side of the cerebrum, then _____. What processes might be interrupted?

5. If the ganglia associated with the thalamus and hypothalamus are damaged, then _____. What might be the resulting effects?

6. If the exterior of the cerebrum is "gray" due to densely packed cell bodies, then _____. Why is the exterior of the spinal cord "white?"

7. If an individual loses consciousness due to a sharp twist of the head, then _____. What area of the brain probably has been affected?

8. If a spinal tap is ordered for a patient, then _____. What type of material will be collected and from where will it be taken?

9. If an individual loses coordination of subconscious movement, then _____. What area of the brain is most likely involved?

LABELING

Fig. 14-2 Flow chart of the nervous system.

Complete the organizational flow chart of the human nervous system.

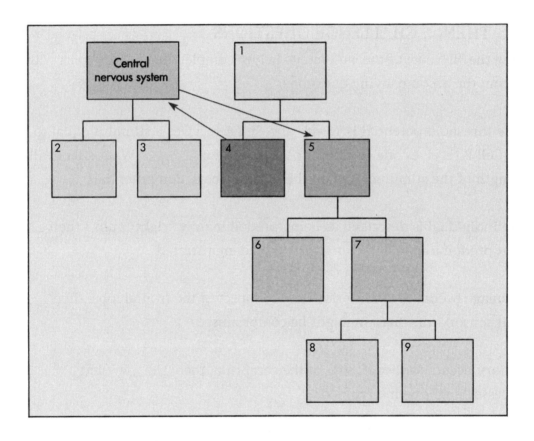

Fig. 14-3 Human brain.

Label the diagram of the human brain below.

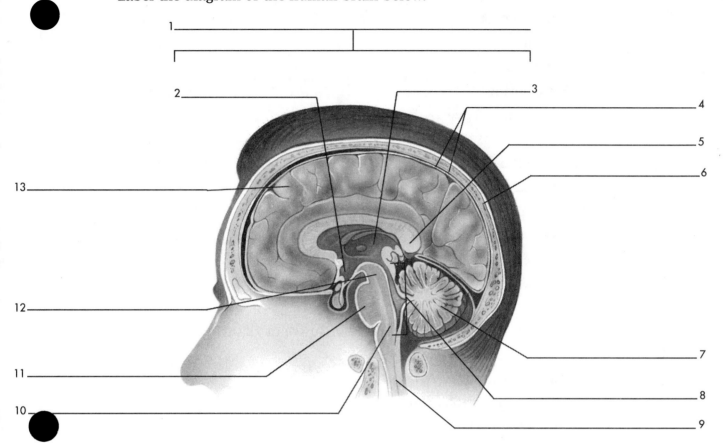

MULTIPLE CHOICE QUESTIONS

_____ 1. Which one of the following abilities is most assoicated with the left side of the brain?

a. Spatial relationships

b. Musical ability

c. Logical activities

d. Control of the left side of the body

e. Artistic ability

_____ 2. Alzheimer's disease seems to be associated with the excess accumulation of which of the following?

   a. Amyloid protein
   b. Myelin
   c. Dopamine
   d. Epinephrine
   e. None of the above

_____ 3. Which of the following would NOT be found in the brainstem?

   a. Medulla
   b. Midbrain
   c. Pons
   d. Cerebellum
   e. None of the above

_____ 4. Which of the following is the MOST inclusive of all others in the list?

   a. Somatic nervous system
   b. Parasympathetic nervous system
   c. Autonomic nervous system
   d. Peripheral nervous system
   e. Sympathetic nervous system

_____ 5. Which of the following would NOT be considered one of the main regions of the brain?

   a. Cerebrum
   b. Cerebellum
   c. Corpus callosum
   d. Brainstem
   e. Diencephalon

_____ 6. Which of the following would be found in the diencephalon?

   a. Midbrain
   b. Reticular formation
   c. Pons
   d. Corpus callosum
   e. Thalamus

7. Which of the following statements is true of the hypothalamus?

   a. It functions to connect the two cerebral hemispheres.

   b. It is an area of the brain that inhibits sensory input.

   c. It is the seat of logical reasoning.

   d. It is the verbal communication center.

   e. It is the center for homeostasis for the body.

8. Which of the following is mismatched?

   a. Motor area of the cerebral cortex - parietal lobe

   b. Auditory - temporal lobe

   c. Sensory area of cerebral cortex - occipital lobe

   d. Olfaction - frontal lobe

   e. Vision - occipital lobe

9. Proportionally speaking, which area of the sensory and motor regions of the cerebral cortex has the greatest area devoted to it?

   a. Nose

   b. Hand and fingers

   c. Eye

   d. Trunk

   e. Genitals

10. Which of the following is NOT one of the lobes of the cerebral cortex?

   a. Frontal

   b. Sagittal

   c. Parietal

   d. Occipital

   e. Temporal

11. Which of the following best describes the ventricles of the brain?

   a. Lobes

   b. Nerve tracts

   c. Meninges

   d. Spaces

   e. Convolutions

_____ 12. Which part of the brain is responsible for deep-seated drives such as sex, hunger, thirst, anger, and pleasure?

    a.   Medulla

    b.   Pons

    c.   Limbic system

    d.   Corpus callosum

    e.   Hypothalamus

SHORT ANSWER QUESTIONS

1.   What is the general function of the medulla?

2.   What are the meninges and where would they be found?

3.   What are the four major parts of the brain?

4.   How does the brain differentiate among sensory nerve impulses?

CRITICAL QUESTIONS

1.   Some spinal injuries cause paralysis while others result in a loss in sensation in specific areas. How would you explain these varying results?

2.   Most motor neurons are covered with myelin. What do you think the adaptive advantages of this structural arrangement is?

3.   What would happen if your body was deprived of sensory input?

4.   Explain why spinal cord injuries result in varying levels of permanent damage.

TELLING THE STORY

Develop a text that demonstrates your understanding of the nervous system. Focus mainly on the organization and function of specific regions. Use as many terms from the chapter as you can. If you have difficulty with any concepts as you write, go back and review them.

CHAPTER REVIEW ANSWERS

MATCHING

1. T	5. H	9. J	13. P	17. N
2. B	6. I	10. A	14. C	18. F
3. L	7. R	11. M	15. G	19. Q
4. D	8. S	12. O	16. K	20. E

IF.... THEN.... CHALLENGE QUESTIONS

1.   The resulting action potentials will have the same force.
2.   The characteristics exhibited would be predominantly those of enhanced spatial orientation, artistic and musical ability, and emotional expression.
3.   The sense of smell would be effected.
4.   The left side of the cerebrum controls speech, writing, logical thought, and mathematical abilities.
5.   The result might be the loss of some types of subconscious control.
6.   The white exterior would be due to myelinated nerve fibers.
7.   The reticular formation in the brainstem would have been damaged.
8.   Cerebrospinal fluid would be taken from the lower portion of the CNS.
9.   The cerebellum would be damaged in some way.

LABELING

Fig. 14.2

1. peripheral	4. sensory	7. autonomic nervous system
2. brain	5. motor	8. sympathetic nervous system
3. spinal cord	6. somatic nervous system	9. parasympathetic system

Fig. 14.3

1. Diencephalon	6. Skull	11. Pons
2. Hypothalamus	7. Cerebellum	12. Midbrain
3. Thalamus	8. Brainstem	13. Cerebrum
4. Meninges	9. Spinal cord	
5. Corpus callosum	10. Medulla	

MULTIPLE CHOICE QUESTIONS

1. C	4. D	7. E	10. B
2. A	5. C	8. C	11. D
3. B	6. E	9. B	12. C

SHORT ANSWER QUESTIONS

1. In general terms, the medulla is the site where nerve tracts pass from the spinal cord to the higher levels of the brain. Additionally, it is the center of vital reflex actions involving respiration rate, heart rate, and other responses such as sneezing, coughing, and vomiting.

2. The meninges are membranes which surround the central nervous system. They serve to protect the brain and spinal cord by acting as barriers, providing space for vascularization, and providing a space for cerebrospinal fluid to circulate. There are three layers of membrane: the dura mater, the arachnoid layer, and the pia mater.

3. The four major parts of the brain are the cerebrum, brainstem, cerebellum, and diencephalon.

4. The brain is able to interpret the action potential as a specific sensory input based on the source of the information being received, the frequency of the impulses, the number of neurons involved, and by recognizing to which area of the brain the action potential is sent.

CRITICAL QUESTIONS

1. Certain injuries only affect limited areas of the brain. Observations of impared functions are often used to relate the type of damage incurred with the functions of the specific areas of the brain that have been injured.

2. Myelination not only provides a protective barrier for the axon, it also allows impulses to travel much faster. The adaptive effect is an organism that can transmit impulses quickly and coordinate movement with great agility and control.

3. Stimulation is an essential component of maintaining perceptive ability. If an individual is exposed to an environment where stimulation is not received, relaxation occurs. If continued for extended periods of time, the individual may not be able to receive or respond to stimuli appropriately.

4. Depending on the location of the injury, the nerve tracts of the spinal cord attach to different regions of the peripheral nervous system and to different

regions of the brain. Any disruption in the tract will cause the communication between the control center and the activity center to be cut off.

CHAPTER 15

SENSES

KEY CONCEPTS OVERVIEW

1.   Sensory Communication

To better understand the mechanisms associated with sensory communication, it is important to conceptualize what it is. When information is transmitted in the nervous system, it moves in the form of an action potential. When you hear the term *sensation*, you should think about a stimulus of sufficient magnitude being received through sensory receptors causing an impulse to move along sensory neurons toward the brain. This impulse travels through a very organized pathway to the CNS and is then routed to different areas of the brain to be interpreted. The interpretation or *perception* of this impulse is dependent on which areas of the brain receive it. The resulting awareness or effect becomes a conscious sensation that you realize.

The body has numerous sensory receptors dispersed throughout. Their specific locations are related to the type of reception they can receive. Sensory receptors provide the body with three types of information: (1) information regarding the body's internal environment, (2) information regarding the body's position in space, and (3) information regarding the external environment. It then becomes evident that the location of receptors is directly related to which of these tasks they perform. Changes in internal environment are received by receptors located deep within the body, such as the walls of organ structures and blood vessels. Changes in external environment are received by more surface type receptors. Those that provide position information are located within the muscles, tendons, joints, and inner ear.

2. Receptors

Sensory receptors are usually neurons that occur individually or in clusters within sensory organs. They function by collecting and transmitting information from various environmental stimuli. The sensory receptors that monitor the body's internal environment are simple nerve endings. Their job is to receive information regarding internal body temperature, the pH of body fluids, the presence of certain chemicals such as carbon dioxide, blood pressure, and pain. Remember that this information will be received in various parts of the brain so the necessary regulation and control mechanisms can be maintained.

The receptors that sense the body's positions in space are called **proprioceptors.** These tiny structures are found deep in muscle tissue and recognize the degree of contraction within the muscles. The receptors themselves are actually specialized muscle cells called **muscle spindles.** (See Figure 15.2, P. 308 of your text)

Each spindle has the ending of a sensory neuron, called a stretch receptor, wrapped around it. Stretching the muscle stimulates this nerve ending. There are also proprioceptors in the tendons and in the tissue surrounding the joints. They too are activated when stretching occurs. Other receptors are gravity and motion sensitive. These are located in the inner ear and help maintain balance, relative position, and equilibrium.

The sensory receptors that monitor the body's external environment are usually located at or near the surface of the body. These receptors are classified as either general receptors or special sense receptors. The **general receptors** monitor information regarding touch, pressure, pain, and temperature. The **special sense** receptors detect smell, taste, sight, hearing, and balance. Due to the nature of their role in the nervous system, the special sense receptors are very complex structures. Let's review their functions and characteristics.

- Smell: The sense of smell is called the olfactory sense and smell receptors are referred to as the olfactory receptors. These receptors are neurons located in tissue within the roof of the nasal cavity. Dendrites extend from the cell bodies of these neurons. The ends of these dendrites are ciliated and as atmospheric gases are dissolved in the mucus that bathes these neurons, the gas molecules bind to special receptor chemicals on the cilia causing the neuron to depolarize. The information travels in the form of an action potential to the olfactory region of the cerebral cortex. Additionally, this information is transmitted to the limbic system allowing you to interpret the smell and construct a "feeling" or sense about it.

- Taste: The sense of taste occurs because of the close to 9,000 **taste buds** located on tiny projections on the tongue called **papillae**. Each taste bud consists of 30 to 80 receptor cells that have cilia capable of sensing chemicals that dissolve in saliva after entering the mouth. These specialized cilia allow us to differentiate among sweet, salty, sour, and bitter tastes.

- Sight: The eye is a specialized sensory organ that is able to detect various wavelengths of light. Special cells called rods and cones, which are located in the retina, absorb light. Most of the cones are concentrated in a region of the retina called the **fovea**. There are three kinds of rod cells that absorb three different wavelengths of light-red, blue, and yellow wavelengths. **Rod cells** are able to function in dim light only, while **cone cells** function in bright light and are able to detect color. The absorption of light by these cells results in a process known as **hyperpolarization**. During hyperpolarization, the sodium channels close instead of opening. This results in the interior of the cells becoming even more negatively charged. Adjacent neurons are stimulated by this event. Each cone cell is attached to a **bipolar cell** and each of these are attached to a ganglion cell. It is the ganglion cell that transmits the visual stimulus information to the cerebral cortex through the **optic nerve**.   (See Figure 15-7, P.312 of your text)

- Hearing : The ear consists of three parts: the outer ear, middle ear, and inner ear. The **outer ear** is composed of the **pinna** and the **auditory canal**. The **tympanic membrane** or eardrum of the middle ear transmits sound waves to the small bones in the ear (**malleus, incus,** and **stapes**) that in turn conduct the waves to the inner ear by way of the **oval window**. The inner ear is a fluid-filled labyrinth of channels within the temporal bone on the sides of the skull. There are two large fluid-filled canals (the **vestibular canal** and the **tympanic canal**) within the **cochlea** or coiled entrance to the inner ear. These canals are separated by a small duct containing a structure called the **organ of Corti** that contains the receptor **hair cells**. Vibrations from found waves activate the hair cells and conduct nerve impulses through the **auditory nerve** to the specialized regions of the cerebral cortex. The louder the sound, the more hair cells are activated. The pitch of the sound depends on the wavelength or frequency of the sound waves. Different hair cells are activated by different frequencies.

- Balance : Balance and equilibrium are maintained by sense organs within the **vestibule** of the inner ear. Inside the vestibule are two chambers known as the **utricle** and **saccule** and three **semicircular canals.** These structures contain hair cells that project into a gelatinous material that contains calcium carbonate particles called **otoliths.** Otoliths always exert a pull on the hair cells in the direction of gravitational attraction. Movement relative to gravity initiates a stimulus and an impulse is then transmitted. Other hair cells located at the base of the semicircular canals (**ampula**) are projected into a gelatinous material called the **cupula.** Movement of these cells results from rotational changes in direction. Because each of the three semicircular canals is oriented in a different plane, movement in any plane can be detect. (See Figure 15-10, P. 317 and Figure 15-11, P. 318 of your text)

CHAPTER REVIEW ACTIVITIES

CONCEPT MAPPING

Develop a concept map that demonstrates your understanding of the sensory processes. Try to incorporate as many of the key terms as possible and select linking phrases that demonstrate how you congitively connect one component to another.

MATCHING

Select the term from the list provided that best relates to the following statements. Decide on a term BEFORE looking through the list, then search the list for the one you selected. If you are unsure of the correct answer or chose an incorrect response, review the related concept in Chapter 15.

_____ 1.   This term refers to the projected part or "flap" of the ear.

_____ 2.   This is the opening in the center of the iris through which light passes.

_____ 3.   This receptor is located within skeletal muclses, tendons, and the inner ear and gives the body information about the position of its parts relative to each other and to the pull of gravity.

_____ 4.   This is the collective term for the hair cells, the supporting cells of the basilar membrane, and the overhanging tectorial membrane in the cochlea.

_____ 5.   These three fluid-filled structures in the inner ear detect the direction of the body's movement.

_____ 6.   This is the central region of the retina and the location of the highest concentration of cones in the eye.

_____ 7.   This inch-long channel receives sound waves funneled from the pinna and carries them directly to the eardrum.

_____ 8.   This tough outer layer of connective tissue covers and protects the eye.

_____ 9.   These receptors located within the retina function in bright light and detect color.

_____ 10. This is the larger of two membranous sacs within the vestibule that functions in the maintenance of bodily equilibrium and coordination.

_____ 11. This rounded, transparent portion of the eye's outer layer permits light to enter the eye.

_____ 12. This bone in the middle ear resembles a stirrup.

_____ 13. These sensory neurons are sensitive to any stretching that occurs in the tissues in which they are located.

_____ 14. These projections on the tongue contain the taste receptors.

_____ 15. This is the smaller of two membranous sacs within the vestibule that functions in the maintenance of bodily equilibrium and coordination.

_____ 16. These light receptors are located within the retina and function in dim light and detect white light only.

_____ 17. This structure connects the middle ear with the nasopharynx and functions to equalize the air pressure on both sides of the eardrum.

_____ 18. This result event occurs when sodium gates are closed and the internal environment of a cell becomes increasingly more negative.

_____ 19. These neurons detect smells when different airborne chemicals bind with receptor chemicals.

_____ 20. This is the entrance to the inner ear.

TERMS LIST

a. Organ of Corti	h. Pupil	o. Papillae
b. Stretch receptors	i. Utricle	p. Fovea
c. Oval window	j. Hyperpolarization	q. Sclera
d. Pinna	k. Proprioceptor	r. Eustachian tube
e. Cones	l. Rods	s. Auditory canal
f. Saccule	m. Cornea	t. Olfactory receptors
g. Stapes	n. Semicircular canals	

LABELING

Label the diagrams below.

Fig. 15-1A Muscle spindle.

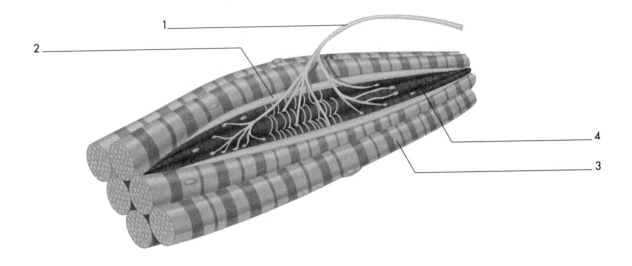

Fig. 15-1B Sensory receptors of the skin

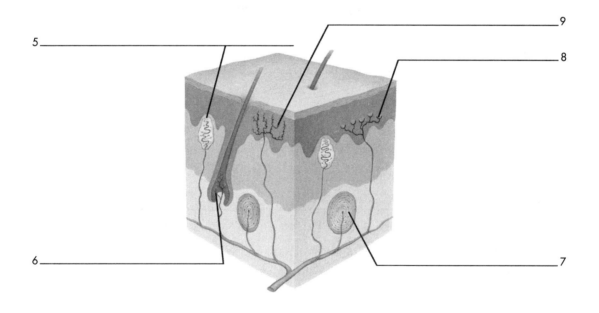

Fig. 15 -2 Inner ear.

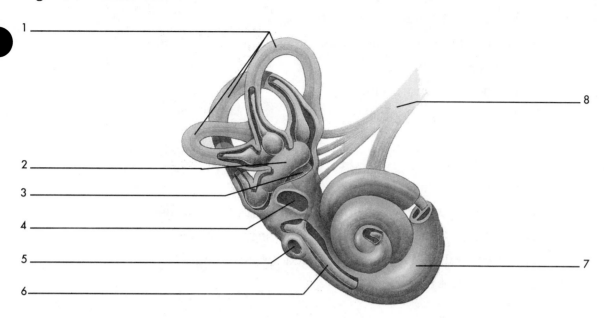

MULTIPLE CHOICE QUESTIONS

_____ 1. Which of the following is NOT a membrane structure?
      a. Eardrum
      b. Oval window
      c. Basilar membrane
      d. Round window
      e. All of the above are membranes

_____ 2. Which of the following is NOT considered one of the general senses?
      a. Pressure
      b. Touch
      c. Balance
      d. Pain
      e. Temperature

_____ 3. Which of the following pairs is mismatched?

　　a. Sight - electromagnetic radiation

　　b. Smell - chemoreceptor

　　c. Hearing - mechanoreceptor

　　d. Balance - thermoreceptor

　　e. Taste - chemoreceptor

_____ 4. Hyperpolarization is caused by which of the following?

　　a. Enhanced influx of sodium

　　b. Closed potassium gates and increased positive charges

　　c. Closed chlorine gates and increased negative charges

　　d. Closed sodium gates and increased negative charges

　　e. Cells cannot hyperpolarize

_____ 5. Which of the following monitors body temperature?

　　a. Cerebrum

　　b. Cerebellum

　　c. Hypothalamus

　　d. Pons

　　e. Medulla

_____ 6. Which of the following describes a receptor?

　　a. It is the first part of a reflex arc.

　　b. It is attached to a dendrite.

　　c. It is essential for the initiation of a sensory impulse.

　　d. A and B

　　e. A, B, and C

_____ 7. Where does nearsightedness cause an image to be focused?

　　a. Directly onto the retina

　　b. In front of the retina

　　c. At the blind spot

　　d. Behind the retina

　　e. On the optic nerve

_____ 8. Which of the following is detectable by specific taste buds?
   a. Sour
   b. Sweet
   c. Bitter
   d. Salty
   e. All of the above

_____ 9. What are otoliths?
   a. Calcium carbonate particles that slide within a gelatinous matrix
      When the head is moved
   b. Limestone formations in the cochlea
   c. Pull on cilia of hair cells and bend them
   d. A and C
   e. B and C

_____ 10. Which of the following processes occurs first?
   a. Rhodopsin absorbs a photon of light.
   b. Molecular shape of rhodopsin changes separating retinal and
      opsin.
   c. Neurotransmitters are released from rod cells.
   d. A rod is hyperpolarized.
   e. A nerve impulse reaches the brain.

_____ 11. Light does not pass through which of the following on its way to being
      converted into an impulse in the optic nerve?
   a. Sclera
   b. Cornea
   c. Aqueous humor
   d. Lens
   e. Retina

_____ 12. Which of the following is mismatched?
   a. Fovea - sight
   b. Olfactory receptor - smell
   c. Papilla - hearing
   d. Utricle - balance
   e. Ossicle - hearing

SHORT ANSWER QUESTIONS

1. Differentiate between sensation and perception.

2. List the three types of receptors.

3. What three color pigments are used to perceive and distinguish color?

4. What is the difference between general and special sense receptors?

5 What accounts for loudness in hearing?

CRITICAL QUESTIONS

1. Amputees often report a condition known as phantom limb pain. How do you explain this distortion in perception?

2. What would be the advantage of having a large number of proprioceptors?

3. Biofeedback is often used as a therapy for sufferers of chronic pain. What is the theory behind the use of this treatment?

4. Sarah went to the ophthalmologist complaining that her vision had changed considerably. She no longer sees color, everything appears in shades of gray. She also lacks fine detail and bright illumination. Suggest a possible problem Sarah might have.

TELLING THE STORY

Develop a text that demonstrates your understanding of senses and the structures that facilitate the reception and transmission of sensory information. Focus mainly on the organization and function of specific regions. Use as many terms from the chapter as you can. If you have difficulty with any concepts as you write, go back and review them.

CHAPTER REVIEW ANSWERS

MATCHING

1. D	5. N	9. E	13. B	17. R
2. H	6. P	10. I	14. O	18. J
3. K	7. S	11. M	15. F	19. T
4. A	8. Q	12. G	16. L	20. C

LABELING

Fig. 15.1A   Fig.15.1B

1. Nerve fiber	5. Meissner's corpuscle	8. Merkel's disk
2. Spindle sheath	6. Root hair plexus	9. Free nerve endings
3. Skeletal muscle	7. Pacinian corpuscle	
4. Muscle spindles		

Fig. 15.2

1. Semicircular canals	4. Oval window	7. Cochlea
2. Utricle	5. Round window	8. Cochlear nerve
3. Saccule	6. Cochlear duct	

MULTIPLE CHOICE QUESTIONS

1. E	4. D	7. B	10. A
2. C	5. C	8. E	11. A
3. D	6. E	9. D	12. C

SHORT ANSWER QUESTIONS

1. The brain must first receive an action potential from the peripheral nervous system to interpret its meaning. Sensations are impulses that travel along sensory neruons that must be routed to the appropriate part of the brain so that they can be perceived. The perception is dependent on the destination of the impulse in the brain.

2. The three types of receptors are: (1) those that receive information regarding the body's internal environment, (2) those that receive information regarding the body's position in space, and (3) those that receive information regarding the body's external environment.

3. The three colors are red, blue, and green.

4. General sensory receptors are simple mechanical receptors scattered throughout the body that detect touch, pressure, pain, and temperature. Special sensory receptors are more complex and are located in specific places within the body. The special senses are sight, sound, taste, smell, and balance.

5. While the intensity of an impulse does not affect the intensity of the response, the frequency does. The brain is able to intepret differences in sound wave patterns. Different patterns cause the fluids in the inner ear to vibrate differently. Louder sound waves cause greater vibrations of the basilar membrane than can be interpreted by the brain. Repeated exposure to such intense sounds can cause permanent hearing damage.

CRITICAL QUESTIONS

1. Phantom pain is associated with the sensory neural network within the central nervous system. Normally these receptors are activated by some type of stimulus, internal or external. However, in the case of amputation, these receptors are sometimes activated independently causing the central nervous system to interpret the information as though it were receiving input from the peripheral nervous system.

2. The more proprioceptors a person has, the more aware that individual is of the exact position of his/her body. A greater number of proprioceptors should enable better coordination and better motor performance. This might be exemplified in the performance of an elite athlete.

3. Since it is the brain that actually perceives pain, there are numerous ways to prevent that information from resulting in an individual realizing the pain. One of these is the process of biofeedback. This technique allows the sufferer to decrease the sensitivity to the pain by conscious control.

4. Sarah probably has vision that is now restricted to the rods only. Damage to cones results in a low illumination, no color visual perception.

●CHAPTER 16

PROTECTION, SUPPORT, AND MOVEMENT

KEY CONCEPTS OVERVIEW

1.   Protection, Support, and Movement

Movement is an essential characteristic of all living things. It occurs in a variety of ways as a result of the integration of many different structural aspects of an organism. Coordinating such movement is a very complicated physiological process. This chapter focuses on the integrated efforts of skin, bones, and muscles to produce movement.

The tissues associated with movement are very diverse and perform more than one function. For example, the skin is a very dynamic organ that flexes when your body moves, protects the body from the external environment, receives sensory information, and aids in controlling the body's temperature. This multifunctional characteristic is not limited to the skin. Bones and muscles also serve the body in various ways.

2.   Skin

The skin, called **integument**, is the largest organ of the body and constitutes about 15% of the total body weight. Because of the highly specialized cells that make up this tissue, the skin is able to perform numerous functions. Remember that on average one square centimeter of skin contains approximately 200 nerve endings, 10 hair follicles, 100 sweat glands, 15 oil glands, 3 blood vessels, 12 heat receptors, 2 cold receptors, and 25 pressure-sensing receptors.

Skin provides a waterproof, protective barrier against the external environment and provides the first line of defense against any foreign objects or microbes.

(See Figure 16-1, P. 325 of your text)

It contains a protein pigment called **melanin** that offers a protective mechanism against ultraviolet radiation and gives skin its color. It also provides a surface for receiving sensory information. Many general sensory receptors are located within the skin and they acquire information about the body's external environment. Skin is flexible and compensates for body movement. It stretches when joints bend and accommodates growth. Additionally, skin helps in controlling the body's internal or **core** temperature by means of blood flow, sweat, and hair.

3.   Bones

Bones provide a protective and supporting framework for the body. These dense structures of varying size, shape, and arrangement provide a scaffold for the attachment of skeletal muscles to act against when they contract in order to cause movement. Bone tissue is also the site for blood cell formation and a storage site for minerals. The internal matrix of bones contain a protein material called **collagen,** which provides the flexibility and strength needed for bone structure. The hardness of bone is due to deposits of calcium salts.

Bone-producing cells are called **osteoblasts.** They function by producing new bone material and configuring it in concentric circles called **lamellae.** Narrow channels called **haversian canal** provides space for blood vessels to reach and nourish bone cells surrounded by their hardened environment.

There are two types of bone in the human body. This design allows the skeleton to be strong, yet not too heavy. **Spongy bone** is found in nearly all bones. The spaces within spongy bone contain **red bone marrow,** which is the site for blood cell production. **Compact bone,** which provides the dense exterior of bones and surrounds most of the spongy bone material, gives the strength to the skeleton. Within the central core of the long bones made of compact bone there is a soft, fatty connective tissue called **yellow marrow.**

The skeletal system is actually divided into two sections: the **axial skeleton,** consisting of the trunk bones (vertebrae, ribs, and skull) and the **appendicular skeleton,** which consists of the **pelvic girdle** and the **pectoral girdle** and their respective appendages. The point where two bones come together is called a **joint** or **articulation.** These bones are held together by a dense connective tissue called **ligaments.** There are different types of joints within the skeleton that dictate the type of movement that can occur at that site. These include the **hinge, ball-and socket, pivot, gliding, saddle,** and **ellipsoid** joints. Those joints that are freely movable have spaces between the articulating surfaces and contain a lubricating material called **synovial fluid.** These joints are referred to as **synovial joints.** There are some joints that are totally immovable or only slightly moveable. For example, the bones of the skull are imovable and the site of articulation is called a **suture.**

4.   Muscles

Muscles provide the contractile material for movement to occur. They are attached to bones by connective tissue called **tendons.** One end of a muscle is attached to a site on a stationary bone (**origin**) and the other end is attached to a site on a moveable bone (**insertion**). This arrangement gives the muscle an anchor and and lever to move. Sometimes muscles work in opposition to one another to stabilize the movement. The opposing pair of muscles are referred to as **antagonists.**

Skeletal muscle cells must be designed to provide a means for contraction or shortening. Whole muscles are composed of numerous bundles of muscle cells or **fibers.** Within each fiber there are bundles, called **myofibrils,** that contain an organized arrangement of thick and thin protein filaments. The thick filament is called **myosin** and the thin filament is called **actin.** These filaments are arranged in such a way that they appear as dark and light bands when light is passed through them. This banding effect makes the skeletal muscle appear **striated** or stringy. Closer look at the arrangement within a muscle fiber reveals that these filaments are also organized into repeating units for contraction. One contractile unit is called a **sarcomere** and these units run the length of the fiber.

Contraction results from the interaction between actin and myosin. Actin is pulled inward toward the center of the sarcomere by molecular alterations on myosin. As actin slides inward, the muscle cell shortens and contraction occurs. This very complex process is dependent on the amount of cellular energy available to fuel the process and generally, the voluntary nerve impulse that signals the initiation of the contractile process. When a nerve impulse reaches a motor end plate, the neurotransmitter (acetylcholine) causes depolarization in the muscle cell membrane. The **T tubules** within the membrane transmit this action potential to each myofibril and cause the release of calcium ions from the sarcoplasmic reticulum. In order for contraction to be initiated, the site for myosin and actin interaction (**cross bridge binding sites**) must be exposed. Calcium ions bind to a molecule called **troponin** that swings the obstructive **tropomyosin** molecule out of the way and exposes the binding site. Now the myosin cross bridge can bind to the actin and initiate contraction. (See Figure 16-15, P. 342 in your text)

CHAPTER REVIEW ACTIVITIES

CONCEPT MAPPING

Develop a concept map that demonstrates your understanding of the movement, protection, and support as they relate to skin, bones, and muscles. Try to incorporate as many of the key terms as possible. Be careful to organize the map to reflect the main themes stated above.

MATCHING

Select the term from the list provided that best relates to the following statements. Decide on a term BEFORE looking through the list, then search the list for the one you selected. If you are unsure of the correct answer or chose an incorrect response, review the related concept in Chapter 16.

_____ 1. These cords of dense connective tissue attach muscles to bones.

_____ 2. This part of the appendicular skeleton is made up of the clavicles and the scapulae.

_____ 3. This term is used to collectively describe the skin, hair, and nails of the human body.

_____ 4. This type of connective tissue provides attachment for muscles, serves as a storage site for minerals, and produces blood cells.

_____ 5. These repeating bands of actin and myosin myofilaments appear between two Z lines in a muscle fiber.

_____ 6. This is the thicker of the two protein myofilaments within a muscle fiber that is involved in the contractile process.

_____ 7. This is the freely movable joint in which a space exists between the articulating bones.

_____ 8. This immovable type of joint is found between bones in the skull.

_____ 9. This is the synapse between a motor neuron and a skeletal muscle cell.

_____ 10. This is the central column of the skeletal system that consists of the bones of the trunk and head.

_____ 11. This narrow channel runs parallel to the length of a bone and contains blood vessels and nerves.

_____ 12. This thin protein myofilament is involved in the contractile process.

_____ 13. This part of the appendicular skeleton is composed of the coxal bones.

_____ 14. This type of bone provides spaces for blood cell formations and assures that the human skeleton is not too heavy.

_____ 15. This ion is necessary for exposing the binding site for contraction to occur.

_____ 16. This protein material protects against ultraviolet radiation and gives color to the skin.

_____ 17. This site is where two bones join or where bones and cartilage join.

_____ 18. This portion of the human skeleton consists of the appendages and the supporting girdles for these appendages.

_____ 19. These structures on a muscle cell membrane transmit action potentials from nerve cells to deep within the muscle cell.

_____ 20. This soft, fatty connective tissue is found in the center of compact bone.

_____ 21. This dense bone tissue is configured in concentric rings.

_____ 22. These are bone forming cells.

_____ 23. These bundles of dense connective tissue hold bone to bone.

TERMS LIST

a. Sarcomere	i. Ligament	q. Bone
b. Actin	j. Axial skeleton	r. Yellow marrow
c. T-tubules	k. Tendons	s. Haversian canals
d. Compact bone	l. Sutures	t. Pectoral girdle
e. Synovial joint	m. Pelvic girdle	u. Osteoblasts
f. Calcium	n. Appendicular skeleton	v. Integument
g. Myosin	o. Articulation	w. Melanin
h. Spongy bone	p. Neuromuscular junction	

IF.... THEN....CHALLENGE QUESTIONS

Using the "IF" part of the statements below, complete the "THEN" part with guidance from the accompanying questions.

1. If an impulse is to be transported across a neuromuscular junction, then _____. What must be released to bridge the gap?

2. If contraction results in the shortening of sarcomeres, then _____. What configural changes take place to allow for shortening to occur?

3. If you compare the difference between the type of joint in the shoulder and the elbow, then _____. What types of joints are found at each site?

4. If ligaments attach bone to bone, then _____. What is the role of a tendon?

5. If you experience pain in the anterior portion of the lower leg following vigorous exercise involving running, jumping, or quick stops and starts, then _____. What type of condition might you develop and why?

6. If the body is going to move, then _____. What will it use to fuel this process?

7. If the external environmental temperature becomes very cold, then _____. What happens to the vascular beds in the skin?

8.   If bone tissue contains living cells called osteoblasts, then _____. What function do these cells serve during bone growth or repair?

9.   If a joint is considered to be a synovial joint, then _____. What characteristics might you expect for this joint type?

10.   If actin is a thin myofilament that moves inward during contraction, then _____. What is the thick myofilament and what does it do?

LABELING

Fig. 16.1

   The diagram below illustrates the architecture of a muscle. Label the parts.

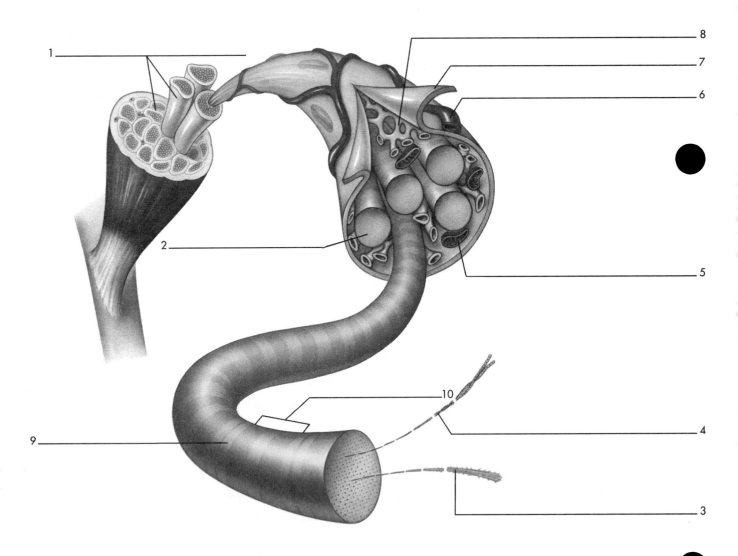

Dennehy Fig. 16-1

MULTIPLE CHOICE QUESTIONS

_____ 1. Which of the following bones is located on the medial (toward the middle) aspect of the lower arm?
   a. Humerus
   b. Ulna
   c. Radius
   d. Tibia
   e. Phalange

_____ 2. Which of the following bones is found in the upper portion of the lower limb?
   a. Tibia
   b. Fibula
   c. Humerus
   d. Femur
   e. Clavicle

_____ 3. Which of the following bones would be found in the pectoral girdle?
   a. Scapula
   b. Clavicle
   c. Pelvis
   d. A and C
   e. A and B

_____ 4. What is the largest organ of the body?
   a. Skin
   b. Liver
   c. Heart
   d. Brain
   e. Skeletal muscles

_____ 5. Vertebrae have which type of joint?

   a. Hinge

   b. Saddle

   c. Gliding

   d. Pivot

   e. Ball-and-socket

_____ 6. Which of the following would NOT be found in the appendicular skeleton?

   a. Clavicle

   b. Vertebrae

   c. Femur

   d. Metatarsal

   e. Fibula

_____ 7. Which of the following ions is released from the sarcoplasmic reticulum when a threshold level action potential is received by a muscle cell?

   a. Potassium

   b. Sodium

   c. Magnesium

   d. Hydrogen

   e. Calcium

_____ 8. Collectively both arms, including wrists and hands, contain how many bones?

   a. 80

   b. 40

   c. 25

   d. 75

   e. 38

_____ 9. What does the release of the housed ion from the sarcoplasmic reticulum cause?

    a. The sarcomere to lengthen

    b. The sarcomere to shorten

    c. An additional action potential to be formed

    d. A and C

    e. B and C

_____ 10. Which of the following would NOT be found in the arm?

    a. Ulna

    b. Fibula

    c. Radius

    d. Humerus

    e. Carpals

_____ 11. Which of the following is a bone forming cell?

    a. Osteon

    b. Osteoclast

    c. Osteocyte

    d. Osteoblast

    e. Haversian cell

_____ 12. Which of the following is NOT true of the skin?

    a. It becomes less elastic as you age.

    b. It serves as the first line of defense.

    c. It contains cells that produce pigment molecules that protect against ultraviolet radiation.

    d. It serves as a general sensory receptor site.

    e. All of the above are true.

SHORT ANSWER QUESTIONS

1. Draw a diagram of a long bone and label the spongy bone, the compact bone, the site for red marrow, and the site for yellow marrow.

2.	List the different types of joints and indicate the type of movement characteristic of each.

3.	Differentiate between the axial and appendicular skeletons.

4.	Explain the value of having both spongy and compact bone.

CRITICAL QUESTIONS

1.	What causes rigor mortis?

2.	Explain why movement of some type is important for most organisms.

3.	Explain how muscles actually move bones.

4.	Some individuals have a condition known as acromegaly. This anomoly causes exaggerated features in adults. Explain how such a condition might occur.

TELLING THE STORY

Develop a text that demonstrates your understanding of protection, support, and movement. Focus on the three systems addressed in Chapter 16. Use as many terms from the chapter as you can. If you have difficulty with any concepts as you write, go back and review them.

CHAPTER REVIEW ANSWERS

MATCHING

1. K	6. G	11. S	16. W	21. D
2. T	7. E	12. B	17. O	22. U
3. V	8. L	13. M	18. N	23. I
4. Q	9. P	14. H	19. C	
5. A	10. I	15. F	20. R	

IF.... THEN.... CHALLENGE QUESTIONS

1. Neurotransmitters must be released from the presynaptic cell and bind to receptors on the postsynaptic cell membrane.

2. Actin will move inward by the mechanics of the myosin crossbridges attached at the active sites. This will result in a shortening of the sarcomere and will be evident by the changed distance between sequenced the Z lines.

3. The shoulder joint, formed by the humerus and the scapula, would be considered a ball-and-socket type joint and the elbow, formed by the humerus, ulna, and radius, would be considered a hinged joint.

4. The tendons would attach muscles to bones at sites called origins and insertions.

5. The condition you might develop is known as shin splints which result from inflammation of the outer covering (perimysium) of the tibia.

6. Sufficient supply of ATP molecules would be necessary to fuel the muscle contractions and many other processes involved in human movement.

7. The vascular structures in the skin would constrict and blood flow would be reduced in these areas. Blood normally flowing in these vessels would be redirected to the vital organs.

8. The osteoblasts are bone forming cells and would initiate the production of bone tissue to repair any damaged areas.

9. Synovial joints are freely movable and they are characterized by having a fluid-filled space between articulating bones covered by hyaline cartilage. The entire joint is encapsulated in a double layer of connective tissue.

10. The myosin filaments would form crossbridges with the actin filaments and pull the actin inward during the contraction process.

LABELING

Fig. 16.1

1. Bundles of muscle fibers or faciculi	5. Mitochondrion	9. I-band
2. Myofibril	6. Capillary	10. A-band
3. Myosin	7. Sarcolemma	
4. Actin	8. Sarcoplasmic reticulum	

MULTIPLE CHOICE QUESTIONS

1. B	4. A	7. E	10. B
2. D	5. C	8. A	11. D
3. E	6. B	9. B	12. E

SHORT ANSWER QUESTIONS

1.   Students should refer to Fig. 16.2 in the text for the answer to this question.

2.   Hinge joint - hinge or swinging movement
Ball and socket - movement in any direction, basically unlimited range
Pivot joint - rotational movement
Gliding joint - sliding in many directions
Saddle joint - movement in right angles
Ellipsoid joint - nearly hinge movement with rotation restricted

3.   The axial skeleton forms the axis or central part of the skeleton while the appendicular skeleton forms the bones of the appendages. The axial skeleton includes the vertebral column, the rib cage, and the bones of the skull. The appendicular skeleton includes the bones of the pectoral girdle, the pelvic girdle, and their appendages.

4.   Spongy bone provides the space necessary for the formation of blood cells and reduces the weight of the skeleton. The compact bone provides a solid, rigid surface for muscle attachment and gives strength to the skeleton.

CRITICAL QUESTIONS

1.   When death occurs, the metabolic process that results in the formation of ATP ceases. The lack of ATP causes the cross bridges to stay attached to the actin

filaments. The resulting effect is contracted muscles. As the chemical degradation of the muscle tissue progresses, the body will eventually relax.

2. The process of movement is necessary for organisms because it provides a means for gathering food, mating and reproducing, and, in the case of some organisms, the ability to escape from predators.

3. Muscles are attached to bones at the origin of fixed bone and at the insertion of the movable bone. When contraction occurs, the muscle shortens. This shortening provides the force necessary to cause the bone to move in the direction of the shortened muscle. The process works physically like that of a lever, having a fulcrum or pivot point, a resistance or lever arm, and an effort or force.

4. Acromegaly occurs because of an excessive secretion of growth hormone. The long bones do not continue to elongate because the growth or epiphyseal plate is closed. The growth hormone stimulates the development of cells and causes exaggerated growth of certain areas. The soft tissue may continue to enlarge and bones become thicker. This can result in an exaggerated size of the face, feet, hands, jaw, nose, and mouth.

CHAPTER 17

HORMONES

KEY CONCEPTS OVERVIEW

1.   Hormones

Control over the body's functions is primarily controlled by the nervous system. However, the endocrine system also plays a role in controlling certain functions. This tandem relationship assures that consistant monitoring, control, and needed changes occur.

Hormones are chemicals produced and secreted by one cell that has an effect on another. These chemical messengers are generally produced by cells that group together to form glands. A gland is said to be **endocrine** if the chemical messengers it produces are transported by the blood. These hormones will be carried by the blood throughout the body. However, only certain cells will be effected by these hormones. These cells are called target cells and they possess receptor sites that can be located on the membrane, in the cytoplasm, and/or in the nucleus. These receptor sites allow the target cells to recognize only specific hormones and no others. For this reason, hormones are considered specialized. Because they lack ducts, endocrine glands are characterized as ductless glands. There are also glands within the body that have ducts through which chemical substances are released and transported. These types of glands are called **exocrine glands.**

There are two types of hormones that work within the human body: peptide hormones and steroid hormones. **Peptide hormones**, composed of amino acids, bind to receptor molecules located on the membrane of target cells. This binding triggers the production of a **second messenger** (the hormone being the first messenger) within the cell. An example of a second messenger would be cAMP. Second messenger works within the target cell to alter the cell's activities. Steroid hormones are lipid-soluble. Because of their chemical nature, they are able to enter target cells and interact with the cell's DNA influencing the production of proteins.

The role of hormones in the body is to regulate the internal environment and assist in maintaining homeostasis. The messages that hormes carry can be grouped into four different categories: regulation, response, reproductive functions, and growth and development.

2. Endocrine Glands

There are 10 different endocrine glands and they collectively make over 30 different hormones. Being aware of the gland and the hormones secreted by each of these glands helps us to better understand how the endocrine system can assist the nervous system to control and regulate cell functions.

The different hormones and the glands that secrete them are summarized in the table that follows.

3. Nonendocrine Hormones

There are several nonendocrine hormones that do not enter the bloodstream. **Local hormones** are intercellular chemical messengers that regulate nearby cells without having to enter the bloodstream to do so. An example is that of the prostaglandins, which are derived from cell membranes in response to injury or tissue damage. They stimulate the contraction of smooth muscle and the dilation and constriction of blood vessels. Additionally, there have been local hormones found within the brain. These chemicals are known as enkephalins and endorphins. Both inhibit the sensation of pain and have also been linked to the euphoric feeling athletes receive during aerobic exercise.

4. Hormone Production

Hormones production is regulated by **feedback loops**. When endocrine glands are stimulated, they secrete their hormones into the bloodstream. The stimulus can come from one of the following sources: (1) direct stimulation by the nervous system, (2) stimulation by a **releasing hormone**, or (3) the concentration level of a specific substance in the bloodstream. Once the hormone is released, it produces a desired effect within a target cell. This effect acts as a new stimulus and initiates another response on the part of the endocrine gland that turns off the hormone production and secretion. This type of homeostatic mechanism is called a **negative feedback loop**. A **positive feedback loop** acts differently. Rather than inhibiting hormone production and secretion, it enhances the process causing more hormone to be released from the gland.

Table 17.1 Endocrine glands and their hormones.

ENDOCRINE GLAND AND HORMONE	TARGET TISSUE	PRINCIPAL ACTIONS
HYPOTHALAMUS		
Releasing hormones	Other endocrine glands	Stimulate the release of hormones by other endocrine glands
POSTERIOR PITUITARY		
Oxytocin	Uterus Mammary glands	Stimulates contraction of uterus and milk production
Antidiuretic hormone (ADH)	Kidneys	Stimulates reabsorption of water by the kidneys
ANTERIOR PITUITARY		
Follicle-stimulating hormone (FSH)	Sex organs	Stimulates ovarian follicle, spermatogenesis
Luteinizing hormone (LH)	Sex organs	Stimulates ovulation and corpus luteum formation in females
Adrenocorticotropic hormone (ACTH)	Adrenal cortex	Stimulates secretion of adrenal cortical hormones
Thyroid-stimulating hormone (TSH)	Thyroid	Stimulates secretion of T_3 and T_4
Growth hormone (GH)	Cartilage and bone cells, skeletal muscle cells	Stimulates division of cartilage and bone cells, growth of muscle cells, and deposition of minerals
Prolactin	Mammary glands	Stimulates milk production
Melanocyte-stimulating hormone (MSH)	Melanocytes	Stimulates production of melanin
THYROID GLAND		
Thyroxine (T_4) and triiodothyrone (T_3)	General	Regulates metabolism
Calcitonin	Bone	Regulates calcium levels in the blood
PARATHYROID GLAND		
Parathyroid hormone (PTH)	Bone, kidney, small intestine	Regulates calcium levels in the blood
ADRENAL CORTEX		
Aldosterone	Kidney	Increases sodium and water reabsorption and potassium excretion
Glucocorticoids	General	Stimulate manufacture of glucose
ADRENAL MEDULLA		
Adrenaline and noradrenaline	Heart, blood vessels, liver, fat cells	Regulate fight or flight response: increase cardiac output, blood flow to muscles and heart, conversion of glycogen to glucose
PANCREAS (ISLETS OF LANGERHANS)		
Insulin	Liver, skeletal muscle, fat	Decreases blood glucose levels by stimulating movement of glucose into cells
Glucagon	Liver	Increases blood glucose levels by converting glycogen to glucose
OVARY		
Estrogens	General, female reproductive organs	Stimulate development of secondary sex characteristics in females, control monthly preparation of uterus for pregnancy
Progesterone	Uterus	Completes preparation of uterus for pregnancy
	Breasts	Stimulates development
TESTIS		
Testosterone	General	Stimulates development of secondary sex characteristics in males and growth spurt at puberty
	Male reproductive structures	Stimulates development of sex organs, structures spermatogenesis

CHAPTER REVIEW ACTIVITIES

CONCEPT MAPPING

Develop a concept map that demonstrates your understanding of hormones and their effect on the body. Try to incorporate as many of the key terms as possible. Try to construct the map as you conceptualize the information. Don't just put one together for the sake of doing so.

MATCHING

Select the term from the list provided that best relates to the following statements. Decide on a term BEFORE looking through the list, then search the list for the one you selected. If you are unsure of the correct answer or chose an incorrect response, review the related concept in Chapter 17.

_____ 1.   This a hormone is produced by the anterior pituitary that works with the thyroid hormones to control normal growth.

_____ 2.   This mechanism causes a gland to slow down or stop the production of its hormone.

_____ 3.   This outer yellowish portion of each adrenal gland secretes a group of hormones known as corticosteroids in response to the hormone ATCH.

_____ 4.   These glands' secretions reach each destination by means of ducts.

_____ 5.   This gonadotropic hormone triggers the maturation of one egg each month in females and initiates sperm production in males.

_____ 6.   This hormone, produced by the hypothalamus but stored and released by the posterior lobe of the pituitary affects the contraction of the uterus during childbirth and stimulates the mammary glands to produce and secrete milk.

_____ 7.   This inner, reddish portion of each adrenal gland secretes the hormones adrenaline and noradrenaline.

_____ 8.   This term refers to any disturbance that affects the body.

_____ 9.   This hormone, produced by the islets of Langerhans in the pancreas, works antagonistically to glucagon by decreasing blood glucose levels.

_____ 10. This mineralcorticoid hormone regulates the level of sodium and potassium ions in the blood, thereby conserving sodium and water and promoting the excretion of potassium.

_____ 11. These hormones, made of amino acids, are unable to pass through the cell membrane, so they bind to a receptor on the membrane of a target cell.

_____ 12. This stimulates the release of an egg in females and the production of testosterone in males.

_____ 13. These four small glands are embedded in the posterior side of the thyroid gland.

_____ 14. These tropic hormones affect the male and female sex organs.

_____ 15. These hormones are produced by the hypothalamus and affect the secretion of specific hormones from the anterior pituitary.

_____ 16. This chemical messenger is secreted and sent by a gland to other cells of the body.

_____ 17. This small gland hanging from the underside of the brain is controlled by the hypothalamus and secretes nine major hormones.

_____ 18. This hormone is produced by the thyroid gland and helps regulate the body's metabolism.

_____ 19. This is the mechanism in which the effect produced by the hormone enhances the secretion of that hormone rather than inhibiting it.

_____ 20. This class of endocrine hormone is made up of cholesterol and is able to pass through the target cell membrane.

_____ 21. This mass of nerve cells regulates the pituitary's secretion of hormones, based on the information it receives from the peripheral nerves and parts of the brain.

_____ 22. These are the three stages of reaction by the body to stress.

_____ 23. This hormone is produced by the adrenal medulla and readies the body to react to stress.

_____ 24. This hormone is secreted by the thyroid and works with parathyroid hormones to regulate the concentration of calcium in the blood.

TERMS LIST

a. Aldosterone	m. Oxytocin
b. Growth hormone	n. Releasing hormone
c. Adrenaline	o. Hypothalamus
d. Steroid hormone	p. Exocrine gland
e. Stress	q. Negative feedback loop
f. Hormone	r. Thyroxin
g. Calcitonin	s. Pituitary gland
h. Adrenal cortex	t. Peptide hormone
i. General adaptation syndrome	u. Gonadotropins
j. Parathyroid glands	v. Follicle stimulating hormone
k. Positive feedback loop	w. Leutinizing hormone
l. Insulin	x. Adrenal medulla

LABELING

The diagram below illustrates the location of glands in the body. Supply the correct names of the glands.

Fig. 17.2 Endocrine glands.

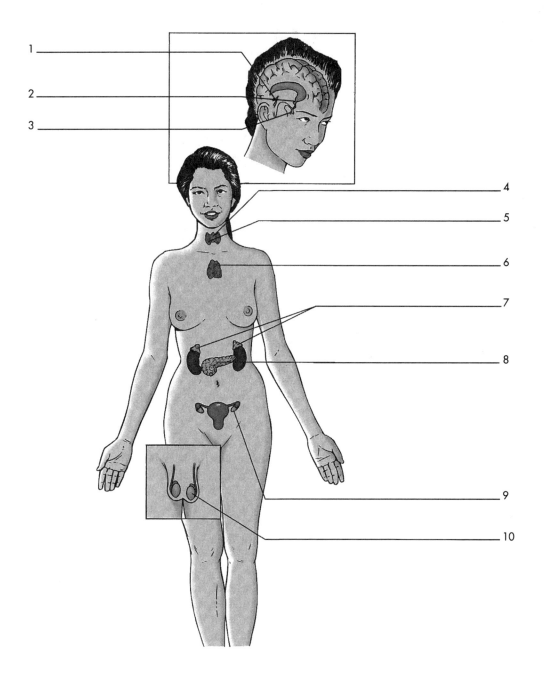

MULTIPLE CHOICE QUESTIONS

1. Which of the following occurs in females as a response to anabolic steroids?
   a. A deepening of the voice
   b. Increase of body hair
   c. Reduction in breast size
   d. Increase in muscle mass
   e. All ofthe above

2. An endocrine gland can be stimulated to secrete its hormone by
   a. Simulation by tropic hormones from the hypothalamus
   b. Trigger from the nervous system
   c. Chemical signals in the blood such as glucose or calcium
   d. Both A and B
   e. A, B, and C

3. Which of the following is NOT one of the inducing hormones?
   a. Follicle-stimulating hormone
   b. Growth hormone
   c. Adrenocorticotropic hormone
   d. Thyroid-stimulating hormone
   e. Lutenizing hormone

4. Ovulation is triggered by which of the following?
   a. A surge of lutenizing hormone
   b. The release of estrogen
   c. The production of estrogen
   d. The release of gonadotropin from the pituitary
   e. The release of follicle stimulating hormone

_____ 5. In males, the production of sperm is triggered by which of the following hormones?
   a. Lutenizing hormone
   b. Estrogen
   c. Follicle-stimulating hormone
   d. Testosterone
   e. A and D

_____ 6. The leading cause of blindness is due to a defect in which of the following?
   a. Adrenal gland
   b. Parathyroid gland
   c. Pancreas
   d. Thyroid gland
   e. Pituitary gland

_____ 7. Which of the following is true of Type I diabetes?
   a. It occurs during childhood.
   b. It may be due to an environmental condition that causes the body's immune system to attack the pancreas.
   c. It is the result of the body becoming resistant to the effects of insulin.
   d. It is characterized by excess insulin.
   e. It does not require daily injections of insulin.

_____ 8. Diabetes increases the risk of all but which one of the following?
   a. Stroke
   b. Blindness
   c. Kidney failure
   d. Heart attack
   e. Emphysema

_____ 9. Which ion is associated with the thyroid hormone?
   a. Calcium
   b. Iodine
   c. Chlorine
   d. Sodium
   e. Iron

_____ 10. The parathyroid hormone regulates which of the following in the blood?
   a. Calcium
   b. Chloride
   c. Iodide
   d. Sodium
   e. Magnesium

_____ 11. Stress reduction is controlled by which of the following?
   a. Thyroid
   b. Parathyroid
   c. Adrenal cortex
   d. Adrenal medulla
   e. Pituitary

_____ 12. Which of the following would be most closely associated with the reception of steroid hormones?
   a. Membranes
   b. Mitochondria
   c. Ribosomes
   d. DNA
   e. tRNA

_____ 13. Increasing the reabsorption of which of the following is controlled by aldosterone?
   a. Sodium
   b. Chloride
   c. Calcium
   d. Potassium
   e. Phosphorus

_____ 14. Which of the following glands is both exocrine and endocrine at the same time?
  a. Thyroid
  b. Pancreas
  c. Adrenal cortex
  d. Parathyroid
  e. Adrenal medulla

_____ 15. How does the pineal gland function?
  a. In immunity
  b. By controlling pigmentation
  c. By regulating the biological clock
  d. As the seat for the major drives such as sex and hunger
  e. To control secretion for sebacceous glands and control of acne

_____ 16. The anterior pituitary is controlled by the hypothalamus through which of the following?
  a. Vasopressin
  b. Insulin
  c. Releasing hormones
  d. cAMP
  e. Thyroxin

SHORT ANSWER QUESTIONS

1. Explain why the definition of hormone has been broadened recently.

2. State two differences between endocrine and exocrine glands.

3. Explain the function of second messenger in chemical signaling.

4. Explain how the body reacts over a prolonged period of stress.

CRITICAL QUESTIONS

1. Endocrine hormones are released in the blood. How is it that they can find and react with specific target cells?

2. Explain how the release of oxytocin results in a positive feedback mechanism. How would this ever be stopped?

3. Why is it that health care professionalS express such concern for stress management?

TELLING THE STORY

Develop a text that demonstrates your understanding of hormonal activity in the body. Focus on the hormones, their affect on cells, and the resulting effects. Use as many terms from the chapter as you can. If you have difficulty with any concepts as you write, go back and review them.

CHAPTER REVIEW ANSWERS

MATCHING

1. B	6. M	11. T	16. F	21. O
2. Q	7. X	12. W	17. S	22. I
3. H	8. E	13. J	18. R	23. C
4. P	9. L	14. U	19. K	24. G
5. V	10. A	15. N	20. D	

LABELING

Fig. 17.2

1. Pineal	5. Parathyroids	9. Ovaries
2. Hypothalamus	6. Thymus	10. Testes
3. Pituitary	7. Adrenals	
4. Thyroid	8. Pancreas	

MULTIPLE CHOICE QUESTIONS

1. E	5. C	9. B	13. A
2. E	6. C	10. A	14. B
3. B	7. A	11. C	15. C
4. A	8. E	12. D	16. C

SHORT ANSWER QUESTIONS

1.   Researchers now know that there are more types of chemical messengers than those described briefly before. The definition now includes chemical messengers such as growth factors, prostaglandins, pheromones, endorphins, etc.

2.   Endocrine glands release hormones directly into the blood and these hormones are carried to target cells. Exocrine glands release hormones into ducts and these hormones move to the site where they will stimulate some activity.

3.   Second messenger carries information from the hormone that has been received on the cell surface to the interior of the cell. This occurs because of the restriction of the initial hormone to the exterior of the cell.

4.   There are three stages of stress that the body will express following prolonged periods of stress: (1) the alarm reaction, (2) resistance, and (3)   exhaustion.

CRITICAL QUESTIONS

1.   Hormones are selective in that they bind only to the receptor sites of the appropriate target cell. Chemically, the receptors and the hormones are similar, much like a lock is to a key.

2.   Oxytocin causes the smooth muscle of the uterus to contract. The resulting contraction signals the release of more oxytocin to cause additional contractions. As this process continues, the contractions are more frequent and stronger. Once birth occurs, contractions subside and eventually diminish and the positive feedback loop is broken.

3.   Continual stimulation and triggering of the adrenal medulla is counterproductive. The resulting physiological toll causes high blood pressure, poor nutritional habits that generally cause vitamin deficiencies, high cholesterol, or lack of appropriate hydration, inability to concentrate or irritability, excessive weight gain or loss, etc. All these conditions place serious stress on the body and can lead to disease or even death.

CHAPTER 18

DNA, GENE EXPRESSION, AND CELL REPRODUCTION

KEY CONCEPTS OVERVIEW

1.   Chromosomes

Gregor Mendel's work (late 1800s) lead to the understanding that certain traits or characteristics associated with living things were passed on from generation to generation. From this finding, Walter Sutton surmised that the information that resulted in these traits was found on molecular structures called **chromosomes**. It wasn't until Joachim Hammerling identified that the most likely place for these structures was the nucleus that scientists really began to embrace the notion that chromosomes contain the information that allows traits to be passed on from parent to offspring. Friedrich Miescher (1869) isolated the molecules that would become known as **deoxyribonucleic acid** or DNA. He found this nucleic acid material in the chromosomes of various cells.

DNA is the hereditary material of cells, and it is responsible for controlling the cell's ability to produce necessary proteins. **Chromosomes** are tightly wound strands of DNA and assorted proteins that become visible during cell division. The somatic (body) cells of humans contain 23 pairs of chromosomes and the gametes (sex cells) contain 23 chromosomes not paired with another set.

2.   Nucleic Acids

P. A. Levene (1920s) discovered that there were two types of nucleic acids that were located within cells. One of these actually contains the genetic information (DNA or **deoxyribonucleic acid**) and the other type helps express that information (RNA or **ribonucleic acid**). **Nucleic acids** have the same basic molecular structure. They contain a phosphate group, 5-carbon sugars, and four types of nitrogen-containing bases. These bases tend to accept hydrogen ions even though the nucleic acid molecule is acidic. These structural units bond together in equal portions to form the building blocks for DNA and RNA. Each unit is referred to as a **nucleotide**. (See Figure 18-3, P. 371 in your text)

The nitrogen bases found in DNA and RNA are **guanine, cytosine,** and **adenine.** The fourth nitrogen base varies for these two types of nucleic acids. DNA contains the nitrogen base **thymine**, while RNA contains **uracil.** These bases are grouped by their structure. The **pyrimidines** (thymine, cytosine, and uracil) are single-ring bases. The **purines** (guanine and adenine) are double-ring bases. In DNA, there are equal amounts of adenine and thymine and equal amounts of guanine and cytosine (A=T and G=C). Therefore there are always equal proportions of purines and pyrimidines.

3.   Deoxyribonucleic Acid

The structure of DNA is of great significance in understanding how coded genetic information is stored. Roaslind Franklin and Maurice Wilkins determined that DNA was shaped helixically and that the diameter of this helix was uniformed. James Watson and Francis Crick used this information to develop the idea that DNA was a double helix in which two complementary strands were held together by hydrogen bonds betweeen two nitrogenous bases, one on each strand. They also determined that the pairing was always the same (A with T, G with C).

Scientists also discovered that just prior to cell division, the hydrogen bonds between the complementary base pairs breaks and causes sections of the molecule to *unzip.* Each of the separated strands is called a **replication fork** and serves as a template for the production of a complementary strand. DNA replication begins with a single, double-stranded DNA molecule and ends with two double-stranded DNA molecules. This assures that the new cells being produced contain the same genetic information as the parent cells.

4.   Genes

A **gene** is defined as a unit of heredity. It is a sequence of nucleotides that codes for the amino acid sequence necessary to produce an enzyme or other protein. George Beadle and Edward Tatum are credited with the initial information on how DNA directs the growth and development of an organism. They proposed the **one gene-one enzyme theory,** which states that the production of a given enzyme is under the control of a specific gene. If the gene mutates or changes in any way, the enzyme will be inappropriately constructed or not made at all. The ramifications for mutations can be serious, depending on the role of that enzyme in the organism.

Proteins are synthesized in the cytoplasm at the ribosomes. Ribosomes are made up of proteins and RNA. There are three types of RNA. **Ribosomal RNA (rRNA)** is the type found in the ribosomes. **Transfer RNA (tRNA)** is found in the cytoplasm and is responsible for transporting amino acids to the ribosomes during protein synthesis and for positioning each amino acid at the correct place in the developing polypeptide chain. **Messenger RNA (mRNA)** brings the information regarding the code for protein synthesis from the DNA in the nucleus to the ribosomes in the cytoplasm.

The process whereby these types of RNA and the ribosome work to synthesize new proteins for the cell occurs in definable stages. The first stage is called **transcription**. During this process, the gene coding is copied by mRNA from DNA. Messenger RNA then exits the nucleus and moves to the ribosomes in the cytoplasm bearing the encoding for the protein to be made. The second stage is called **translation**. During this process, the information being carried by mRNA is "read" or deciphered and results in the appropriate sequencing of amino acids for a specific protein. The genetic information on the mRNA is carried in "packages" of nucleotides called **codons**. One codon is composed of three nucleotides. Each codon corresponds to a gene on the DNA and codes for one amino acid. There are approximately 20 different amino acids and 64 possible codon combinations. This suggests that some codons code for the same amino acid and some are signals to stop reading the encoding.

Once mRNA reaches the ribosome it configures itself in such a way that only the codon is exposed. Transfer RNA binds to mRNA at its **anticodon**, a loop on the molecule that consists of the complementary base pairs to mRNA. On the other end, tRNA carries an amino acid that is specific to its anticodon. In this way, tRNA can supply the amino acids that are coded for on mRNA. Once a stop codon is reached, it is recognized by proteins called **special release factors**, that cause the release of the newly formed polypeptide.

Scientists recognize the importance for cells to "know" <u>when</u> to make proteins as well as <u>how</u> to make them. Eukaryotic cells have proteins that interact with some of the nucleotide sequences on the code. These specific nucleotides are called **regulatory sites** and they control the transcription process. This means that genes can be turned on (positive control) or turned off (negative control) depending on the nature of the cell. Eukaryotes usually have positive control because their genes are generally inactive and must be turned on.

5. Cell Division, Growth, and Reproduction

Cell division occurs to facilitate cell growth, repair, and reproduction. These processes allow organisms to progress through stages. These stages are referred to as the **life cycle**.

Figure 18-1 Human life cycle

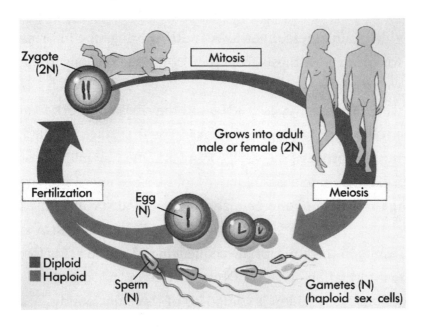

The fusion of **gametes** or sex cells during **sexual reproduction** results in an offspring that is composed of genetic information from both parents. These sex cells are formed by a process of cell division called **meiosis**. Meiosis is a process of cell division that produces four cells from one parent cell. Each sex cell produced in this process contains half the amount of genetic information of that of the parent cell and is referred to as a **haploid**. During fertilization, haploid cells from both parents fuse and produce a cell containing double the haploid amount of genetic information (one from each parent). This new cell is referred to as **diploid**. All other cells in an organism divide by a process called **mitosis**. This process produces two cells that are identical to the original parent cell.

Plants and animals both use the process of mitosis to grow. However, the diploid phase in animals predominates. In plants, the haploid phase is the **gametophyte generation** and the diploid phase is the **sporophyte generation**. Haploid plants produce gametes by mitosis, whereas diploid plants produce spores by meiosis. The life cycle of a plant consists of the alternation of the growth and reproduction of each of these plant types.

Single-cell eukaryotes reproduce by mitosis. This process does not enhance growth but does increase the population numbers of these types of cells. This process is called **asexual reproduction**.

6.   Mitosis

Fig. 18.2 Cell cycle.

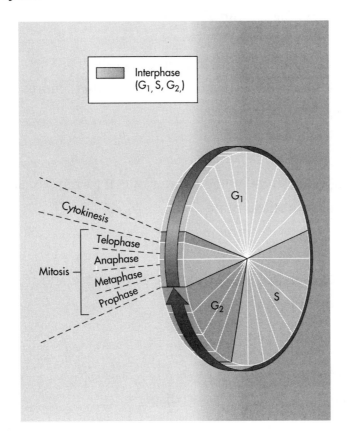

Mitosis is a continuous sequence of events that occurs just after interphase and results in the division of the chromosomes that duplicated during the S phase of interphase. To understand when mitosis occurs and the resulting effects this process has, it is important to understand the **cell cycle**, or life cycle of a cell. The cell cycle controls the timing of mitosis. This cycle is composed of the following phases: interphase (G1, S, and G2), mitosis (prophase, anaphase, metaphase, and telophase), and cytokinesis. **Interphase** is a very active phase that involves cell growth and metabolism. During interphase, **G1** is the growth phase between the last mitotic division and the replication of chromosomes for the next mitotic division. The **S** (synthesis) phase involves the replication of the cell's chromosomes. The **G2** phase is a preparatory phase for mitosis in which the chromatin condenses and other structures are replicated and mobilized for cell division.

Mitosis is the actual process in which cells divide to produce daughter cells. This process of cell division involves four phases which are described below.

Prophase

- Division and movement of centrioles (animal cells) to opposite sides of nucleus
- Asters assembled (animal cells)
- Chromatin condenses
- Nucleolus and nuclear membrane disappear
- Spindle fibers assembled and attach to kinetochores of the centromeres

Metaphase

- Chromosomes line up on equator of cell forming a circular disk called the **metaphase plate**

Anaphase

- Sister chromatids separate at the centromere and move to opposite poles of the cell

Telophase

- Cytokinesis
- Chromatin decondenses and disperses
- Nucleolus and nuclear membrane reappear
- Aster disassembles (animal cells)
- Spindle fibers disappear
- Cell pate (plant cells) forms

When mitosis is complete, the cell undergoes the process called **cytokinesis**. In this process the cytoplasm divides. In animal cells, a cleavage furrow forms at the location of the metaphase plate and microtubules cleave the cell into two. In plant cells, vesicles manufacture the new cell wall by forming a partition called the **cell plate**.

7. Meiosis

In order to facilitate the process of sexual reproduction, plants and animals must have a way to produce the necessary sex cells. **Meiosis** is the process whereby sex cells are formed. Meiosis is similar to mitosis in that it can be divided into the same general phases. The major difference in the process is that meiosis requires two divisions, and the chromatin material is not replicated between the two divisions. The result of meiosis is the development of four haploid cells (gametes or spores) from a diploid cell. The stages of meiosis are as follows:

Prophase I

- Nucleolus and nuclear membrane disappear
- Homologous chromosomes pair and line up side-by-side, a process called **synapsis**
- The homologous chromosomes exchange genetic material in a process called **crossing over**
- Spindle fibers assemble; attach to kinetochore of each homologous pair

Metaphase I

- Homologous chromosomes align along the metaphase plate

Anaphase I

- Homologous chromosomes move to opposite poles; sister chromatids do not separate

Telophase I

- Nuclear membrane and nucleolus reestablish
- Chromosomes gather at both poles

Prophase II

- Spindle fibers assemble; attach to kinetochore of centromere
- Nucleolus and nuclear membrane disappear

Metaphase II

- Chromosomes line up along the metaphase plate

Anaphase II

- Sister chromatids separate at the centromere and move to opposite poles of the cell

Telophase II

- Nucleolus and nuclear membrane reappear
- Spindle fibers disappear

CHAPTER REVIEW ACTIVITIES

CONCEPT MAPPING

Develop a concept map that demonstrates your understanding of how genes are expressed in organisms. Since there is a large amount of information in this chapter, make sure to organize your thoughts before you attempt to construct your map. Use as many of the key terms as possible, but make sure they fit into your conceptualization appropriately. Try not to "force" a term into the map if it doesn't fit.

MATCHING

Select the term from the list provided that best relates to the following statements. If you are unsure of the correct answer or chose an incorrect response, review the related concept in Chapter 18.

_____ 1. In this type of cell division the gametes or spores are produced.

_____ 2. These small round structures are found in the endoplasmic reticulum of cells or within the cytoplasm that serve as the site for protein synthesis.

_____ 3. This is the physical division of the cytoplasm of a eukaryotic cell.

_____ 4. These sequences of three nucleotide bases in transcribed mRNA code for specific amino acids.

_____ 5. This complex of DNA and protein makes up the chromosomes of eukaryotes.

_____ 6. These single-ring compounds are components of nucleotides.

_____ 7. This unit of heredity is formed of a sequence of nucleotides that codes for the amino acid sequence of an enzyme or other proteins.

_____ 8. This term describes a somatic or body cell that contains the appropriate number of paired chromosomes for that particular organism.

_____ 9. In eukaryotic cells, these specific nucleotide sequences control the transcription of genes.

_____ 10. This is the diploid phase of a plant life cycle that alternates with the haploid phase.

_____ 11. These double-ring compounds are components of nucleotides.

_____ 12. This is a term for the life cycle of a cell.

_____ 13. This term describes a sex cell that contains half the amount of hereditary material of the original parent cell.

_____ 14. This is the process during prophase I of meiosis in which homologous chromosomes line up side-by-side, initiating the process of crossing over.

_____ 15. This type of RNA, found in cellular cytoplasm, transports amino acids to the ribosomes and correctly positions them for protein synthesis.

_____ 16. This is the second step of gene expression in which the mRNA directs the synthesis of a polypeptide.

_____ 17. This process of cell division produces two identical cells from an original parent cell.

_____ 18. This is another term for sex cells.

_____ 19. This is the haploid phase of a plant life cycle that alternates with the diploid phase.

_____ 20. This is the single unit of nucleic acid, consisting of a 5-carbon sugar bonded to a phosphate group and a nitrogen base.

_____ 21. This is the first step in the process of protein synthesis and gene expression in which a gene is copied into a strand of mRNA.

_____ 22. This type of RNA brings information from the DNA within the nucleus to ribosomes in the cytoplasm.

_____ 23. This term describes the union of a sperm and an egg.

_____ 24. This is the portion of a tRNA molecule with a sequence of three base pairs complementary to a specific mRNA sequence.

a.	Synapsis	m.	Cell cycle
b.	tRNA	n.	Transcription
c.	Pyrimidines	o.	Ribosomes
d.	Nucleotide	p.	Regulatory site
e.	Anticodon	q.	Gametophyte
f.	Cytokinesis	r.	Meiosis
g.	Gametes	s.	Gene
h.	Diploid	t.	Chromatin
i.	mRNA	u.	Sporophyte
j.	Codons	v.	Haploid
k.	Purines	w.	Fertilization
l.	Translation	x.	Mitosis

THE CONTRIBUTIONS OF SCIENTISTS

Complete the table below.

Scientists	Organisms or Materials used	Techniques or investigations	Results
Mendel			
Hammerling			
Levene			
Chargaff			
Hershey and Chase			
Watson and Crick			

LABELING

Label the diagram with the correct terms for the numbered structures.

Fig. 18.3 Human chromosome structure.

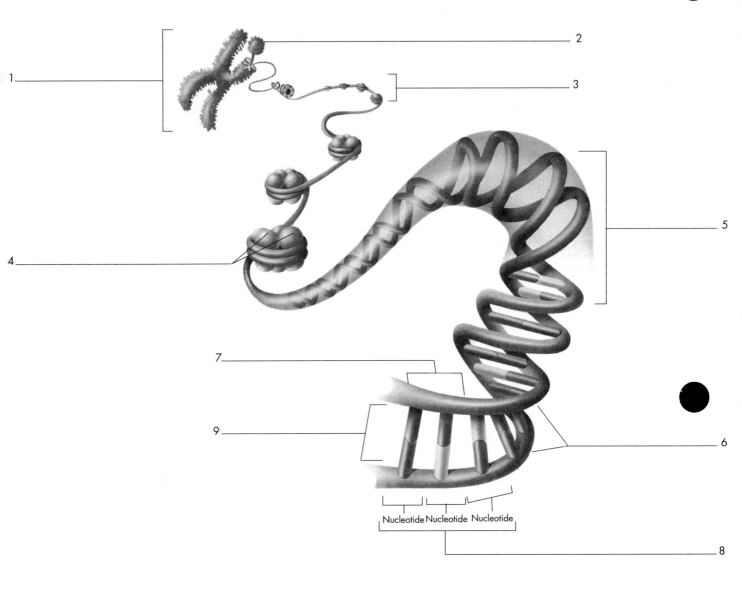

Nucleotide Nucleotide Nucleotide

MULTIPLE CHOICE QUESTIONS

_____ 1. Which of the following statements is TRUE?

a. DNA is only found in human cells.

b. The complexity of an organism does not determine the number of chromosomes it has.

c. The DNA in one chromosome is organized into discrete packages of genes that are isolated from each other.

d. DNA in a chromosome can form a helix or a single, tight coil.

e. Throughout the entire lifetime of a cell, the DNA remains condensed in individual bodies known as chromosomes.

_____ 2. The fusion of gametes from opposite sexes of the same species results in which of the following?

a. Asexual reproduction

b. A haploid cell

c. Meiosis

d. A zygote

e. The formation of additional gametes

_____ 3. How does meiosis differ from mitosis?

a. The cells produced are genetically different from each other.

b. The number of chromosomes in the resulting cells is less than that of the original parent cell.

c. It requires two divisions to be complete.

d. Synapsis or pairing of homologous chromosomes occurs.

e. All of the above

_____ 4. Spindle fibers are composed of which of the following?

a. Microfilaments

b. Actin and myosin

c. Microtubules

d. Uracil

e. Strands of DNA

_____ 5. Which of the following statements is FALSE?

  a. Meiosis always leads to gamete formation and sexual reproduction.

  b. Meiosis converts a diploid cell into four haploid cells.

  c. The gametophyte phase of a plant life cycle is haploid.

  d. All of the cells produced by mitosis are genetically identical.

  e. All of the above are true.

_____ 6. During which stage of interphase are the chromosomes replicated?

  a. S

  b. G1

  c. Prophase

  d. D

  e. G2

_____ 7. During meiosis, the homologous chromosomes separate during which of the following phases?

  a. Metaphase

  b. Prophase I

  c. Prophase II

  d. Telophase I

  e. Anaphase I

_____ 8. What is the first stage of mitosis?

  a. Telophase

  b. Cytokinesis

  c. Prophase

  d. Metaphase

  e. Anaphase

_____ 9. Which of the following describes a kinetochore?

  a. It is a microtubule organizing center.

  b. It is a duplicated chromosome consisting of sister chromatids.

  c. It is a condensed mass of chromatin.

  d. It is the site in a centromere where spindle fibers attach

  e. None of the above

10. Which of the following describes the biological significance of crossing over?
   a. It makes sexual reproduction possible.
   b. It distinguishes meiosis from mitosis.
   c. It allows chromosomes to replicate.
   d. It allows for new combinations of genes.
   e. It separates traits inherited from your mother and father.

11. During which of the following stages of mitosis do sister chromatids separate, become chromosomes, and begin to move apart?
   a. Anaphase
   b. Interphase
   c. Metaphase
   d. Prophase
   e. Telophase

12. Which of the following occurs during the pairing of bases in nucleic acids?
   a. Purines pair with purines
   b. Pyrimidines pair with pyrimidines
   c. Purines pair with pyrimidines in both DNA and RNA
   d. Purines pair with pyrimidines in DNA only
   e. None of the above

13. Which of the following should always pair with cytosine in DNA?
   a. Adenine
   b. Guanine
   c. Thymine
   d. Uracil
   e. Other cytosines

14. Which of the following bases is found in DNA, but not RNA?
   a. Thymine
   b. Cytosine
   c. Uracil
   d. Adenine
   e. Guanine

15. If the template DNA molecule reads AGTCACCCT, which of the following would describe the appropriate sequence of nitrogen bases on messenger RNA?

    a. AGGGTGACT

    b. TCAGTGGGA

    c. UCAGUGGGA

    d. AGTCACCCT

    e. There is no way to tell.

16. In Hershey and Chase's experiment, an isotope of which of the following was used to tag the nucleic acid part of the virus?

    a. Carbon

    b. Potassium

    c. Nitrogen

    d. Phosphorus

    e. Sulfur

17. When Beadle and Tatum exposed *Neurospora* to x-rays, the radiation directly affected which of the following?

    a. The media

    b. Enzymes

    c. Genes

    d. Metabolic pathway

    e. The ability of the organism to grow.

18. How many different codon combinations are there?

    a. 4

    b. 20

    c. 36

    d. 64

    e. Infinite number

_____ 19. Approximately how many amino acids are there?

    a. 5

    b. 20

    c. 64

    d. Thousands

    e. Infinite number

_____ 20. Which of the following persons determined the structure of DNA?

    a. Beadle and Tatum

    b. Watson and Crick

    c. Chargoff

    d. Mendel

    e. Franklin and Wilkins

SHORT ANSWER QUESTIONS

1. Using the three molecular components of a nucleotide, diagram how they combine with one another.

2. What is the significance of mRNA?

3. How many different amino acids could be specified if the codons were only two nucleotides long?

4. What are the two 5-carbon sugars found in nucleic acids?

5. What is the difference in cytokinesis between plant and animal cells?

6. Draw and label the cell cycle and make the segments appropriately sized.

CRITICAL QUESTIONS

1.   What would the ramifications be for protein synthesis if one nucleotide was deleted when mRNA was reading DNA?

2.   In the early days of genetics, why were scientists more likely to consider proteins to be the genes rather than the nucleic acids?

TELLING THE STORY

Develop a story that demonstrates your understanding of DNA and gene expression in the body. Use as many terms from the chapter as you can. If you have difficulty with any concept as you write, go back and review the specific section in Chapter 18.

CHAPTER REVIEW ANSWERS

MATCHING

1. R	6. C	11. K	16. L	21. N
2. O	7. S	12. M	17. X	22. I
3. F	8. H	13. V	18. G	23. W
4. J	9. P	14. A	19. Q	24. E
5. T	10. U	15. B	20. D	

CONTRIBUTIONS OF SCIENTISTS

SCIENTISTS	ORGANISMS USED	TECHNIQUES, INVESTIGATIONS	RESULTS
Mendel	Peas	Crossing a variety of peas that differed in only one trait	Traits are inherited in discrete packets of information
Hammerling	*A. mediterranea* *A. crenulata*	Removed the cap and feet to determine where genetic information was stored	Heriditary information was stored in the foot of *Acetabularia* and is necessary for the foot to regenerate
Levene	Nucleic acids	Chemical analysis	Determined that there are two types of nucleic acids and that they are nearly identical in structure
Chargaff	DNA from a variety of organisms	Compared nucleotide composition of DNA	DNA composition is specific to a species, ratio of A =T and C=G
Hershey and Chase	Viruses and bacteria	Tagged DNA and protein, found DNA to be reason cells produced more virus particles	DNA, not protein, is the hereditary material
Watson and Crick	DNA	X-rays and models	DNA is a double helix, rungs are nitrogen bases, backbone is alternating sugar and phosphate groups

LABELING

Fig. 18.3

1. Human chromosome	4. Histone proteins	7. Nucleotide
2. Supercoil within chromosome	5. Gene region	8. Nucleotide strand
3. Chromatin	6. DNA	9. Nitrogen base

MULTIPLE CHOICE QUESTIONS

1. B	6. A	11. A	16. E
2. D	7. E	12. C	17. C
3. E	8. C	13. B	18. D
4. C	9. D	14. A	19. B
5. A	10. D	15. C	20. B

SHORT ANSWER QUESTIONS

1. Students should refer to Figure 18.3, p. 371 in the text.

2. Messenger RNA is necessary for transcribing the code from DNA. Since DNA is contained within the nucleus and protein synthesis occurs at the ribosomes in the cytoplasm, mRNA can move in and out of the nucleus to the cytoplasm to facilitate this location issue.

3. Only 16.

4. Ribose and deoxyribose

5. In animal cells a cell furrow develops between the two nuclei, and the two cells are pinched into two new daughter cells. In plants, cells are surrounded by a rigid cell wall. A new cell wall is produced from tiny vesicles probably derived from the Golgi complex. A structure is produced in the middle of the cell called the cell plate. It will extend between the two nuclei and form a new cell wall.

6. Students should refer to Fig. 18.1 in the Key Concept Review.

CRITICAL QUESTIONS

1.   The deletion of one nucleotide would cause all of the codons following the deletion to be read inappropriately by tRNA. The amino acids would be inappropriately sequenced. Since proteins are specific for each species, any protein synthesized must be compatible with the organism making it.

2.   So much more information was known about proteins during this time. Proteins were made from the combinations of 20 different amino acids and nucleic acids were composed of only four types of nucleotides. It was logical to hink that such a complex process as inheritance involved the more complex compound.

CHAPTER 19

PATTERNS OF INHERITANCE

KEY CONCEPTS OVERVIEW

1. Inheritance

Although it is commonly accepted that organisms inherit characteristics from the parent generation, this was not always the case. **Hippocrates** surmised that offspring inherited traits from "particles" given off by all parts of the parents' bodies. During the time of **Anton van Leeuwenhoek** it was thought that either the sperm or the egg contained a minute whole human being. The advent of the cell theory led scientists to begin to explain how new organisms developed and grew from individual cells of parent organisms. By the late 1800s, scientists concluded that the cell's nuclear material was the physical bond that linked generations of organisms.

2. Gregor Mendel's Work

Some 25 years prior to discovering the hereditary significance of the nuclear material, **Gregor Mendel** began a study involving generations of garden peas. He discovered the fundamental laws governing inheritance by doing detailed genetic experiments on pea plants and keeping exact mathematic records of all the crosses he performed. The pea plants were used because the flowers on individual plants produced both male and female gametes and could therefore fertilize themselves. This process is known as **self-fertilization**. Following generations of self-fertilization, some of the plants exhibited the exact traits of the parent plants. These plants are known as **true-breeding**. Mendel however noted that the anatomy of the flower allowed him to remove the stamen (male part) from one flower and pollinate the pistil (female part) of another flower. This process is called **cross-fertilization**. Mendel's success was due to the fact that he concentrated on seven contrasting physical traits that he noted from the true-breeding varieties. His work was designed to focus on one trait at a time.

Mendel crossed a true-breeding plant having white flowers with a true-breeding plant having purple flowers. These plants represented the **parental (P) generation.** Their hybrid offspring were called the **first filial (F$_1$) generation.** All the offspring in this generation had purple flowers. It was this realization that disproved the concept of blending inheritance because no light purple flowers were produced. Mendel referred to the form of a trait that was expressed in the F$_1$ generation as **dominant** and the alternative form not expressed as **recessive.** Mendel then wondered what became of the white flowers. He crossed two of the F$_1$ generation to get the **second filial (F$_2$) generation.** It was in the offspring of the F$_2$ generation that he found a **3:1 ratio** of purple to white flowers.

Mendel determined that each individual inherits what he called a "factor" for each trait. And these factors are given to the offspring by each parent. We know these "factors" as genes and know that they are molecules of DNA. Mendel's P generation had two identical factors (white flower in one and purple flower in the other). Since these traits were identical, the plants were called **homozygous** for that trait. Each parent contributed their factor to the hybrid offspring. The offspring were therefore **heterozygous** for that trait.

Dominant genes are expressed whenever they are present. The recessive gene is hidden and only expressed in the absence of the dominant gene. The two alternative forms of a gene are called **alleles.** Mendel used a capital letter to denote the dominant allele and a lower case letter to denote the recessive allele. Since the white flower was recessive and homozygous, they could be represented pp and the homozygous dominant purple flower could be represented PP. When PP and pp are crossed, their offspring will all receive one dominant and one recessive trait (Pp) and all will be purple. The expression of the trait (how it looks) is called the **phenotype.** A cross between a Pp and Pp (F$_2$) generation will yield a different combination of alleles for the offspring. A **Punnett square** is used to demonstrate the gametes and potential offspring ratio. On the square, the gametes are written on the top and side and the genetic combinations for the offspring within the boxes of the square.

Fig. 19.1 Punnett square.

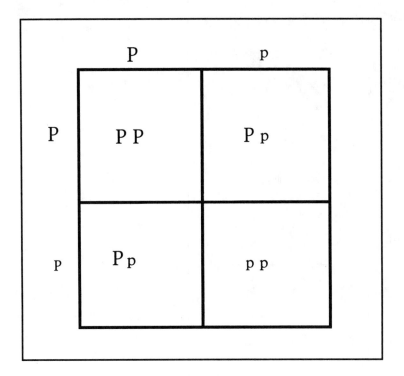

While three of the plants will appear purple and one white, note that the genetic make-up or **genotype** is different. The phenotypic ratio is 3:1, but the genotypic ratio is 1:2:1.

If two traits are involved, the cross is referred to as a **dihybrid cross.** If a true-breeding plant with smooth seeds (S) and yellow seeds (Y) is crossed with a plant with wrinkled (s) seeds and green (y) seeds, the cross would result in the following possibilities of offspring:

Fig. 19.2 Dihybrid cross

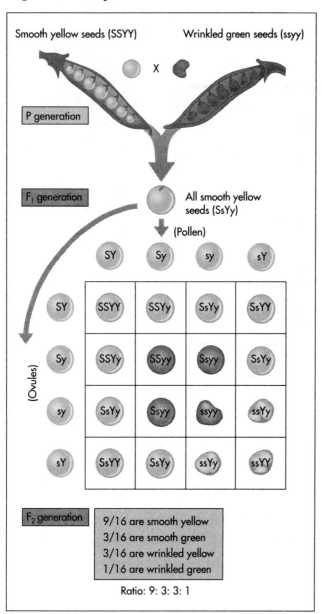

As you can see, the phenotype ratio is found to be 9:3:3:1.

3. Mendel's Laws

As a result of his work, Mendel proposed three laws related to inheritance. First was the **law of dominance**. This law states that the dominant gene will be expressed if it is present. The second one is the **law of segregation** which states that if an individual is heterozygous for a particular trait, there is a 50-50 chance that either allele for that trait will be passed on to the gamete. The third law is the **law of independent assortment.** This law states that the distribution of the alleles for the first gene does not affect the distribution of the alleles for the second gene.

The laws of genetics follow the laws of probability. Although Mendel could demonstrate these physical traits, he was unable to explain the reasons for his observations. We now know that genes are located on chromosomes within the nucleus of cells. Mendel's laws were rediscovered in 1900 by three independent researchers (Correns, DeVries, and Tschermak) who worked out Mendel's principles of heredity prior to knowing about his work. A graduate student at the time, Walter Sutton proposed that the genetic material was located on the chromosome. This was finally demonstrated by T. H. Morgan who found the first gene linked to a specific chromosome - the X chromosome.

CHAPTER REVIEW ACTIVITIES

CONCEPT MAPPING

Develop a concept map that demonstrates your understanding of inheritance. Focus on the work and findings of Mendel and others that followed him. Use as many of the key terms as possible, but make sure they fit into your conceptualization appropriately.

MATCHING

Select the term from the list provided that best relates to the following statements. If you are unsure of the correct answer or chose an incorrect response, review the related concept in Chapter 19.

_____ 1. This is the total set of genes that constitutes an organism's genetic makeup.

_____ 2. This is the hybrid offspring of the parental generation.

_____ 3. ' This form of a trait will be expressed in a hybrid offspring or in any cross if it is present.

_____ 4. Each member of a factor pair containing information for an alternative form of trait occupies corresponding positions on paired chromosomes.

_____ 5. This branch of biology deals with the principles of heredity.

_____ 6. This term refers to an individual who has two identical alleles for a particular trait.

_____ 7. This term refers to plants that after generations of self-fertilization, produce offspring consistently identical to the parent with respect to certain defined characteristics.

_____ 8. This simple diagram provides a way to visualize the possible combinations of genes in a cross and illustrates the expected offspring ratios.

_____ 9. These are distinguishing characteristics or features.

_____ 10. This form of a trait recedes or disappears entirely in a hybrid offspring.

_____ 11. This is the outward appearance or expression of an organism's gene.

_____ 12.   This the the name of the generation of offspring from first generation hybrid parents.

_____ 13.   This term refers to an individual who has two different alleles for a trait.

_____ 14.   This term refers to the product of two organisms that differ from one another in two traits.

_____ 15.   These are the units of transmission of hereditary information.

_____ 16.   This is the product of two organisms that differ from one another in a single trait.

TERMS LIST

a. Gene	g. Genotype	m. Homozygous
b. Phenotype	h. Dihybrid	n. Alleles
c. Dominant	i. Genetics	o. Punnett squares
d. Recessive	j. Monohybrid	p. F_2 generation
e. Heterozygous	k. F_1 generation	
f. True-breeding	l. Traits	

GENETIC PROBLEMS

1.   In one species of plants, yellow-flowered plants crossed with yellow-flowered plants always produed yellow-flowered offspring. On the other hand, when blue-flowered plants were crossed with blue-flowered plants, some of the offspring exhibited blue flowers and otheres produced both blue flowers and yellow flowers. Which one of the genes for colored flowers would you think would be dominant and why?

2.   If tall (D) is dominant to short (d), give the F_2 genotypic and phenotypic ratios.

3. How many different gametes can be produced by organisms with the following genotypes?
   a. Aa
   b. BbCc
   c. DdEeFf
   d. DdEEff

4. If AaBb is crossed with AaBb, how many true-breeding varities could be produced? List the genotypes of each.

5. If CcDd is crossed with CcDd, which genotype has the greatest possibility of being expressed among all possibilities?

6. If black (B) is dominant to brown (b) and wire hair (W) is dominant to smooth (w), what would be the genotypic and phenotypic ratios of the F_2 generation of a cross of a homozygous black smooth and homozygous brown wire-haired dog?

7. If tall (D) is dominant to short (d) and purple (P) is dominant to white (p), what are the chances of producing the following offspring?
   a. A tall white offspring
   b. A true-breeding tall white offspring
   c. Offspring with a genotype of ddPp

8. If smooth seed (S) is dominant to wrinkled (s), what is the genotype of two smooth-seeded plants that produced an offspring with wrinkled seeds?

9. If 10% of the genes in a population are recessive (a) and this gene is sex-linked, what is the frequency that males and females in the population will express it?

10. In watermelons, if striped (S) is dominant to solid (s) and elongated (E) is dominant to round (e), give the phenotypic results of the cross between SsEE and SsEe.

11. In spaniels, solid colored coat is dominant and spotted coat is recessive. Suppose a true-breeding solid colored (SS) is crossed with a spotted dog, and the offspring (F_1) are interbred.

a. What is the probability that the first puppy born will have a spotted coat?

b. What is the probability that if four puppies are born, all of them will be a solid color?

SHORT ANSWER QUESTIONS

1. Why was Mendel successful in his quest for determining hereditary patterns in peas?

2. Which would be easier to establish as a true-breeding variety -- the dominant or recessive allele? Why?

3. Which would be easier to eliminate from a population -- the dominant or recessive allele?

CRITICAL QUESTIONS

1. Blood typing is used as evidence in paternity suits. Using the following mother and child combinations, determine which blood groups of fathers would be excluded from possibly fathering the child.

Mother's blood type	Child's blood type	Blood type excluding the man (cannot be the father of the child)
B	A	_____
A	AB	_____
O	B	_____
O	O	_____
AB	A	_____

2.   Construct a Punnett square of the following cross. Black (B) is dominant to chestnut (b) and trotting gait (T) is dominant to pacer (t). Cross a black trotter (BBTt) with a black pacer (Bbtt). Summarize the genotypic and phenotypic characteristics.

TELLING THE STORY

Develop a story that demonstrates your understanding of inheritance. Use as many terms from the chapter as you can. If you have difficulty with any concept as you write, go back and review the specific section in Chapter 19.

CHAPTER REVIEW ANSWERS

MATCHING

1. G	5. I	9. L	13. E
2. K	6. M	10. D	14. H
3. C	7. F	11. B	15. A
4. N	8. O	12. P	16. J

GENETIC PROBLEMS

1. Blue alleles are dominant. The blue-flowered plants that produced yellow flowers have a recessive gene for yellow. The yellow flowers always breed true because they are homozygous or have the same alleles for a particular trait.

2.
Genotype	Phenotype
1 DD	3 tall
2 Dd	1 short
1 dd	

3. a. 2 A
   2 a

   b. 4 BC   1 bC
      1 Bc   1 bc

   c. 8 DEF   1 dEF
      1 DEf   1 dEf
      1 DeF   1deF
      1 Def   1 def

   d. 2 DEf
      1 dEf

4. There would be 4 varieties: AABB, AAb, aaBB, aabb

5. CcDd occurs 4/16. The other combinations are CCDD 1/16, CCDd 2/16, CCdd 1/16, CcDD 2/16, Ccdd 2/16, ccDD 1/16, ccDd 2/16, ccdd 1/16.

6.

Genotypic ratio	Phenotypic ratio
1BBWW, 1bbWW	9 Black, wire-haired
2BBWw, 2bbWw	3 Black, smooth-haired
2BbWW, 1BBww	3 Brown, wire-haired
4BbWw, 1bbww	1 Brown, smooth-haired

7.   a. 3/16
       b. 1/16
       c. 2/16

8.   Ss X Ss

9.   Ten percent of the males express the recessive trait because they inherit the dominant trait 90% of the time (aY). Whenever a gene is inherited by the male, it must be expressed because there is no alternative allele available. In females, for a sex linked recessive gene to be expressed it must be in the homozygous condition. The chance of having the genotype aa is 1% (10% X 10%). One percent of the females are homozygous recessive. Eigthy-one percent (90% X 90%) are homozygous dominant while 18% of the females are heterozygous.

10.   Three-fourths of the offspring will be striped and 1/4 will be solid. All the watermelon will be elongated.

11.   The F_1 generation is heterozygous for coat color. In the F_2 generation, the probability of a solid-colored puppy is 3:4, and the probability of a spotted puppy is 1:4.
       a. 1:4
       b. 1:256

SHORT ANSWER QUESTIONS

1.   Mendel focused on one single trait at a time rather than considering all the facets of the phenotype. He kept accurate records and treated his results

mathematically. He also carried the experiments into the second (F$_2$) generation.

2.   The recessive gene is easier to establish as a true-breeding form because if an organism expresses a recessive trait, it must be homozygous for that trait. If an organism expresses a dominant trait, it is possible that the expression of the dominant allele may mask the presence of a recessive allele. The recessive gene may reappear in later crosses and thus the organism would not be in a true-breeding form. It would be necessary to conduct a testcross to determine if an organism expressing a dominant trait is homozygous.

3.   The dominant allele is easier to eliminate because if it is present it will be expressed. The recessive allele on the other hand may be masked and more difficult to determine and to eliminate.

CRITICAL QUESTIONS

1.   a.   B type or O type
       b.   A type or O type
       c.   A type or O type
       d.   AB type
       e.   No group could be excluded

2.

	BT	Bt	BT	Bt
Bt	BBTt	BBTt	BBTt	BBTt
Bt	BBTt	BBTt	BBTt	BBTt
bt	BbTt	Bbtt	BbTt	Bbtt
bt	BbTt	Bbtt	BbTt	Bbtt

Genotype	Phenotype
8 BBTt	12 Black, trotters
4 BbTt	
4 Bbtt	4 Black, pacer

CHAPTER 20

HUMAN GENETICS

KEY CONCEPTS OVERVIEW

1.   Karyotypes

Of the 23 pairs of chromosomes that humans have, one pair is considered the **sex chromosomes** and the other 22 are called autosomal chromosomes or **autosomes.** Autosomes are the same regardless of the gender of the individual. Chromosomes are easiest to study in there most condensed form and are stained to reveal the banding patterns. Then the chromosomes can be photographed, enlarged, and cut out to be arranged in order by their size and shape. Such an arrangement is known as a **karyotype.** Karyotypes can be used to identify genetic disorders caused by the loss of all or part of a chromosome or by the addition of extra chromosomes or fragments of chromosomes. However, they cannot help researchers see changes in single genes. Any permanent changes in the genetic material, known as a **mutation,** could only be seen if it resulted in the alteration of large pieces of a chromosome, whole chromosomes, or entire sets of chromosomes.

2.   Abnormal Chromosomal Conditions

An abnormal number of chromosomes in gametes may occur. This can result when homologous chromosomes fail to separate during synapsis in meiosis. This occurs because of a process called **nondisjunction.** J. Langdon Down was the first to describe the occurrence of nondisjunction of chromosome 21, known as **trisomy 21.** The developmental defect produced by this abnormal chromosomal condition is called **Down syndrome.** Nondisjunction can also occur with the sex chromosomes. The resulting abnormalities are not as severe as when nondisjunction occurs on autosomal chromosomes. Examples of these conditions include: **Triple X females, Klinefelter syndrome, Turner syndrome,** and **XYY males.**

Even though the appropriate number of chromosomes are inherited, a chromosome can be structurally defective. They may break naturally or by the effects of outside agents. Exposure to ionizing radiation or chemicals can cause chromosomal breaks or alterations in the nucleotide structures within molecules of DNA. When broken chromosomes are repaired, aberrations may occur. The aberrations include **duplication, deletion, inversion**, and **translocation**. An example of a genetic disorder that results from chromosomal deletion is **cri du chat syndrome.**

Mutations or changes in a single allele can occur rather than alterations in entire sections of chromosomes. A change in the genetic information of a chromosome that is caused by alterations in the molecules of DNA is called a **point mutation.** The result is a new and permanent allele of a gene. These types of mutations are usually the result of exposure to x-rays, ultraviolet radiation, and certain chemicals.

3.   Cancer

It is known that mutations in the DNA of somatic cells are directly related to the development of cancer. While this is thought of as one disease, it is in fact a large (200 + types) group of diseases. All types of cancer however have four common characteristics: (1) uncontrolled cell growth, (2) loss of cell differentiation, (3) invasion of normal tissues, and (4) **metastasis** or spreading to multiple sites.

Cancer cells occur as a result of a series of step-wise progressive mutations of the DNA in normal somatic cells. The first mutations may involve the **proto-oncogenes,** which are the dormant forms of cancer causing genes or **oncogenes.** These are the signals or **initiatiors** to speed up the growth of cells and decrease their levels of differentiation. The tumor suppressor genes, required for normal functioning of cells, signals cells to slow their growth and increase their level of differentiation. Mutation causes the inactivation of these genes allowing cancerous growth to occur.

For cancer to occur, cells involved must undergo a process, called **promotion,** by which the DNA is further damaged. This continual damage occurs over a long period of time and causes further cell division and growth. Agents that cause initiation and promotion of cancer cells are called complete **carcinogens.** Although few in number, some agents act as the intiator only or the promoter only. Heredity is only an initiator and asbestos is only a promoter.

Following promotion, the DNA damage accumulates and the expression of oncogenes begins. If the damage to the DNA is not sever, most of the cellular components still function normally. A **benign tumor** might result. These are growths of cells that are made up of partially transformed cells and are usually confined to one location, encapsulated from surrounding tissue. While these types of tumors are not life-threatening, they exhibit growth patterns similar to those of cancer cells. The cells making up these types of tumors are said to exhibit **dysplasia.**

If the damage to the DNA is severe, the cells will reach a point where they can not reverse the process and cancer cells result. This is the beginning of the third stage of cancer development called **progression.** During this stage, the transformed cells become less differentiated than benign cells and increase their rate of growth and division without regard for the body's needs. It is at this stage that the cells are able to invade and kill other tissues and metastasize. Tumors produced by this process are considered **malignant.**

4.   Pedigrees

Pedigrees are family "trees" that demonstrate how a trait is inherited within generations of a family. By studying a pedigree, researchers can determine if a trait is dominant or recessive and if it is sex-linked or autosomal. Additionally, carriers of recessive genes can be identified.

5.   Genetic Disorders

In most cases, human genetic disorders are the result of the expression of a recessive gene. Some of these recessive genetic anomolies are **cystic fibrosis, sickle cell anemia,** and **Tay Sachs.** Dominant recessive disorders, although not as common, include **Huntington's disease, Marfan's syndrome, polydactyly, achondroplasia,** and **hypercholesterolemia.**

In addition to simple dominance and sex-linked inheritance, there are other types of inheritance that affect humans. Some traits demonstrate what is termed **incomplete dominance,** in which the heterozygote is intermediate to the homozygote. An example is wavy hair, which is intermediate between curly hair (HH) and straight hair (H^1H^1). In **codominance,** both alleles for a trait are dominant and both characteristics are exhibited in the phenotype. An example are the alleles that code for blood type. The human blood type alleles are also an example of **multiple alleles,** in which there are more than two alleles for a specific trait. In this case there are three: A, B, and O.

Genetic counseling involves identifying couples at risk for producing offspring with genetic defects. During counseling, the family pedigree is studied and statistical analyses are applied to determine the probability that a person is a carrier of a recessive trait. Procedures such as **amniocentesis** and **chorionic villi sampling** allow cells to be extracted and studied for signs of genetic abnormalities. New technologies such as gene therapy may make it possible to treat many genetic disorders prior to birth.

CHAPTER REVIEW ACTIVITIES

CONCEPT MAPPING

Develop a concept map that demonstrates your understanding of human genetics. Try to incorporate as many of the key terms as possible. Try to construct the map as you have conceptualized the information.

MATCHING

Select the term from the list provided that best relates to the following statements. Decide on a term BEFORE looking through the list, then search the list for your answer. If you are unsure of the correct answer or chose an incorrect response, review the related concept in Chapter 20.

_____ 1. This is one of the characteristics of cancer cells that allows the cancer to spread to multiple sites throughout the body.

_____ 2. This term describes cancerous tumors that have the ability to invade and kill other tissues and move to other areas of the body.

_____ 3. These are cancer-causing genes.

_____ 4. This failure of homologous chromsomes to separate after synapsis, results in gametes with abnormal numbers of chromosomes.

_____ 5. This system of more than two alleles for certain genes exhibits either complete dominance or codominance.

_____ 6. This change in the genetic message of a chromosome is caused by alterations of molecules within the structure of the chromosomal DNA.

_____ 7. This term refers to traits in which the alternative forms of an allele are both dominant, and both characteristics are exhibited in the phenotype.

_____ 8. This is a latent form of cancer-forming genes that all people have.

_____ 9. These are cancer-causing substances.

_____ 10. These are permanent changes in the genetic material whether they affect single genes, pieces of chromosomes, whole chromosomes, or entire sets of chromosomes.

_____ 11. This term describes any of the 22 pairs of human chromosomes that carry the majority of an individual's genetic information, but have no genes that determine gender.

_____ 12. These are diagrams of genetic relationships among family members over several generations.

_____ 13. These growths or masses of cells are made up of partially transformed cells, are confined to one location, and are encapsulated from the surrounding tissues.

_____ 14. This is a form of dominance in which alternative alleles are neither dominant over nor recessive to other alleles governing a particular trait.

_____ 15. This genetic disease is associated with Jewish people from eastern and central Europe.

_____ 16. This type of gene is responsible for most human genetic disorders.

_____ 17. This is the name for the chromosomal disorder that causes Down syndrome.

_____ 18. This genetic disease afflicts mostly people of African descent and results in red blood cell deformation.

_____ 19. This is the process of isolating, staining, photographing, and enlarging chromosomes to identify any abnormalities.

_____ 20. This is the process of identifying couples at risk of having children with genetic disorders and presenting them with the probability of passing such traits to their children.

TERMS LIST

a. Codominance	k. Oncogenes
b. Malignant	l. Trisomy 21
c. Karyotyping	m. Benign tumors
d. Recessive	n. Proto-oncogenes
e. Point mutation	o. Nondisjunction
f. Pedigree	p. Sickle-cell anemia
g. Matastasis	q. Mutation
h. Autosomal	r. Multiple alleles
i. Tay Sachs	s. Incomplete dominance
j. Genetic counseling	t. Carcinogens

MULTIPLE CHOICE QUESTIONS

_____ 1. Which of the following is characterized by a webbed neck, sterility in females, and a broad chest?

a. Klinefelter syndrome

b. Down syndrome

c. Turner syndrome

d. Cri du chat

e. Trisomy 13

_____ 2. An individual with Klinefelter syndrome has a chromosome composition recorded as which of the following?

a. XO

b. XXX

c. XXY

d. XYY

e. YY

3. An individual with Turner syndrome has a chromosome composition recorded as which of the following?

a. XO
b. XXX
c. XXY
d. XYY
e. YY

4. Which of the following is NOT a chromosomal aberration?

a. Inversion
b. Trisomy
c. Deletion
d. Duplication
e. Translocation

5. Which of the following involves the exchange of genetic material between two chromosomes?

a. Inversion
b. Trisomy
c. Deletion
d. Duplication
e. Translocation

6. What was the first genetic disorder to be treated by gene therapy?

a. Severe combined immune deficiency
b. Huntington's disease
c. Cystic fibrosis
d. Sickle cell anemia
e. Multiple sclerosis

7. In a pedigree chart, how is an affected male represented?

a. An open circle
b. A closed, darkened circle
c. An open square
d. A closed, darkened square
e. A circle with a dot in it

_____ 8. From whom does a son inherit his sex-linked traits?

  a. His mother

  b. His father

  c. Both parents

_____ 9. From whom does a daughter inherit her sex-linked traits?

  a. Her mother

  b. Her father

  c. Both parents

_____ 10. In a pedigree chart, how is a female carrier is represented?

  a. An open circle

  b. A closed, darkened circle

  c. An open square

  d. A closed, darkened square

  e. A circle with a dot in it

_____ 11. Which of the following is controlled by a recessive allele?

  a. Polydactyly

  b. Marfan's syndrome

  c. Achondroplasia

  d. Huntington's disease

  e. Sickle cell anemia

_____ 12. Which of the following has the greatest frequency of occurance among the ethnic group that expresses it?

  a. Sickle cell anemia

  b. Tay-Sachs disease

  c. Cystic fibrosis

_____ 13. Which of the following crosses would produce all three phenotypes in incomplete dominance?

  a. Red carnations with red carnations

  b. Red carnations with pink carnations

  c. Pink carnations with pink carnations

  d. Pink carnations with white carnations

  e. White carnations with white carnations

14. If a child has type O blood, the father could not have which of the following blood types?

    a. A

    b. B

    c. AB

    d. O

    e. Rh

15. Which of the following blood types do NOT have surface antigens?

    a. A

    b. B

    c. AB

    d. O

    e. All of them have surface antigens

16. Which of the following blood types is considered the universal donor?

    a. A

    b. B

    c. AB

    d. O

17. Cri du chat syndrome is an example of which type of chromosomal mutation?

    a. Deletion

    b. Inversion

    c. Translocation

    d. Nondisjunction

    e. Duplication

18. Which type of mutation below occurs when two simultaneous breaks in a chromosome cause the loss of a segment of that chromosome?

    a. Deletion

    b. Inversion

    c. Translocation

    d. Nondisjunction

    e. Duplication

_____ 19. Which of the following is TRUE regarding a person with Down syndrome?

    a. They exhibit short stature and oriental-like folds in the eyelids.

    b. They usually have a thickened, fissured tongue.

    c. The disorder results if an egg had two number 21 chromosomes.

    d. The disorder results if the sperm had two number 21 chromosomes.

    e. All of the above

_____ 20. Which sex chromosomal abnormality displays a karyotype of XXX?

    a. Turner's syndrome

    b. Klinefelter syndrome

    c. Down syndrome

    d. Triple X males

    e. Triple X females

DISORDERS AND TRAITS

Match the disorders listed below with one of the following:

    a. Dominant disorder

    b. Recessive disorder

    c. Sex-linked disorder

    d. Recessive trait

    e. Dominant trait

1. Duchenne muscular dystrophy	6. Fragile X syndrome	11. Polydactyly
2. Brown eyes	7. Sickle cell anemia	12. Common baldness
3. Huntington's disease	8. Short fingers	13. Tay-Sachs disease
4. Hemophilia	9. O blood factor	14. Marfan's syndrome
5. Widow's peak	10. Albinism	15. A or B blood factor

SHORT ANSWER QUESTIONS

1. What are some of the ways that chromosomes aberrations or point mutations can occur?

2. Explain nondisjunction.

3. Why does the incidence of Down syndrome increase with the age of the mother?

4. Why are some genetic disorders ethnically related?

5. What are some of the limitations of gene therapy?

6. List the warning signs of cancer.

CRITICAL QUESTIONS

1. Why is it that an individual possessing a Y chromosome will always be male regardless of the number of X chromosomes present (as in the example of Klinefelter syndrome)?

2. Why are there general dietary recommendations to reduce the risk of cancer?

3. . Does having a benign tumor remove all risk of developing cancer?

TELLING THE STORY

Develop a text that demonstrates your understanding of human genetics. Focus on inheritance patterns, chromosomal abnormalities, and the genetic anomolies that can result. Use as many terms from the chapter as you can. If you have difficulty with any concepts as you write, go back to Chapter 20 in your text and review them.

CHAPTER REVIEW ANSWERS

MATCHING

1. G	6. E	11. H	16. D
2. B	7. A	12. F	17. L
3. K	8. N	13. M	18. P
4. O	9. T	14. S	19. C
5. R	10. Q	15. I	20. J

MULTIPLE CHOICE QUESTONS

1. C	6. A	11. E	16. D
2. C	7. D	12. A	17. A
3. A	8. A	13. C	18. A
4. B	9. C	14. C	19. E
5. E	10. C	15. D	20. E

DISORDERS AND TRAITS

1. C	6. C	11. A
2. E	7. B	12. D
3. A	8. E	13. B
4. C	9. D	14. A
5. E	10. C	15. E

SHORT ANSWER QUESTIONS

1. Chromosome aberrations and point mutations can occur as a result of exposure to ionizing radiation. Energy from UV radiation is absorbed by some molecules in a chromosome and can be responsible for genetic mutations. Chemicals that produce damage to DNA directly include LSD, marijuana, cyclamates, and certain types of pesticides.

2. Nondisjunction is the failure of homologous chromosomes to separate. Nondisjunction results in gametes that have two copies of one chromosome or lack the chromosome altogether. The final result after fertilization may be trisomy or an abnormal number of sex chromosomes. Examples of resulting anomolies are Down syndrome, Klinefelter syndrome, and Turner's syndrome.

3. Females develop all of their eggs by the fifteenth week of gestation. The number of eggs declines throughout the life of the female. The older an egg is that becomes fertile, the greater chance that the meiotic process will result in some abnormality.

4. Since reproduction within a single ethnic population generally remains in that population, the offspring produced will have the characteristics of that ethnic group. Should a mutation arise in this population, the gene is retained in the population and can result in disorders associated with that particular ethnic group.

5. The absolute location of a gene on a chromosome is very difficult to determine. Many have yet to be found. If it is found, the next challenge is to isolate it. The means of delivering the gene as a treatment poses a challenge. Then once the gene is delivered, the challenge is to induce it or "turn it on" to begin rectifying the initial problem.

6. The warning signs of cancer include the following:
   - Sores that do not heal
   - Unusual bleeding
   - Changes in a wart or mole
   - A lump or thickening in any tissue
   - Persistent hoarseness or cough
   - Chronic indigestion
   - A change in bowel or bladder functions

CRITICAL QUESTIONS

1. The determining factor for testicular development is found on the Y chromosome only.

2. Researchers suggest that 90% of all cancers are environmentally induced, not inherited. Eating a well-balanced diet and avoiding foods that have been linked to cancer will decrease the risks of getting cancer. Nutritional balance gives cells the appropriate minerals, fuels, and vitamins to perform all necessary metabolic functions.

3.   Any type of tumor warrants heightened awareness of the potential for malignancy to occur. A lump or thickening in any tissue should be seen by a physician and monitored closely.

CHAPTER 21

SEX AND REPRODUCTION

KEY CONCEPTS OVERVIEW

1. Sexual Reproduction

    As one of the characteristics of all living things, reproduction is one of the most fundamental functions. After all, should this process cease in a species, the perpetuation of organisms in that species would cease. Although reproduction is fundamental, it is not necessary to maintain the life of an individual organism.

    There are different ways organisms reproduce, some are as simple as the division of a parent cell into two identical daughter cells, and some are much more complex. **Sexual reproduction** is the process in which the sex cells or gametes of males and females fuse to form a fertilized egg or **zygote**. These sex cells are produced in specialized organs called **gonads**. The male gonads are responsible for the production of highly motile sex cells called **spermatozoon** or **sperm**. The female gonads produce sex cells called **eggs** or **ooctyes**.

2. Male Reproductive System

    The reproductive system of males yields haploid cells that carry one-half of the hereditary information necessary to produce a new individual organism. The production of the male haploid sex cell occurs in the **testes** or **testicles**. These structures are located outside the body cavity in a thin-walled sac called the **scrotum**. The purpose of such an arrangement is to assure the proper temperature (less than body temperature) for sperm development. Each testis contains about 750 feet of very tightly coiled tubes called **seminiferous tubules**. Hundreds of millions of sperm cells develop within the lining of these tubules each day during a process called **spermatogenesis**. This process usually begins at the onset of puberty by the release of **follicle-stimulating hormone (FSH)** from the pituitary gland. Another hormone released by the pituitary, the **leutinizing hormone (LH)**, regulates the secretion of testosterone. **Testosterone** is the male sex hormone responsible for the development and maintenance of secondary sex characteristics. The release of both of these hormones is regulated by a negative feedback mechanism.

Spermatogenesis begins when the diploid **spermatogonia**, cells that give rise to the sperm cells, undergo meiosis to produce haploid **spermatids**. The spermatids then move to a long, coiled tube located on the posterior side of each testis called the **epididymis** where they mature into spermatozoa (sperm cells). Mature sperm cells are stored in the epididymis and in a tube that connects it to the urethra called the **vas deferens**. The flagellated sperm are propelled through the vas deferens to the urethra during **ejaculation**. Along the pathway, accessory glands (**seminal vesicles, prostate gland**, and **bulbourethral glands**) add fluid to the sperm. This fluid, called **semen**, nourishes and protects the sperm.

The external genitals of males are composed of the scrotum and the **penis**. The penis is made up of three cylinders of erectile tissue that allows it to become erect when the cavities within these layers of tissue fill with blood. This results when neural stimulation causes the vascular tissue within the penis to dilate. Physical stimulation is usually required for semen to be ejaculated from the penis.

3.   Female Reproductive System

In the female reproductive system, the process of producing female gametes is called **oogenesis** and takes place in structures called the **ovaries**, which are located in the pelvic cavity. Each ovary contains **primary oocytes**, which have the potential to develop into mature egg cells or **secondary oocytes. Follicular cells** surround these oocytes and collectively they make up tissue known as a **follicle**.

The process of meiosis that produces the eggs occurs before birth in females. In other words, females have all the primary ooctyes (between 1million to 2 million) they will ever have at birth. As the female reaches puberty many of these cells have degenerated and left fewer that half the original number as viable cells. The primary oocytes are derived from diploid cells (**oogonia**). During fetal development, these oogonia undergo meiosis. However, the process is halted at the prophase state of meiosis I. Once a female reaches puberty, the follicle-stimulating hormone (FSH) is released monthly stimulating the development of 20 to 25 follicles. These follicles begin to produce **estrogen**, which triggers the thickening of the uterine lining or **endometrium**. This process prepares the uterus for implantation of a fertilized egg. After approximately 7 days, most of the developing follicles die, but one continues to mature. Rising levels of estrogen signal the pituitary gland to release LH and additional FSH. Increasing levels of leutinizing hormone trigger the process of ovulation. During **ovulation**, the follicle ruptures and releases the secondary oocyte, which travels through the **uterine tubes** to the **uterus**. This oocyte must be fertilized within 24 hours of entering the uterus or it will die. Fertilization by a sperm cell will stimulate the continuation of the second meiotic division. Fertilization also stimulates the production of **human chorionic gonadotropin (HCG)**, which is the hormone used in pregnancy tests. Positive detection of HCG means a positive pregnancy test. Additionally, HCG maintains the corpus luteum so that estrogen and progesterone will continue to be secreted throughout the pregnancy.

Following ovulation, the follicle that ruptured and released the oocyte continues to develop into the corpus luteum ("yellow body"), which secretes progesterone and estrogen. These hormones are necessary to continue preparing the uterine lining for implantation of a fertilized egg. Should the secondary oocyte not be fertilized, the corpus luteum will begin to decrease production of these hormones and the uterine lining will break down and be shed. This process is called **menstruation**. The monthly development and shedding of the outer endometrial layer is known as the **menstrual cycle**. The maturation and release of an egg on a monthly basis is known as the **ovarian cycle**. These two cycle together are referred to as the female **reproductive cycle**.

The female external genitals are collectively called the **vulva** and include the **mons pubis, labia majora and minora, clitoris**, and **vagina**. The **mammary glands** are secondary sex characteristics that begin to develop in females at puberty. They develop as milk-producing glands for lactation. The hormone **prolactin** stimulates the production of milk and the hormone **oxytocin** causes milk letdown.

3. The Sexual Response

Like most physiological processes, sexual response occurs in phases that vary in length, results, and requirements. There are four basic phases associated with sexual response and activity in humans. These phases are **excitement, plateau, orgasm**, and **resolution**.

4. Contraception and Birth Control

The reasons for and results of sexual intercourse in humans may not be associated with procreation. Couples wishing to minimize the possibility of pregnancy have many methods available to them. Birth control can be achieved with varying degrees of effectiveness employing methods such as abstinence, the rhythm method, withdrawal, condom use, diaphragm, cervical cap, spermicide, oral contraceptives, implants, intrauterine devices, vasectomy, and tubal ligation.

5.   Sexually Transmitted Diseases

It is very well understood that sexual activity can result in the transmission of certain infectious agents. The most common infectious agents are viruses and bacteria. Currently 25% of the American population has some type of STD. Since these diseases are highly communicable, precautions should always be used when sexual contact occurs. Abstinence is the only way to totally protect against STD and the risks increase as sexual activity and numbers of partners increase.

CHAPTER REVIEW ACTIVITIES

CONCEPT MAPPING

Develop a concept map that demonstrates your understanding of sexual reproduction. Try to incorporate as many of the key terms as possible. Develop the map as you conceptualize the topic and choose linking phases that clarify how you see one idea connecting to another.

MATCHING

Select the term from the list provided that best relates to the following statements. Decide on a term BEFORE looking through the list, then search the list for your answer. If you are unsure of the correct answer or chose an incorrect response, review the related concept in Chapter 21.

_____ 1. This is a sac of skin in which the testicles are housed.

_____ 2. This rubber dome is inserted immediately before intercourse to cover the cervix and prevent the entry of sperm into the uterine tubes.

_____ 3. This period marks the permanent cessation of menstrual activity in a woman, usually between the ages of 50 and 55.

_____ 4. This gland surrounds the male urethra and adds a milky alkaline fluid to the semen.

_____ 5. This is the monthly process by which an egg is produced and released by the ovary.

_____ 6. These male and female reproductive organs produce the sex cells.

_____ 7. This is a long, coiled tube that sits on the posterior side of the testes and in which sperm mature.

_____ 8. These various hormones develop and maintain the female reproductive structures, such as the ovarian follicles, the uterine lining, and the breasts.

_____ 9. These are sex cells.

_____ 10. This small mass of erectile and nervous tissue in the female genitalia responds to sexual stimulation.

_____ 11. These male gonads are where sperm production occurs.

_____ 12. This is the inner lining of the uterus.

_____ 13. These two longitudinal folds of skin run posteriorly from the mons in the exterior genitals of females.

_____ 14. These tightly coiled tubes within each testis are each approximately 750 feet in length and are where sperm cells develop.

_____ 15. This is the female organ in which a fertilized egg can develop.

_____ 16. This muscular tube brings urine from the bladder to the outside and carries semen to the outside of the body during ejaculation.

_____ 17. This is the collective term for the external genitals of a female.

_____ 18. This is the monthly sloughing off of the flood-enriched lining of the uterus when pregnancy does not occur.

_____ 19. This set of tiny accessory glands lying beneath the prostate secretes an alkaline fluid into the semen.

_____ 20. This is another term for the male reproductive cell.

_____ 21. This hormone triggers ovulation.

_____ 22. This is the hormone, along with estrogen, that is secreted by the corpus luteum.

TERMS LIST

a. Bulbourethral glands	l. Scrotum
b. Menopause	m. Clitoris
c. Progesterone	n. Ovulation
d. Prostate	o. Luteinizing hormone
e. Uterus	p. Menstruation
f. Testes	q. Vulva
g. Endometrium	r. Spermatozoan
h. Diaphragm	s. Estrogens
i. Gonads	t. Gametes
j. Urethra	u. Epididymis
k. labia	v. seminiferous tubules

MULTIPLE CHOICE QUESTIONS

_____ 1. Which of the following structures is first to be involved in sperm production?

a. Urethra

b. Seminiferous tubules

c. Epididymis

d. Vas deferens

e. Penis

_____ 2. Which of the following is severed to produce sterility in males?

a. Urethra

b. Seminiferous tubules

c. Epididymis

d. Vas deferens

e. Penis

3. Which of the following statements is TRUE of the uterus?

   a. It is attached to the oviducts and to the vagina.

   b. It is involved in placental development.

   c. It has a layer of smooth muscle to facilitate contraction.

   d. It is the place where the fertilized egg will develop.

   e. All of the above

4. Which of the following is stimulated by FSH?

   a. Development of follicles

   b. Development of the endometrium

   c. Onset of menstrual flow

   d. A and B

   e. B and C

5. Prior to implantation of the fertilized egg in the uterus, secretions from which of the following would be necessary?

   a. Pituitary gland

   b. Corpus luteum

   c. Vagina

   d. B and C

   e. A and B

6. Which of the following relates best to the prostate gland?

   a. A vasectomy is performed to remove it.

   b. It adds a milky alkaline fluid to semen.

   c. It is necessary for an erection to occur.

   d. It is the site for spermatogenesis.

   e. B and D

7. Which of the following hormones is secreted by the pituitary to stimulate the production of testosterone?

   a. FSH

   b. Progesterone

   c. LH

   d. GRH

   e. HCG

_____ 8. The beginning of menstruation occurs in response to which of the following?
   a. Decreased progesterone levels
   b. Increased estrogen levels
   c. Rupture of the ovarian follicle
   d. Changes in body temperature
   e. Release of FSH from the pituitary

_____ 9. Which of the following occurs during ovulation?
   a. The polar body divides into two ooctyes.
   b. The secondary oocyte breaks away from the ovary.
   c. Oogonia
   d. Estrogen concentrations are at their lowest levels.
   e. Uterine lining is shed.

_____ 10. Each month, approximately how many primary ooctyes begin to develop within an ovary?
   a. Thousands
   b. Millions
   c. Hundreds
   d. Twenty-five
   e. Tt varies significantly from month to month.

_____ 11. Which of the following birth control methods interfere with implantation?
   a. Cervical cap
   b. Intrauterine devices
   c. Diaphragm
   d. Implants
   e. Sponge

_____ 12. Where does the process of fertilization occur?
   a. Uterus
   b. Vagina
   c. Vulva
   d. Oviduct
   e. Ovary

13. Sperm is stored in which of the following structures?
   a. Urethra
   b. Penis
   c. Epididymis
   d. Seminiferous tubules
   e. Testes

14. Which of the following hormones is involved in the female orgasm?
   a. Oxytocin
   b. Adrenalin
   c. FSH
   d. LH
   e. Estrogen

15. What is the percent of failure rate of the rhythm method of birth control?
   a. 2%
   b. 20%
   c. 10%
   d. 50%
   e. 0%

16. Which of the following terms are synonomous with intercourse?
   a. Coitus
   b. Copulation
   c. Orgasm
   d. B and C
   e. A and B

17. Which of the following structures in the female is homologous to the tip of the penis in males?
   a. Labia minora
   b. Labia majora
   c. Clitoris
   d. Vagina
   e. Mons pubis

_____ 18. Which of the following structures is considered to be the first accessory gland to add fluid to the sperm?

a. Seminal vesicles

b. Bulbourethral glands

c. Prostate gland

d. Scrotum

e. Testicles

HUMAN MENSTRUAL CYCLE

Answer the following questions regarding the information provided in the graphic illustration below.

Fig. 21.1 Menstrual cycle.

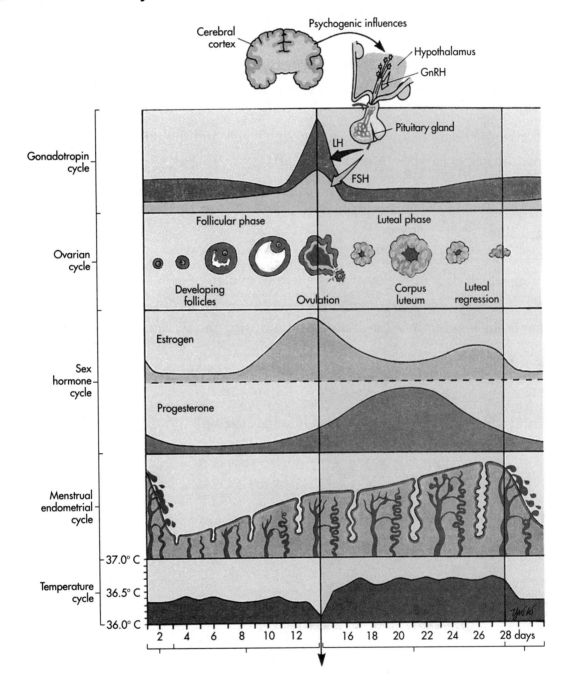

1. Approximately how many days does the follicular phase last?

2. Approximately how long is the menstrual phase?

3. During what days (0-28) do the levels of estrogen begin to rise? What is the condition of the endometrium during this rise?

4. During which days are the levels of progesterone the highest? What is occurring in the ovarian cycle at this time?

5. Describe the relationship between estrogen levels and the release of gonadotropic hormones from the pituitary.

6. What changes in the endometrial lining can be associated with luteal regression?

7. During which days of the ovarian cycle does ovulation occur?

8. What is the condition of the endometrium during the secretory phase?

9. What happens to the body temperature during ovulation?

10. During which phase does the corpus luteum develop?

LABELING

Label the following diagrams.

Fig. 21-2 Male reproductive structures.

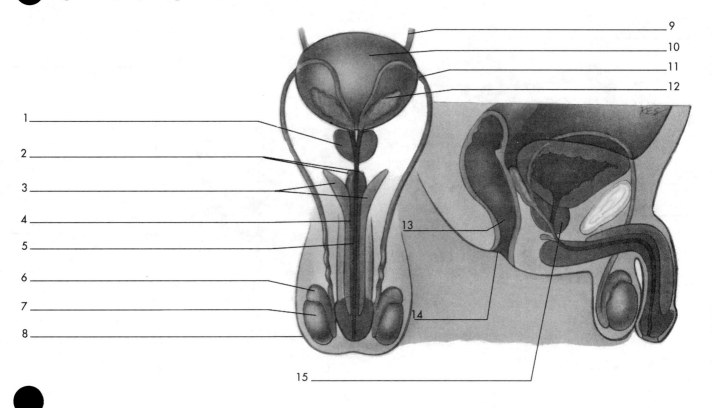

Fig. 21.3 Testicle.

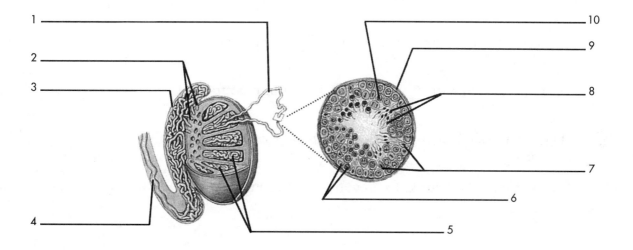

Fig. 21.4 Female reproductive structures.

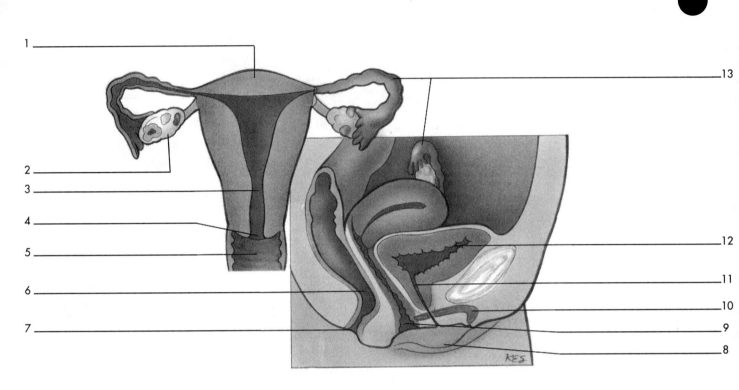

SHORT ANSWER QUESTIONS

1. Differentiate between the male sex organ and copulatory organ.

2. Describe the development of a functional female gamete.

3. List and describe the four stages of sexual intercourse.

4. Outline the steps involved in spermatogenesis.

5. List some methods used to prevent conception.

CRITICAL QUESTIONS

1. What would the effects be if a testis fails to descend within the scrotal sac?

2. The process of in vitro fertilization occurs outside the body of the female. How can fertilization be verified?

3. Explain why the risk of genetic abnormalities in the fetus is more likely as females age.

TELLING THE STORY

Develop a text that demonstrates your understanding of sex and reproduction in humans. Focus on similarities and differences in female and male reproductive systems. Discuss your understanding of fertilization and the events that lead up to the production of gametes. Use as many terms from the chapter as you can. If you have difficulty with any concepts as you write, go back to Chapter 21 in your text and review them.

CHAPTER REVIEW ANSWERS

MATCHING

1. L	7. U	13. K	19. A
2. H	8. S	14. V	20. R
3. B	9. T	15. E	21 O
4. D	10. M	16. J	22. C
5. N	11. F	17. Q	
6. I	12. G	18. P	

MULTIPLE CHOICE QUESTIONS

1. B	4. A	7. C	10. D	13. C	16. E
2. D	5. E	8. A	11. C	14. A	17. C
3. E	6. B	9. B	12. D	15. B	18. A

HUMAN MENSTRUAL CYCLE

1. 14 days

2. 5 to 7 days

3. Levels of estrogen begin to rise at approximately day 9 or at the onset of the proliferative phase
   The endometrium is beginning to thicken.

4. During the secretory phase or luteal phase, the levels of progesterone are the highest. The corpus luteum is forming at this time.

5. At the height of LH and FSH levels in the blood, estrogen levels are at their peak. This is in conjunction with the process called ovulation.

6. The beginning of the endometrial shedding or menstruation occurs in conjunction with the luteal regression.

7. Ovulation occurs around day 14.

8. The endometrium is at its thickest level in preparation for the implantation of a fertilized egg.

9. It decreases to approximately 36^0 C.

LABELING

FIG. 21.2

1. Prostate gland	6. Epididymis	11. Vas deferens
2. Bulbourethral glands	7. Testis	12. Seminal vesicle
3. Corpus cavernosum	8. Scrotum	13. Rectum
4. Corpus spongiosum	9. Ureter	14. Anus
5. Urethra	10. Bladder	15. Prostate gland

FIG. 21.3

1. Seminiferous tubule	6. Spermatogonia
2. Efferent ductules	7. Sertoli cells
3. Epididymis	8. Mature sperm cells
4. Ductus deferens	9. Basement membrane
5. Lobules	10. Spermatocyte

FIG. 21.4

1. Uterus	6. Rectum	11. Urethra
2. Ovary	7. Anus	12. Bladder
3. Cervical canal	8. Labia	13. Fallopian tube
4. Cervix	9. Vagina	
5. Vagina	10. Clitoris	

SHORT ANSWER QUESTIONS

1. The testes are the sex organs because they are responsible for the production of sperm cells. The penis is the copulatory organ because it is the male organ used in sexual intercourse.

2. During embryonic development, the female possesses all the primary oocytes she will ever produce. The number declines significantly by puberty to about 400,000. The primary oocytes are halted in meiosis I during prophase. With the onset of puberty, one or sometimes two of the follicles develop and the secondary oocyte will complete meiosis I. The follicle ruptures releasing the secondary oocyte to travel to the oviduct where it will complete the second meiotic division. The female gamete is then ready for potential fertilization.

3. The four stages of sexual intercourse are:
    Excitement phase - sexual activity that precedes intercourse
    Plateau phase - a period that intensifies the physiological changes that wereinitiated during the excitement phase
    Orgasm - a series of reflexive muscular contractions
    Resolution phase - slow paced return to normal physiological states

4. Diploid spermatogonia develop into primary spermatocytes that undergo meiosis I. After the first meiotic division, the secondary spermatocytes are produced and undergo the second meiotic division to produce haploid spermatids. The spermatids undergo maturation to become motile, functional sperm.

5. Contraceptive methods include abstinence, rhythm method, withdrawal, condoms, diaphragm, cervical cap, sponge, spermicides, oral contraceptives, implants, intrauterine devices, vasectomy, tubal ligation.

CRITICAL QUESTIONS

1. In the scrotal sac, the temperature is lower since the sac is outside the pelvic cavity. The muscles in the scrotum can contract to bring the testes closer to the body if it is necessary to raise the temperature or can relax to lower them and reduce their temperature. If a testis does not descend into the scrotum, it will remain in the pelvic cavity at normal body temperature, which is too warm for normal sperm cell development. This can result in sterility due to the lack of sufficient viable sperm.

2. The process of fertilization initiates meiosis in the secondary oocyte. The production of the second polar body would be an indication that fertilization had occurred.

3. As females age, the viability and number of eggs declines. Since the eggs are much older, the risk of mutation or damage is greater and the result could be genetic abnormalities in the fetus.

•CHAPTER 22

DEVELOPMENT BEFORE BIRTH

KEY CONCEPTS OVERVIEW

1.   Fertilization

Fertilization is defined as the union of a male gamete (sperm) and a female gamete (egg). The number of sperm produced is very high to assure that at least one reaches a viable egg. Males ejaculate more than 200 million sperm cells at once because they do not live long after leaving the male reproductive system. Due to the acidic conditions in the vagina and the downward current of female secretions, the sperm cells have to contend with less than favorable conditions in order to reach the egg. Timing really is everything in this process. Successful fertilization generally occurs when the sperm and egg are released at about the same time.

Following ejaculation into the vagina, sperm swim through strands of **mucin**, which is a special protein secreted by females at the time of ovulation. Sperm use this material to move toward the uterus. The motile process wears away inhibitor molecules stored on the head of the sperm. Exposed enzymes are necessary for the sperm to break down the outer protective layers on the egg's surface. Penetration of a sperm into the oocyte causes surface transformations on the egg that prevent other sperm from entering. When the sperm penetrates the secondary oocyte, the second meiotic division will be completed. At this time a mature ovum and a second polar body are produced. Following the fusion of the sperm and egg cell nuclei, a new cell or **zygote** is formed. These events take place in the upper regions of the oviduct. (See Figure 22-2 A&B, P. 469 and Figure 22-3 A&B, P 469 in your text)

2.   Early Development

Throughout the 2 weeks following fertilization, the developing cell mass is called the **pre-embryo**. Cells mitotically divide to form this undifferentiated cell mass. As the pre-embryo moves through the oviduct toward the uterus, cell division continues. The resulting cell mass, now 3 days after fertilization, consists of 16 cells and is referred to as a **morula.** It is the morula that reaches the end of the oviduct and the entrance to the uterus. After 2 days in the uterus, the free moving morula develops into a hollow cell ball called a **blastocyst** or **blastula.** This structure consists of an **inner cell mass** that will eventually differentiate into specific cells and an outer **trophoblast** that develops into embryonic membranes that will protect and nuture the developing embryo.

Fig. 22.1 Blastula.

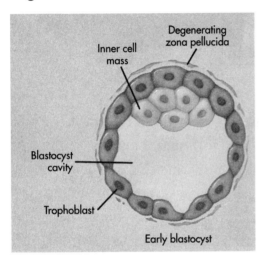

After 7 days, the blastocyte implants (**implantation**) into the uterine wall. At this time human chorionic gonadotropin (HCG) hormone is secreted to maintain the corpus luteum so that it continues to secrete progesterone. Progesterone prevents the loss of the uterine lining.

In the second week of development, the **three primary germ** layers develop: the **ectoderm, endoderm,** and **mesoderm.** The ectoderm will eventually give rise to the outer layer of skin, the nervous system, and portions of the sense organs. The endoderm will give rise to the lining of the digestive tract, the digestive organs, the respiratory tract, the lungs, the urinary bladder, and the urethra. The mesoderm will give rise to the skeleton, muscles, blood, reproductive organs, connective tissue, and the inner layer of the skin.

During the same 2 weeks of development, the **extraembryonic membranes** form. These membranes provide structure and nutrition. The **chorion,** or outermost membrane, is the primary part of the **placenta.** The placenta assures that nutrients, gases, and waste will be moved appropriately between the mother and the embryo. The placenta is attached to the embryo by the **umbilical cord,** which develops from the **body stalk,** the **yolk sac**, and the **allantois.** Arteries and veins in the umbilical cord are developed from the allantois. Waste and oxygen depleted blood are transported from the placenta through the arteries in the umbilicus.

3.   Third through Eighth Week of Development

During the third and fourth week a thin, protective membrane, called the **amnion,** grows around the embryo. Inside this membranous envelope is the **amniotic fluid,** which protects the fetus by helping to support it, aiding in temperature control, and giving the fetus freedom to move.

At the third week of development what was referred to as the pre-embryo is now called the **embryo.** Many important changes take place during this time. For example, certain cells within the inner cell mass divide and form a three-layered embryo. This process in known as **gastrulation.** The **notochord** develops at the head end of the primitive streak and completes the gastrulation process. The heart of the embryo begins to develop. The cardiovascular system is actually the first system that is fully functional in the embryo. There is also the formation or identification of mesodermal bodies called **somites,** which ultimately give rise to most of the axial skeleton, the associated muscles, and most of the dermis.

Sometime within that third week of development, the formation of the **neural groove** occurs. This process, called **neurulation,** results in the development of a hollow nerve cord that subsequently develops into the brain, spinal cord, and other related structures. The embryo is now called a **neurula** and this stage signals the process of **tissue differentiation**. Tissues begin to differentiate or specialize through signals produced by certain cells. This process, called **induction,** is controlled by certain cells that act as "on and off" switches for the genes of neighboring cells.

The fourth week of development is noted by a change in the fetal posture. At this time the fetus begins to curl as folds are formed at the head and tail ends. The **branchial (gill) arches** develop on the sides of the head region. These arches eventually develop into the middle ear, eustachian tubes, thymus, parathyroid, and the tonsils.

During the fifth week, the limbs begin to appear, developing from the **limb buds.** A nose appears as tiny pits and the brain grows very rapidly resulting in a disproportioned head on the embryo. At this time the fetus is about 8 millimeters long.

The sixth week is marked by the appearance of fingers and ears. The trunk straightens somewhat, but the head remains bent forward. The liver and digestive tracts continue to grow.

4.   The Last Trimester

This is a time of growth rather than development. Labor, the series of events leading to birth, is initiated by hormonal changes occurring in the fetus. These hormonal changes cause the placenta to release **prostaglandins** that cause the secretion of **oxytocin** from the mother's pituitary gland. Oxytocin causes the uterus to contract, which induces the release of more oxytocin. This continuous cycle is an example of a **positive feedback loop.** Working together, prostaglandins and oxytocin trigger waves of contractions in the uterine walls forcing the fetus downward. Eventually, these contractions result in the birth of the infant. Contractions continue after the birth to expel the placenta and its associated membranes. The umbilical cord is severed and the baby begins its life outside the womb.

The first breath taken by the baby induces the lungs to function for the first time. The circulatory system adjusts to the new source of oxygen by the closure of the **foramen ovale** and the **ductus arteriosus.**

CHAPTER REVIEW ACTIVITIES

CONCEPT MAPPING

Develop a concept map that demonstrates your understanding of human development up to the birthing process. Use as many of the key terms as possible, but make sure you really understand how they fit into the theme of the map. Try not to "force" a term into the map if it doesn't fit.

MATCHING

Select the term from the list provided that best relates to the following statements. If you are unsure of the correct answer or chose an incorrect response, review the related concept in Chapter 22.

_____ 1. This term is used to describe the stage of prenatal development after 8 weeks.

_____ 2. This extraembryonic membrane gives rise to the umbilical arteries and veins as the umbilical cord develops.

_____ 3. This term describes the development of a hollow nerve cord, which later develops into the brain, spinal cord, and related structures, such as the eyes.

_____ 4. This term describes the vagina, through which the fetus is expelled during birth.

_____ 5. This is the embedding of the developing blastocyst into the posterior wall of the uterus approximately 1 week after fertilization.

_____ 6. In this stage of development the pre-embryo consists of about 16 densely clustered cells.

_____ 7. This is the union of a female and male gamete.

_____ 8. This is the process by which some cells switch on and off the genes of neighboring cells.

_____ 9. During this process by groups of inner cells mass migrate, divide, and differentiate into three primary germ layers from which all the organs and tissues of the body develop.

_____ 10. This germ layer forms the outer layer of skin, the nervous system, and portions of the sense organs.

_____ 11. This extraembryonic membrane facilitates the transfer of nutrients, gases, and wastes between the embryo and the mother's body.

_____ 12. This early stage of development in a vertebrate occurs when cells begin to move and results in the shaping of the new organism.

_____ 13. This structure forms the midline axis along which the vertebral column develops in all vertebrate animals.

_____ 14. This membranous sac surrounds the food yolk and produces blood for the embryo until its liver becomes functional.

_____ 15. This flat disc of tissue grows into the uterine wall, through which a mother supplies the offspring with food, water, and oxygen.

_____ 16. This primary germ layer gives rise to the digestive tract lining, the digestive organs, the respiratory tract, the lungs, the urinary bladder, and the urethra.

_____ 17. This outer ring of cells of the blastocyst will give rise to most of the extraembryonic membranes as the embryo develops.

_____ 18. This term refers to the developing cell mass formed by the zygote as it begins to divide by mitotic cell division.

_____ 19. During this period of pregnancy, the fetus grows to about 1 1/4 pounds and about 1 foot in length.

_____ 20. During this stage of development, the pre-embryo is a hollow ball of cells.

_____ 21. This new cell contains intermingling genetic material from both the sperm and egg cells.

_____ 22. During this period of pregnancy, the fetus reaches about 3 1/2 inches in length and growth is very rapid.

_____ 23. This is a mature egg cell.

_____ 24. This process of bone formation, begins at the eighth week of fetal development and continues to approximately the age of 18 years.

_____ 25. This primary germ layer differentiates into the skeleton, muscles, blood, reproductive organs, connective tissue, and the innermost layer of the skin.

_____ 26. These structures form from the trophoblast and provide nourishment and protection.

TERMS LIST

a. Yolk sac	j. Allantois	s. Fertilization
b. Chorion	k. Notochord	t. Second trimester
c. Birth canal	l. Ovum	u. Trophoblast
d. Fetus	m. First trimester	v. Mesoderm
e. Zygote	n. Endoderm	w. Ossification
f. Blastocyst	o. Neurulation	x. Pre-embryo
g. Ectoderm	p. Gastrulation	y. Induction
h. Morula	q. Placenta	z. Implantation
i. Morphogenesis	r. Extraembryonic membrane	

LABELING

Label the diagram with the correct terms for the numbered structures.

Fig. 22.2 Release of egg.

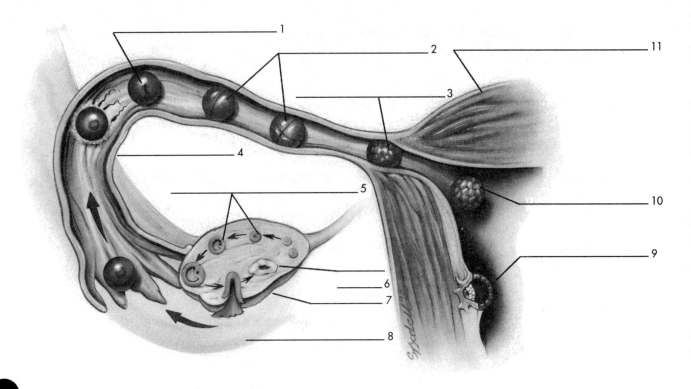

Fig. 22.3 Embryo development.

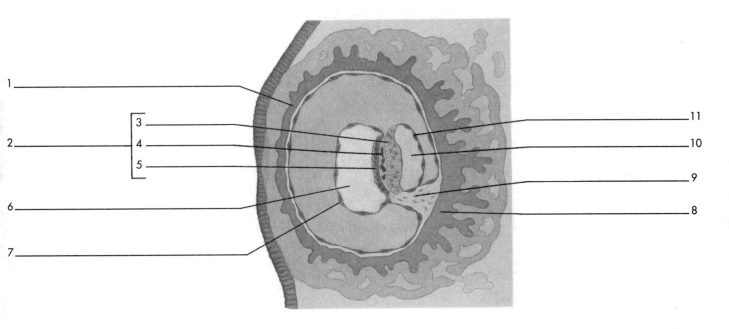

IF.... THEN.... CHALLENGE QUESTIONS

Using the "IF" part of the statements below, complete the "THEN" part with guidance from the accompanying questions.

1. If the amnion develops during this time, then _____. What week of development is it?

2. If a hollow nerve cord develops and the eyes can been seen, then _____. What week of development is it?

3. If progesterone and estrogen levels drop, then _____. What effects would result?

4. If the blastocyst implants in the uterine lining, then _____. What hormone would be secreted and what effects does this hormone have?

5. If a pre-embryo becomes caught in the folded inner lining of the uterine tube, then _____. What is the resulting condition?

6. If the fetus is about 12 inches long, then _____. What is the period of pregnancy?

7. If an ear can be seen on the side of the head, then _____. What week of development is it?

8. If insufficient production and/or secretion of oxytocin happens, then _____. What happens during the birthing process?

MULTIPLE CHOICE QUESTIONS

_____ 1. Gentle rocking hastens the development of which of the following
   a. Cerebrum
   b. Cerebellum
   c. Foramen ovale
   d. Heart
   e. Lungs

_____ 2. Where does embryonic development begin?

    a. Uterus

    b. Follicle

    c. Oviduct

    d. Ovary

    e. Cervix

_____ 3. For what period of time does the yolk in a human egg provide food?

    a. 2 days

    b. 1 week

    c. 2 weeks

    d. 1 month

    e. 9 months

_____ 4. Which of the following is true of the protein mucin?

    a. It forms the mucus plug over the cervix.

    b. It is in one of the enzymes used in the formation of the placenta.

    c. It is released during the menstrual flow.

    d. It forms tracts for sperm to travel through the vagina to the oviduct.

    e. It controls the way the egg divides during cleavage.

_____ 5. Which of the following words is derived from the Latin word for mulberry?

    a. Morula

    b. Morphogenesis

    c. Menstruation

    d. Maturation

    e. Menopause

_____ 6. Which of the following structures is produced by invagination?

    a. Gastrula

    b. Blastula

    c. Morula

    d. Neurula

    e. Blastocyst

_____ 7. In which of the following stages does implantation occur?

a. Gastrula

b. Blastula

c. Morula

d. Neurula

e. Blastocyst

_____ 8. The blastocyst secretes which of the following substances?

a. Human chorionic gonadotropin

b. Luteinizing hormone

c. Follicle-stimulating hormone

d. Estrogen

e. All of the above

_____ 9. Which of the germ layers form last in the development process?

a. Ectoderm

b. Mesoderm

c. Endoderm

d. Blastoderm

e. Epiderm

_____ 10. Which of the following is the extraembryonic membrane that functions as a shock absorber?

a. Allantois

b. Amnion

c. Chorion

d. Yolk sac

c. Mucin

_____ 11. Which of the following is the extraembryonic membrane associated with the placenta?

a. Allantois

b. Amnion

c. Chorion

d. Yolk sac

c. Mucin

_____ 12. Which of the following stages is the latest stage in embryonic development?

a. Gastrula

b. Blastula

c. Morula

d. Neurula

e. Blastocyst

_____ 13. Which of the following systems becomes functional in the embryo first?

a. Cardiovascular

b. Digestive

c. Pulmonary

d. Endocrine

e. They all become functional at the same time

_____ 14. Which of the following structures is NOT derived from the gill arches?

a. Middle ear

b. Lungs

c. Tonsils

d. Eustachian tube

e. Thymus

_____ 15. Which of the following features is NOT an embryonic feature in all chordates?

a. Gill arches

b. Notochord

c. Tail

d. Dorsal nerve cord

e. Both B and C

_____ 16. Which of the following occurs first?

a. The mother secretes oxytocin from pituitary.

b. There is pressure from the baby's head against the cervix.

c. The placenta secretes prostaglandins.

d. The uterus contracts and the cervix stretches.

e. There are changes in fetal hormone levels.

SHORT ANSWER QUESTIONS

1. What is the difference between identical and fraternal twins?

2. What means might be used to determine if a fetus has a genetic defect?

3. What is the function of the chorionic villi?

4. List the four extraembryonic membranes.

5. Which is responsible for the onset of labor - the mother or the fetus? Explain.

6. When is the embryo first appropriately referred to as a fetus?

CRITICAL QUESTIONS

1. Why are cell division and growth alone not sufficient for development?

2. What is it about the process of fertilization that triggers the development process?

3. The formation of the three distinct germ layers is actually the result of cell to cell interaction. Why is this so?

TELLING THE STORY

Develop a story that demonstrates your understanding of early human development. Use as many terms from the chapter as you can. If you have difficulty with any concept as you write, go back and review the specific section in Chapter 22.

CHAPTER REVIEW ANSWERS

MATCHING

1. D	7. S	13. K	19. T	25. V
2. J	8. Y	14. A	20. F	26. R
3. O	9. P	15. Q	21. E	
4. C	10. G	16. N	22. M	
5. Z	11. B	17. U	23. L	
6. H	12. I	18. X	24. W	

LABELING

Fig. 22.2

1. First mitosis	5. Developing follicles	9. Implantation
2. Cleavage	6. Corpus luteum	10. Blastula
3. Morula	7. Ovary	11. Uterus
4. Fertilization	8. Ovulation	

Fig. 22.3

1. Chorion	5. Endoderm	9. Body stalk
2. Germ layers	6. Yolk sac	10. Amniotic cavity
3. Ectoderm	7. Extraembryonic membrane	11. Amnion
4. Mesoderm	8. Chorionic villi	

IF.... THEN.... CHALLENGE QUESTIONS

1. Between the 3rd and 4th
2. 4 1/2 weeks
3. The uterine lining would be shed.
4. HCG would be secreted and the corpus luteum would produce estrogen and progesterone.
5. Ectopic pregnancy
6. About the sixth month or the end of the second trimester
7. Twelveth week
8. Contractions of the uterus would not occur or would be slowed and less forceful.

MULTIPLE CHOICE QUESTIONS

1. B	5. A	9. C	13. A
2. C	6. A	10. B	14. B
3. B	7. E	11. C	15. D
4. D	8. A	12. D	16. E

SHORT ANSWER QUESTIONS

1. Fraternal twins are called dizygotic twins because they are produced from the fertilization of two different eggs. They have no more biological characteristics alike than any other siblings. Identical twins are called monozygotic twins and are produced when a single zygote splits and both halves develop to maturity. Identical twins are same sex and have identical genotypes.

2. Embryonic cells floating in the amniotic fluid or cells from the chorionic villi can be karyotyped. It is then possible to examine the chromosomes to locate any physically observable aberrations or changes in chromosome number.

3. The chorionic villi dramatically increases the surface area where the embryo comes in contact with the mother.

4. The four extraembryonic membranes are the allantois, amnion, chorion, and yolk sac.

5. The fetus initiates the changes that cause prostaglandins to be released from the placenta and oxytocin from the pituitary.

6. The human embryo is first appropriately called a fetus at the end of the second month or ninth week of pregnancy. At this time, the embryo definitely may be identified as human. Prior to this time, the embryo resembles that of many other vertebrate animals.

CRITICAL QUESTIONS

1. Cell division and growth alone are not sufficient in the development process because it is essential that cells differentiate at some point and begin to develop into tissues with distinct functions and structures.

2. Fertilization assures that a complete set of genetic material is available in the zygote. Prior to this, only individual haploid cells existed and they are unable to complete any requirements for cell division and differentiation without a complete complement of genetic material.

3. The three distinct layers of tissue are initially developed when cells can recognize their own kind. The proximity of one cell to another occurs after cells have migrated to their specific sites. This closeness promotes differentiation and the resulting three germ layers.

CHAPTER 23

THE SCIENTIFIC EVIDENCE FOR EVOLUTION

KEY CONCEPTS OVERVIEW

1.   Development of Darwin's Theory of Evolution

In 1831, Charles Darwin set sail on an adventure that would change the way the science world would view the origin of species. At the age of 22, he took the opportunity to serve as a naturalist aboard H.M.S. Beagle and for the next 5 years, he would sail around the world. Most of the voyage took place in the southern hemisphere. Darwin began his trip convinced, as most people were at the time, that species did not change over time. Darwin believed that **species** are a population of organisms that interbreed freely in the wild and do not interbreed with other populations. Following his 5 year adventure, Darwin returned to write about his observations. He began to formulate a theory that integrated what he had seen with his knowledge of geology, population biology, and fossil records. By 1838, Darwin was proposing a very different view about the diversity of life on Earth.

Some of the observations Darwin made were key to the development of his theory on evolution. He noted the similarity between fossils and living organisms in the same geographic area. He began to surmise that the fossilized structures were distantly related to the present day organisms. He also noted that even though the climates of geographical areas such as Australia, Chile, and South Africa were much the same, they were populated by plants and animals that were similar but not the same species. This suggested to Darwin that there may be other factors in addition to or other than climate that caused the **diversity** in plants and animals.

The volcanic Galapagos Islands held several clues to the questions posed by Darwin. The organisms he found there resembled the organisms found on the mainland of South America, but interestingly they also showed variation from one Galapagos island to another. From these observations, Darwin concluded that the

Galapagos organisms were related to ancestors that came to the islands from the mainland.

All of the observations and conclusions Darwin made over his 5 year journey were summarized into the following theories:

1. Organisms of the past and present are related to one another.
2. Factors other than, or in addition to, climate play a role in the development of plant and animal diversity.
3. Members of the same species often change slightly in appearance after becoming geographically isolated from one another.
4. Organisms living in oceanic islands often resemble organisms found living on a close mainland.

Darwin's theories were constructed from the data he gathered on his journey and his reflections on their significance in light of what was understood about geology, breeding of domesticated plants and animals, and population biology. One important contribution of geologists was the indication that the earth was much older than the 6,000 to 10,000 years suggested by the literal translation of the Bible. This allowed enough time for evolution to occur. Evidence suggested that both the physical world and the organisms living on it had changed over the course of time. Artificial breeding of pigeons revealed how much variation could be produced by **artificial selection**; that is, when humans choose which organisms would survive and reproduce. Artificial selection is based on the natural selection that all organisms exhibit and can be used to select for characteristics deemed to be valuable or useful.

Darwin, interested in the study of **populations**, read the works of Thomas Malthus, an economist is the late 1700s. His *Essay on the Principles of Population* demonstrated that populations of organisms tended to increase geometrically, while food supplies grew arithmetically. From this Darwin realized that in nature species are subjected to limits on the size fo their populations.

2.   Natural Selection

Noting that environmental factors could influence which organisms in a population survived, Darwin began to offer explanations as to why this occurred. He made associations between the process of artificial breeding and reproduction within natural populations. He indicated that some of the naturally occurring variations within a population provided those individuals which possessed them a better chance for survival. Additionally, these surviving organisms could breed and pass these favorable traits on to subsequent generations. Darwin called this process, in which organisms having adaptive traits survive in greater numbers than those without such traits, **natural selection**. This driving force of change is often referred to as "**survival of the fittest**." Through differential survival and reproduction, a population will undergo change through time. This change is called **evolution**.

Perhaps the most noted organisms Darwin observed were the finches of the Galapagos Islands. These birds resembled their mainland relatives, but had been on the islands long enough (perhaps 2 million years) to have evolved into 13 different species. This process of rapid proliferation into many species is called **adaptive radiation**. This occurs when a population of a species changes as it is dispersed within a series of different habitats within a specific region. The birds invaded different habitats, found little competition for necessities, adapted to features they encountered, and through time became very different. In some of these populations, the changes were so drastic that interbreeding was no longer possible. This process of forming a new species during evolution is called **speciation**. The morphological features that enabled the birds to successfully become established were modifications of their beaks. These modifications resulted in new food sources and a reduction in competition between the finches. The most striking finch Darwin encountered was the woodpecker finch. If there had been woodpeckers living on the island prior to this, they would have been more efficient and it would be unlikely that a woodpecker finch could have competed and survived.

3.   Publication of Darwin's Theory

Charles Darwin worked for some 20 years patiently amassing evidence and refining his argument for his theory of evolution. Concurrently with Darwin, a naturalist named Alfred Wallace had been studying the animals in the South Pacific. He too developed a theory of evolution by natural selelction and, in 1858, Wallace sent his paper to Darwin for review. Both Darwin and Wallace published a joint paper as co-discoverers of the **theory of evolution by natural selection**. However, Darwin is usually given the credit for this theory because he published his book <u>On The Origin of Species</u> in 1859. The book created considerable debate over the definition of the term "species" and why, despite ongoing selection, species continue to possess a seemingly inexhaustible supply of variation. He explained that since species continue to evolve, they cannot be defined. Although the theory of evolution by natural selection underwent modification and revision and encountered both support and opposition, it is regarded as one of the cornerstones of biology.

4.   Testing the Theory

Fossil records provide evidence to support the theory of evolution. **Fossils** are any indication that an organism lived in the past. When organisms died they were deposited and preserved in **sedimentary rocks** in layers or **strata**. Fossils are distributed in a world wild pattern called "faunal succession." Information from this pattern provided support for the construction of the geological time scale by as early as the mid 19th century. Fossils found in the same strata are presumed to have been formed at the same time. Fossils can be dated by where they occur in the strata and by **radioactive dating**. This direct method of dating fossils became available in the late 1940s and enables scientists to assign real time dates to the geological time scale. Radioactive dating depends on naturally occurring isotopes of certain elements found in the rock. Many different isotopes are used in this process. One of the most useful isotopes is carbon 14, which has a 5730 year **half life** (the speed of radioactive decay). Any organic molecule contains carbon. This carbon, in living organisms, is in a constant ratio of $^{12}C/^{14}C$, which is the same as that found in the atmosphere and in plants that carry on photosynthesis. When an organism dies, the amount of stable ^{12}C remains constant, while the ^{14}C undergoes radioactive decay at predictable rates. A simple ratio can then be set up to determine the accurate age of the the fossil. Fossils older than 50,000 years have too little ^{14}C remaining to measure the precise age. These fossils may be dated with the isotope thorium-230, which has a half life of 75,000 years.

Until the time of Darwin, most held the belief that the Earth was approximately 6,000 to 10,000 years old. Radioactive dating of rocks has suggested that the geological age of the Earth is about 4.6 billion years.

Comparative studies of anatomy demonstrate a strong support for evolution. Scientists look for common ancestory by comparing the similarity in some structures of organisms. If organisms are derived from the same ancestor, they should possess similar structures with some modifications from adaptations to their environment. When structures are of the same evolutionary origin, but now have different structures and functions, they are said to be **homologous**. Homologous structures that are present in an organism in a diminished size but are no longer useful are called **vestigial organs**. In contrast to these homologous structures, similar structures often evolve within organisms that have developed from different ancestors. These are called **analogous** structures and they have a similar form and function. The presence of analogous structures shows how natural selection can lead to similar but not identical anatomical structures. This change over time is called **convergent evolution**.

Embryologists determined that various groups of organisms, although different as adults, possessed early developmental stages that were very similar. These similar developments suggest to scientists that similar genes are at work during the early stages of development in related organisms. Over time new instructions are added to the old and the results produce organisms that are different from one another.

Molecular biologists are able to study the progressive evolution of organisms by observing their DNA. Organisms more distantly related to one another will have a greater number of changes in their DNA than those more closely related. The presence of vestigial structures shows that change has occurred through time.

5.   Evolutionary Tree

An evolutionary tree is constructed to demonstrate the patterns of relationships among major groups of organisms. These graphs constructed from the analysis of molecular differences among organisms are similar to those constructed from anatomical data. These comparisions yield information that can be used to estimate the rates at which evolution is occurring in different groups of organisms.

6.   Mode and Tempo of Evolution

In 1972, Nile Eldredge and Stephen Jay Gould produced a theory called **punctuated equilibrium** that states that new species arise suddenly and rapidly as small subpopulations of a species split from their original population. This is in contrast to Darwin's theory, subsequently termed **phyletic gradualism**, which asserts that new species occur slowly and gradually as an entire species changes over time. While both theories have played a role in evolution, among those lineages studied, the punctuated pattern appears the most often.

CHAPTER REVIEW ACTIVITIES

CONCEPT MAPPING

Develop a concept map that demonstrates your understanding of the theory of evolution. Use as many of the key terms as possible, but make sure you really understand how they fit into the theme of the map. Try to construct the map to reflect how <u>you</u> see the terms relating to each other.

MATCHING

Select the term from the list provided that best relates to the following statements. If you are unsure of the correct answer or chose an incorrect response, review the related concept in Chapter 23.

_____ 1. This is a record of dead organisms.

_____ 2. These naturally occurring, inheritable traits present in a population of organisms confer reproductive advantages to the portion of the population that possesses them.

_____ 3. This is a graphic representation of the pattern of relationships among major groups of organisms.

_____ 4. This is the phenomenon by which a population of a species changes as it is dispersed within a series of different habitats within a region.

_____ 5. These changes over time, among different species of organisms having different ancestors, resulted in similar structures and adaptations.

_____ 6. This is when a population of organisms interbreed freely in the wild and does not interbreed with other populations.

_____ 7. This theory that states that new species arise suddenly and rapidly as small subpopulations of a species split from the populations of which they were a part.

_____ 8. This is the process in which organisms with adaptive traits survive in greater numbers than organisms without such traits.

_____ 9. This term describes a part or organ of an organism that has a similar form and function to a part or organ of another organism, but possesses different evolutionary origins.

_____ 10. This term describes genetic change in a population of organisms over generations.

_____ 11. This term is often used for the bones of different vertebrates that now differ in structure and function, although they have the same evolutionary origin.

_____ 12. This is the process by which new species are formed during the process of evolution.

_____ 13. This theory asserts that new species develop slowly and gradually as an entire species changes over time.

_____ 14. This is the type of rock in which fossils are found.

_____ 15. This term describes the process where breeders select for desired characteristics.

_____ 16. This term is used to describe the different characteristics that occur naturally from one individual to another in a population.

_____ 17. This is another way to refer to natural selection.

_____ 18. This individual was the co-developer of the theory of evolution by natural selection.

_____ 19. These organs or structures are present in an organism in a diminished size but are no longer useful.

TERMS LIST

a. Convergent evolution	h. Analogous	o. Adaptive radiation
b. Natural variation	i. Vestigial organ	p. Artificial selection
c. Homologous	j. Sedimentary	q. Speciation
d. Evolution	k. Natural selection	r. Species
e. Fossil	l. Phyletic gradualism	s. Survival of the fittest
f. Wallace	m. Evolutionary tree	
g. Adaptations	n. Punctuated equilibrium	

MULTIPLE CHOICE QUESTIONS

_____ 1. In which of the following populations would evolution by natural selection NOT be occurring?
a. Those near extinction
b. Those with fairly stable organism numbers
c. Those where competition among species is rare and resources are plentiful
d. Those having extreme variations in adaptive characteristics of organisms
e. None of the above

_____ 2. The type of selection exhibited by domesticated pigeons is known as which of the following?
a. Artificial selection
b. Adaptive radiation
c. Homologous selection
d. Natural selection
e. Survival of the fittest

_____ 3. Darwin beleived that which of the following features influenced the evolution of the finches to the greatest extent?
a. Feet
b. Wing size
c. Brain
d. Beak
e. Coloring

_____ 4. The ground finches feed on which of the following?
a. Worms
b. Insects
c. Seeds
d. Cacti
e. Fruits

_____ 5. Which of the following types of Darwin's finches was the most unususal?
   a. Vactus finches
   b. Warbler finches
   c. Vegetarian tree finches
   d. Large insectivorous tree finches
   e. Woodpecker finches

_____ 6. Who was the person who convinced Darwin to publish his research paper on evolution?
   a. Mendel
   b. Malthus
   c. Huxley
   d. Wallace
   e. Beagle

_____ 7. Which of the following is least likely to have been preserved as a fossil?
   a. An insect
   b. A horse
   c. A shark's tooth
   d. A jellyfish
   e. A coral

_____ 8. Which of the following is the most widely used radioactive isotope in the dating of fossils over 50,000 years old?
   a. Thorium
   b. Carbon
   c. Nitrogen
   d. Argon
   e. Fluorine

9. Radioactive carbon is formed when which of the following is exposed to cosmic radiation?
   a. Oxygen
   b. Carbon
   c. Nitrogen
   d. Water vapor
   e. Argon

10. The proportion of radiocarbon to stable carbon is least in which of the following?
    a. Young animals
    b. Marine algae
    c. Land plants
    d. Atmosphere
    e. Organisms that have been dead for a long time

11. Darwin and Wallace are both known to have developed the theory of evolution by natural selection independent of each other, yet Darwin is credited with this theory. Which of the following best explains the reason for this?
    a. Darwin was the first to publish the findings in a book.
    b. Darwin had traveled farther from Europe to gather his data.
    c. Darwin had a more respected reputation in the science community.
    d. The evidence Darwin presented was more extensive and believable.
    e. None of the above

12. Which of the following is NOT homologous to the others?
    a. Leg of a dog
    b. Arm of a human
    c. Wing of a bat
    d. Wing of a butterfly
    e. Flipper of a sea lion

_____ 13. The reason for branches to occur on an evolutionary tree would best be explained by which of the following?
   a. All members of a group die
   b. A new species evolves
   c. A common ancestor
   d. A different geographic region
   e. A species reappears from extinction

_____ 14. The functional resemblance between wings of a butterfly and the flipper of a sea lion is an example of which of the following?
   a. Vestigial organs
   b. Homology
   c. Analogy
   d. Artificial selection
   e. Adaptation

_____ 15. Which of the following would be required for adaptive radiation to occur?
   a. Speciation
   b. Geographic isolation
   c. Reproductive isolation
   d. Different environments
   e. All of the above

_____ 16. When speciation results, which of the following is the first way isolation between organisms occurs?
   a. Geographically
   b. Genetically
   c. Reproductively
   d. Homologously
   e. Mechanically

SHORT ANSWER QUESTIONS

1. List some examples of vestigial organs in humans.

2. List the theories that Darwin proposed from the observations he made on his 5 year journey.

3. Explain the difference between exponential progression and arithmetic progression.

CRITICAL QUESTIONS

1. Write a scenario that describes the effects of geographic isolation.

2. Based on the information you have about Darwin's research, what do think were his hypotheses and how were they derived?

3. Explain why there are so many different types of dogs that look very different if they all belong the the same species.

4. Develop an argument that might be presented by a punctuationalist in regard to the rate of evolutionary change.

TELLING THE STORY

Develop a story that demonstrates your understanding of the scientific evidence of evolution. Try to relate the known facts with the theory as it is presented. Use as many terms from the chapter as you can. If you have difficulty with any concept as you write, go back and review the specific section in Chapter 23.

CHAPTER REVIEW ANSWERS

MATCHING

1. E	5. A	9. H	13. L	17. S
2. G	6. R	10. D	14. J	18. F
3. M	7. N	11. C	15. P	19. I
4. O	8. K	12. Q	16. B	

MULTIPLE CHOICE

1. C	5. E	9. C	13. C
2. A	6. D	10. E	14. C
3. D	7. D	11. A	15. E
4. C	8. A	12. D	16. A

SHORT ANSWER QUESTIONS

1. Appendix, wisdom teeth, coccyx or tail bone, muscles that facilitate wiggling of the ears

2. a. Organisms of the past and present are related to one another
    b. Factors other than, or in addition to, climate play a role in the development of plant and animal diversity.
    c. Members of the same species often change slightly in appearance after becoming geographically isolated from one another.
    d. Organisms living in oceanic islands often resemble organisms found living on a close mainland.

3. In an exponential progression, a population increases as its number is multiplied by a constant factor. In arithmetic progression, the elements increase by a constant difference, as in the example progression 2, 6, 10, 14, etc.

CRITICAL QUESTIONS

1. Answers will vary. An example might look like the following: An irrigation ditch is built across a field inhabited by field mice. Following a 10-year study of changes in the mouse population on either side of the ditch, a researcher

notes that the mice on the east side have lighter colored hair that those on the west side. The vegetation on the east side is composed of lighter colored shrubs and grasses and that the vegetation is less dense. Other than this difference, the habitats appear to be the same.

2.   Darwin gathered an enormous amount of information from his own observations and experiments and from corresponding with other naturalists doing the same. Although the information he had pertained to only a small sample of all living organisms, his analysis of this information lead him to develop some generalizations about all life forms. These generalizations were constructed in the form of hypotheses. They reflected his deductions about patterns in nature. Darwin's hypotheses might have looked like the following:

a)   All organisms tend to produce more offspring than necessary to replace those that die.

b)   Over time, there are relatively constant numbers of individuals within a species.

c)   Not all the individuals within a species are alike.

d)   Some of the characteristics exhibited by offspring are passed to them from the parents.

3.   By choosing organisms that naturally exhibit a particular trait and then breeding that organism with another of the same species exhibiting the same trait over time, breeders are able to produce dogs having the desired trait. This artificial breeding process has been the cause of so many different types and characteristics of dogs which are of the same species.

4.   A punctuationalist would argue that new species arise suddenly and rapidly as small subpopulations of a species split from the original population. His or her idea of rapid change means that the period of time during which speciation occurs is short with respect to the period of time of stasis. Stasis is small variations that do not lead to speciation. The term "short" is relative and should be thought of in geologic time.

CHAPTER 24

THE EVOLUTION OF THE FIVE KINGDOMS OF LIFE

KEY CONCEPTS OVERVIEW

1.   Theories About the Origin of Simple Organic Molecules

So where and how did life originate? Although this is one of the most fundamental questions in biology, the complete answer may not ever be known. However, scientists are able to analyze many factors that are known and develop a likely explanation. One of the factors that is agreed on is that the environment of primitive Earth was very harsh, subject to severe ultraviolet radiation from the sun, and subjected to strong electrical storms and volcanic eruptions. These conditions provided opportunity for elements and simple compounds of the atmosphere to react with one another and form new and more complex molecules that captured energy in the bonds that held them together.

At the University of Chicago (1953), Stanley Miller and Harold Urey designed an experiment to test the effects of primitive Earth's atmosphere. They wondered if the complex molecules in their model "atmosphere" would dissolve in water vapor and fall into their model "ocean." They constructed an apparatus that consisted of two connected chambers, one representing the primitive ocean and the other representing the primitive atmosphere. The atmosphere chamber contained methane, ammonia, water vapor, and hydrogen gas. They placed electrodes within the chamber that would discharge simulated lightening periodically. After a week, the environment of the ocean chamber was chemically analyzed. Notably, 15% of the carbon originally present in the methane gas had been converted into more complex carbon compounds. Twenty different organic compounds were found. The process of change from gases to the complex organic compounds is known as **biochemical evolution**. Each step in biochemical evolution represents a level of increased complexity and stability.

Others performed the experiment using other compounds that might have been present in the primitive atmosphere. Some even tried using a variety of energy sources. Each experiment yielded the same outcome: newly formed complex organic molecules such as amino acids, various sugars, purine and pyrimidine bases, and fatty acids. From this host of experiments, scientists developed the **primordial soup theory**, which states that life arose in the primitive seas or in smaller lakes as complex organic molecules formed from the simple molecules in the ancient atmosphere.

More recently other scientists have changed their thinking about the primordial soup theory. Computerized construction of the primitive atmosphere, designed by James Walker at the University of Michigan, suggested that the primary components of the atmosphere were carbon dioxide and nitrogen gases, which could not have formed complex organic molecules. Others now think that life might have arisen in hydrothermal vents that could have supplied the necessary energy and nutrients needed for life. This concept is supported by the hydrothermal vents that maintain communities of living organisms today.

2.   Theories About the Origin of Simple Organic Molecules

Despite all the experimentation, scientists have not been able to generate life from nonliving material. At some point the synthesis of important macromolecules had to have occurred. In order for this to occur, **dehydration synthesis** must take place. And this could have occurred if **condensing agents**, those molecules that combine with water and release energy, were present. If these molecules were not present, heat and evaporation could also have promoted dehydration synthesis and polymer production.

Sidney Fox (1950) used heat to produce polymers, called proteinoids, from dry mixtures of amino acids. These proteinoids acted like enzymes because they increased the rate at which organic reactions occurred. If these reactions were sequenced appropriately, they could be considered the beginning of metabolic systems. However, these metabolic systems have to be organized within a cellular structure and function independently to be considered "living."

More recently, scientists have been looking to the RNA molecule for answers. RNA may have functioned as a template for other polynucleotides. Eigen theorized that RNA assembled enzymes from the amino acids in the primordial soup. The presence of such enzymes however, furthered the understanding that such chemicals required encasement within a membrane to be organized for life functions. The origin of membranes is still a point of discussion. No scientific consensus exists on how much time lapsed between biochemical evolution and biological evolution. And no one really understands how evolution came about.

3.   The History of Life

When we try to conceptualize the time course of Earth, it is difficult to imagine 4.6 billion years. So scientists have divided the time from the formation of Earth until the present day into five major time periods or **eras**. These time periods are further subdivided into shorter time units called **periods**. The periods of the Cenozoic era are subdivided into **epochs**. While these units of time are not consistent lengths, they do give us an idea of relative time or how long each geological time unit is in relation to the others.

The following information is constructed to summarize each of the five major time periods within the Earth's development and key events within each.

Archean Era

During the Archean era (4.6 billion to 2.6 billion years ago) the first forms of life began to change. One of the most important events was the development of the ability to carry out photosynthesis. Later, the availability of oxygen from photosynthesis allowed the development of aerobic respiration. The single-celled organisms, similar to the present-day **cyanobacteria**, are examples of organisms that played key roles in the evolution of life as it is known today. These bacteria grew in colonies and they left fossilized structures referred to as **stromatolites**. Stromatolites are found earliest in the from of layered mats and appear as tall prominent columns later. They date back some 3.5 billion years.

Proterozoic Era

The Proterozoic era extended from 2.5 billion to 544 million years ago. The most notable characteristic of this era is the formation of a stable oxygen-containing atmosphere. The first eukaryotic organism appeared on Earth during this time. Within stromatolite fossils found in California, much more complex microfossils than that of cyanobacteria were discovered. The current explanation for the evolution of eukaryotes is called the endosymbiotic theory. This theory states that larger forms engulfed some small prokaryotes, called **endosymbionts**, that persisted inside the body of the eukaryotes. These symbiotic forms of life functioned as organelles. Those with appropriate enzymes functioned as mitochondria while those with appropriate pigments became chloroplasts. The cell membranes enfolded and surrounded the organelles.

Paleozoic Era

The Paleozoic era denotes a time between 544 million and 250 million years ago. This time is marked by the abundance of many multicellular fossils. This era is divided into six shorter periods. You may want to review these in Table 24-1 in the text. The oldest period within this era is the **Cambrian period**, which ended about 505 million years ago. This period represents the point in evolution where all the main phyla of organisms existed except the chordates and land plants. Any fossils found in strata older than these are referred to as **Precambrian**.

The **Ordovician period** is marked by the evolution of wormlike aquatic animals from ancient flatworms and are the earliest forms of **chordates**. Chordates are distinguished by a single hollow nerve cord along the back, a rod-shaped notochord, and pharyngeal arches and slits.

The **Silurian period** began 430 million years ago and is marked by changes in the notochord. Rather than a rod-shaped structure, it was replaced with a more flexible arrangement of sequenced pieces. These organisms, eventually developed into the first vertebrates - the **ostracoderms**.

The **Devonian period** is defined as the time of the evolution of fishes. The first jawed fishes that evolved were the **placoderms,** which are ancestors of today's bony fish. Early amphibians arose during this time. They had fish-like bodies, short, stubby legs, and lungs.

The **Carboniferous period** was the time for the giant fern forests and the formation of fossil fuels. It is named for the great coal deposits formed during this time. The fungi also invaded the land during this period and became important in their role as decomposers.

The **Permian period** (285 million to 250 million years ago) is marked by very severe drought and extensive glaciation. During this period, the cone-bearing plants (conifers) originated.

Mesozoic Era

This time occurred between 250 million and 65 million years ago. During this time, the reptiles that had evolved from earlier amphibians became dominant. This era is noted for the adaptive radiation of terrestrial plants and animals that had been established earlier.

The Mesozoic era is divided into three periods: **Triassic, Jurassic,** and **Cretaceous.** The primitive dinosaurs and mammals arose during the late Triassic period. The large dinosaurs and first birds appeared in the Jurassic period. Flowering plants arose in the Cretaceous period, which ended abruptly perhaps as a result of the impact of a giant meteorite. The impact of the meteorite produced a thick cloud that blotted out the sun, killed plants, and produced cold temperatures that would be particularly deadly to cold-blooded animals. The idea that a meteor led to the big extinction that terminated the Mesozoic era is called the **Alvarez hypothesis.** Support for this hypothesis comes from iridium deposited some 65 million years ago in the strata.

Earth has shown a great number of changes over time. The most dramatic involves the movement of land masses over the surface of the earth. The crust floats on a sea of molten semiliquid rock. The theory of **plate tectonics** supports the idea of **continental drift.** Cracks between massive plates are the sites for undersea rifts and earthquake zones and volcanoes.

Cenozoic Era

The onset of the Cenozoic era was marked by the mass extinction of the Cretaceous period. This era has two periods - the **Tertiary** and the **Quaternary,** which are subdivided into many epochs. You can review these in Table 24-1 in your text. During the Cenozoic era, mammals became adundant. Natural selection pressures resulted in many changes in these organisms. Notable important changes were that of **warm-bloodedness** in mammals and various types of reproductive characteristics (**monotremes, marsupials,** and **placental**).

Of the 14 orders of placental mammals, humans belong to the **primates.** Primates are divided into two suborders: **prosimians** and **anthropoids** (of which humans are a member). **Hominoids** are one of four superfamilies of the anthropoid suborder and include the apes and the humans.

The first human fossils date back about 2 million years and belong to the extinct species *Homo habilis*, meaning skillful human. Their ability to make clothing and tools have given scientists reason to believe they were more intelligent than their ancestors and thus considered human. Fossils of another species of humans, *Homo erectus*, have suggested that these humans built shelters, used fire, and probably communicated with some form of language.

CHAPTER REVIEW ACTIVITIES

CONCEPT MAPPING

Develop a concept map that demonstrates your understanding of the Earth's general history and how that lead to the five kingdoms of living things. Use as many of the key terms from Chaper 24 as possible.

MATCHING

Select the term from the list provided that best relates to the following statements. If you are unsure of the correct answer or chose an incorrect response, review the related concept in Chapter 24.

_____ 1. This subphylum of chordates has a dorsal nerve cord surrounded by a vertebral column.

_____ 2. This is the second period of the Mesozoic era in which large dinosaurs dominated the Earth and birds first appeared.

_____ 3. These are the subdivisions of the periods of the Cenozoic era.

_____ 4. These placental mammals have characteristics reflecting an aboreal lifestyle.

_____ 5. This idea states that mitochondria and chloroplasts originated symbiotically from aerobic and anaerobic bacteria respectively.

_____ 6. This is the first period of the Cenozoic era.

_____ 7. These early members of *H. sapiens* had anatomical features that were similar to modern humans.

_____ 8. These warm-blooded vertebrates have hair and the females secrete milk to feed their young.

_____ 9. This is the oldest period within the Paleozoic era, which ended approximately 505 million years ago.

_____ 10. This suborder of primates include the lemurs, indris, aye-ayes and lorsies.

_____ 11. This is the family of hominoids consisting of human beings.

_____ 12. These organisms, which include fishes, amphibians, reptiles, birds, mammals, and humans, evolved during the Ordovician period.

_____ 13. This term describes the development of similar structures having similar functions in different species as the result of the same kinds of selection pressures.

_____ 14. This is the third period of the Mesozoic era in which flowering plants began to appear.

_____ 15. These are shorter time units into which eras are subdivided.

_____ 16. This term describes modern man.

_____ 17. These are banded domes of sediment that form around microbes known as cyanobacteria.

_____ 18. This is the geological time unit scientists use to divide the time from the formation of Earth to present day into five major events.

_____ 19. This is the first period of the Mesozoic era in which small dinosaurs and primitive mammals appeared.

_____ 20. This suborder of primates includes the monkeys, apes, gorillas, chimpanzees, and humans.

_____ 21. This is the second extinct species of hominids whose fossil records date back 1.6 million years and who were adapted to upright walking, making tools, and probably communicating with language.

_____ 22. This is the period from the end of the Tertiary period to the present time.

TERMS LIST

a. Epochs	h. Mammals	o. Eras
b. Stromatolites	i. Cretaceous period	p. Primates
c. Anthropoids	j. Jurassic period	q. Homo sapiens sapiens
d. Vertebrates	k. Chordates	r. Cro-magnons
e. Homo erectus	l. Cambrian period	s. Convergent evolution
f. Prosimians	m. Periods	t. Quaternary period
g. Hominids	n. Endosymbiotic theory	u. Triassic period
		v. Tertiary period

GEOLOGICAL TIME

Geological time can be compared to a calendar year.

Months	Events
January Jan. 1	
February Feb. 10	OLDEST DATED ROCKS
April Apr. 16	
October Oct. 1	

November
  Nov. 1
  Nov. 12
  Nov. 28
	PRIMITIVE HIGHER ANIMALS APPEAR

December
  Dec. 10
  Dec. 26
  Dec. 31 (6 pm)
      (80 sec to midnight)
      (14 sec to midnight)
	END OF DINOSAURS

GEOLOGICAL TIME SCALE

Fill in as many of the events as possible within each time

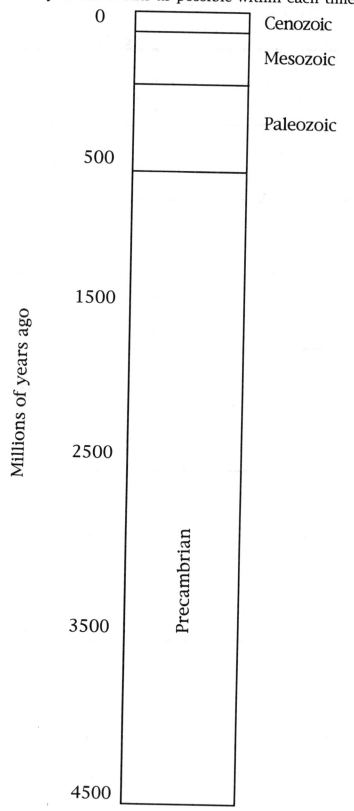

period.

MULTIPLE CHOICE QUESTIONS

_____ 1. Which of the following gases was NOT found in the early atmosphere?
   a. Hydrogen
   b. Carbon dioxide
   c. Water vapor
   d. Nitrogen
   e. Oxygen

_____ 2. Miller and Urey used which of the following as an energy source to produce the complex molecules found in their "primitive ocean?"
   a. Ionizing radiation
   b. Ultraviolet radiation
   c. Heat
   d. Electrical sparks
   e. Microwaves

_____ 3. What did Sidney Fox combine amino acids to form?
   a. Microspheres
   b. Proteinoids
   c. Polypeptides
   d. Coacervates
   e. Protocells

_____ 4. In the early evolution of life, which of the following functioned as enzymes?
   a. Proteinoids
   b. Metal ions
   c. Clay
   d. B and C
   e. A, B, and C

_____ 5. What is the oldest geological era?
   a. Archaen
   b. Proterozoic
   c. Mesozoic
   d. Cenozoic
   e. Paleozoic

_____ 6. What is the most recent geological era?
   a. Archaen
   b. Proterozoic
   c. Mesozoic
   d. Cenozoic
   e. Paleozoic

_____ 7. The oldest known fossil cells are about how many billion years old?
   a. 4.5
   b. 4.0
   c. 3.5
   d. 3.0
   e. 2.5

_____ 8. Until very recently, the first vertebrates were thought to have appeared during which of the following periods?
   a. Cambrian
   b. Ordovician
   c. Carboniferous
   d. Devonian
   e. Permian

_____ 9. The age of fishes was which of the following periods?
   a. Cambrian
   b. Ordovician
   c. Carboniferous
   d. Devonian
   e. Permian

10. What are the stromolites?
    a. Proteinoids that were produced in an anerobic
        environment
    b. Layered mats and columns of fossilized colonial
        cyanobacteria
    c. Microspheres produced by membrane forming
        organic molecules
    d. True cells
    e. Molecules that gave rise to early procaryotes

11. Fossil fuels were formed during which of the following?
    a. Cambrian
    b. Permian
    c. Carboniferous
    d. A and B
    e. B and C

12. Major extinction occurred and conifers appeared at the
    end of which period?
    a. Cambrian
    b. Ordovician
    c. Carboniferous
    d. Devonian
    e. Permian

13. Flowering plants appeared during which of the
    following?
    a. Triassic
    b. Jurassic
    c. Paleocene
    d. Cretaceous
    e. Pliocene

_____ 14. Humans first appeared during which of the following?
   a.   Triassic
   b.   Jurassic
   c.   Paleocene
   d.   Cretaceous
   e.   Pliocene

_____ 15. The first mammals appeared during which of the following?
   a.   Triassic
   b.   Jurassic
   c.   Paleocene
   d.   Cretaceous
   e.   Pliocene

_____ 16. The first birds and large dinosaurs were most prevalent during which of the following?
   a.   Triassic
   b.   Jurassic
   c.   Paleocene
   d.   Cretaceous
   e.   Pliocene

_____ 17. Why do the cyanobacteria differ from the other bacteria?
   a.   Different RNA polymerase molecules
   b.   They have the ability to carry on photosynthesis.
   c.   They require sulfur containing compounds to survive.
   d.   Their cell walls and membranes are structurally different.
   e.   The location of their membrane-bound nuclei

_____ 18. Comparing the course of geologic time to a single calendar year, during which month would life have first appeared?

  a. January
  b. February
  c. March
  d. April
  e. November

_____ 19. The Burgess Shale is located in which of the following?

  a. Mexico
  b. Australia
  c. Zaire
  d. Canada
  e. Brazil

_____ 20. Of the following organisms, which is the most recent on Earth?

  a. Flowering plants
  b. Birds
  c. Conifers
  d. Mammals
  e. Reptiles

SHORT ANSWER QUESTIONS

1. Why was the Proterozoic era once thought to be devoid of life?

2. Explain the theory of endosymbiosis.

3. Define stromatolites.

4. What characterizes the Precambrian period?

5. What is the Burgess Shale?

6. List the five major eras of geologic history in sequence and give their approximate time frames.

CRITICAL QUESTIONS

1. What is the significance of the mass extinctions that occurred at the end of geological periods?

2. What features of the reptiles made them superior to amphibians for terrestrial life?

3. What explanations could you offer for the mass extinction of some species?

TELLING THE STORY

Develop a story that demonstrates your understanding of the evolution of the five kindgoms. Use as many terms from the chapter as you can. If you have difficulty with any concept as you write, go back and review the specific section in Chapter 24.

CHAPTER REVIEW ANSWERS

MATCHING

1. D	7. R	13. S	19. U
2. J	8. H	14. I	20. C
3. B	9. L	15. M	21. E
4. P	10. F	16. Q	22. T
5. N	11. G	17. B	
6. V	12. K	18. O	

GEOLOGICAL TIME

Months	Events
January Jan. 1	Origin of the earth (4,600 bya)
February Feb. 10	OLDEST DATED ROCKS
April Apr. 16	First known life (3,500 mya)
October Oct. 1	Normal oceans and atmosphere (1000 mya)

November
  Nov. 1
  Nov. 12
  Nov. 28
	PRIMITIVE HIGHER ANIMALS APPEAR
	Beginnings of geology (600 mya)
	First land plants (410 mya)

December	
Dec. 10	Supercontinent (200 mya)
Dec. 26	END OF DINOSAURS (65 MYA)
Dec. 31 (6 pm)	Humans appear (2 mya)
(80 sec to midnight)	End of last glacial age (15,000 ya)
(14 sec to midnight)	Birth of Christ (2,000 ya)

GEOLOGICAL TIME SCALE

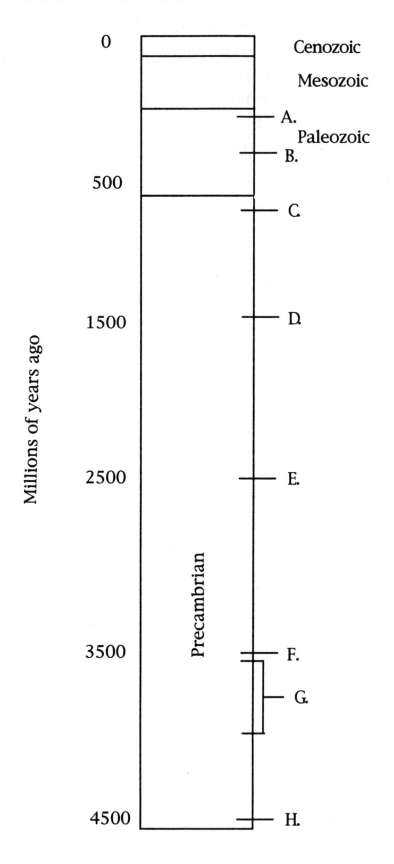

A.	Fossils of fungi	E.	O$_2$ from bacteria
B.	Plants colonize land	F.	Oldest
procaryote			fossil
C.	Oldest animal fossil	G.	Possible origin of life
D.	Oldest eukaryote fossils	H.	Origin of
Earth			

MULTIPLE CHOICE QUESTIONS

1. E	5. A	9. D	13. D	17. B
2. D	6. D	10. B	14. E	18. D
3. B	7. C	11. E	15. A	19. D
4. E	8. B	12. E	16. B	20. A

SHORT ANSWER QUESTIONS

1. Because it was thought that there was little reason to look for the fossils of microorganisms, scientists were not aware of the microfossils.

2. The theory states that bacteria were engulfed or in some way became associated with larger prokaryotes. The cell membranes bent inward and formed a second membrane around the bacteria. The bacteria were symbionts, living within the more advanced cells. Those that had chlorophyll became the eukaryotic chloroplasts and others became the mitochondria.

3. Stromatolites are large masses of colonial cyanobacteria. These formations, found in various parts of the world, date back 3.5 billion years.

4. The Precambrian period ended some 590 million years ago and is associated with the origin of many invertebrate phyla. This is the period that occurred just prior to the Paleozoic era.

5. The Burgess Shale is a geological strata that contains marine animals and plants that lived during the Cambrian period. It is located in British Columbia, Canada.

6.

Archean era	4.6 billion to 2.5 billion years ago
Proterozoic era	2.5 billion to 590 million years ago
Paleozoic era	590 million to 250 million years ago
Mesozoic era	250 million to 65 million years ago
Cenozoic era	65 million years ago to present day

CRITICAL QUESTIONS

1. One obvious feature of the extinctions is that they provide a way to mark the end of a geological period.

2. The three primary preadaptations that suited the reptiles to existence on land were: (1) the evolution of scales that prevented desiccation; (2) the development of the land egg with the four extraembryonic membranes; and, (3) the penis or copulatory organ that negated the need to return to the water for reproduction.

3. Extinction usually occurred in response to major climatic changes. One of the more debated issues is what caused the extinction of the dinosaurs. The Alvarez hypothesis has caused a "winter" that led to massive extinction. When the giant reptiles became extinct, it allowed mammals a better chance to assume dominance and replace them. Another potential explanation might be the breakup of a huge land mass or continent caused by continental drift that resulted in drastic climate changes unsuitable for some species to have survived.

CHAPTER 25

VIRUSES AND BACTERIA

KEY CONCEPTS OVERVIEW

1.   Viruses

   Viruses are infectious agents that cannot reproduce outside a living cell. Due to the limitations for reproduction and the necessity for a host, they are referred to as **obligate parasites**. They are extremely dependent on other organisms and are so at the expense of the host. Viruses were first noted by European scientists in the late nineteenth century. In an attempt to determine the causative agent for tobacco mosaic disease and hoof and mouth disease, they found that the infectious agents were so small that they passed through the fine-pored filter that was being used. Furthermore, no bacteria were isolated from the filtrate. Additionally, it was noted that these agents could multiply only within living cells and therefore lacked the mechanisms for reproduction. Wendell Stanley (1930s) extracted the viral material and identified it as protein. Later others determined that this material also contained RNA.

   Structurally, viruses have a **nucleic acid central core** surrounded by a protein coat or **capsid**, although their overall shape varies from one to another. They are incredibly small and require electron microscopy for adequate viewing.

   The process of viral multiplication is called **replication** and this process can occur in various ways. Some enter a cell, replicate, cause the host cell to burst and release new viruses. This process, called the **lytic cycle**, begins when the virus attaches to a receptor site and injects its nucleic acid into the host cell. The virus gains control of the cell and forces it to produce more viral particles. The cell **lyses** or ruptures and releases the viruses that will begin to infect more cells for the same reasons. Some viruses enter more long-term relationships with the host cells. This type of replication, called the **lysogenic cycle**, begins when the virus enters the cell and instead of reproducing immediately, it becomes incorportated into the host cell's genetic material. An appropriate stimulus will trigger the virus to reproduce and enter the lytic cycle. The infection resulting from this type of replication process is referred to as **latent viral infection. Herpes simplex** causes these types of infections and can cause cold sores or shingles for example.

There is no kingdom into which viruses would be classified. So, scientists have devised a schema for classification based on the host that is infected. The first criterion for classification is the type of organism (plant, animal, or bacteria) invaded. Subsequent charactertists include shape, size, type of nucleic acid, and process of replication.

Certain viral infections can transform normal cells into cancer cells. There are five viruses that have been identified as cancer causing: hepatitis B virus, human T-cell lymphotrophic/leukema virus, human papillomavirus, human cytomegalovirus, and the Epstein-Barr virus. Other viruses are associated with sexually transmitted diseases. There are three such viruses mentioned in your text: genital herpes, genital warts, and acquired immunodeficiency syndrome.

2.   Bacteria

While bacteria and viruses have a "bad" reputation generally associated with being **pathogenic** (causing disease), it is important for you to extend your understanding about the role of bacteria in the world of living things. Bacteria perform many functions that are essential for life. They are natural recyclers of organic compounds, decomposers of organic matter (**heterotrophs**), and useful in the production of certain foods. The green, blue green, and purple bacteria (**photoautotrophs**) carry out photosynthesis, while others known as **chemoautotrophs** derive their energy from ammonia, methane, and hydrogen sulfide.

Bacteria are the oldest, most abundant, and simplest organisms on Earth. They are virtually everywhere. Their simplicity is characterized by the fact that they, as procaryotes, have no membrane-bound organelles nor membrane-bound nucleus. They reproduce asexually by **binary fission**, meaning that one cell divides into two with no exchange of genetic material between cells.

Bacteria are classified in the kingdom Monera. This kingdom is divided into four divisions that differentiate differences in the cell walls of bacteria. Most bacteria are considered **eubacteria** or true bacteria, but there is one section of bacteria different from all the others. These are the **archaebacteria**. The chemical nature of their cell walls and cell membranes allow these organisms to live in places such as the gut of cattle and in land fills where they produce methane gase. Others within this section are associated with deep sea vents or salt marshes.

Despite the "good side" of bacteria, there are negative aspects to these organisms. Many are considered pathogens. Some bacteria infect plants causing **rot** or **wilt**. However fungi, not bacteria, are the primary agents of plant diseases. The bacteria that cause disease are all heterotrophic, using their hosts for food. Upon entering the host, bacteria attach to cells and cause various types of tissue damage. The transmission of bacteria that cause diseases are varied. For example, they can be air borne, in water, in food, on surfaces, in the soil. Some of the diseases associated with bacteria are sexually transmitted, such as **syphilis**, **gonorrhea**, and **chlamydia**.

CHAPTER REVIEW ACTIVITIES

CONCEPT MAPPING

Develop a concept map that demonstrates your understanding of viruses and bacteria. Use as many of the key terms as possible, but make sure you really understand how they fit into the theme of the map. You may want to focus on the structure, reproduction, classification, and impact of viruses and bacteria.

MATCHING

Select the term from the list provided that best relates to the following statements. If you are unsure of the correct answer or chose an incorrect response, review the related concept in Chapter 25.

_____ 1. In this type of asexual reproduction, one cell divides into two with no exchange of genetic material between cells.

_____ 2. In this pattern of viral replication, a virus integrates its genetic material with that of a host and is replicated each time the host cell replicates.

_____ 3. These eubacteria that make their own food by capturing energy from the sun.

_____ 4. A sexually transmitted disease causes a primary infection and inflammation of the urethra in both sexes and of the vagina and cervix of females.

_____ 5. These nonliving infectious agents enter living organisms and cause disease.

_____ 6. These organisms in an ecosystem break down organic molecules of dead organisms and contribute to the recycling of nutrients to the environment.

_____ 7. This chemical layer, rich in protein, lipids, and carbohydrates, is found over the capsid of many viruses.

_____ 8. These eubacteria make their own food by deriving energy from inorganic molecules.

_____ 9. This sexually transmitted disease produces blister-like sores on the genitals.

_____ 10. In this pattern of viral replication, a virus enters a cell, replicates, and then causes the cell to burst, releasing the new viruses.

_____ 11. These organisms cannot produce their own food.

_____ 12. This is a protein coat that covers the nucleic acid core of a virus.

_____ 13. This sexually transmitted disease is caused by the bacterium _T. pallidum_ and produces three stages of infection from localized to widespread.

_____ 14. This is the center of a virus containing a nucleic acid and covered by a protein coat.

TERMS LIST

a. Core	f. Binary fission	k. Chemoautotrophs
b. Envelope	g. Spyhillis	l. Capsid
c. Heterotroph	h. Genital herpes	m. Photoautotrophs
d. Gonorrhea	i. Lysogenic cycle	n. Decomposers
e. Lytic cycle	j. Viruses	

TUMOR VIRUSES

Match the viruses on left with the cancer type they are associated with on the right.

Viruses	Cancer types
1. Epstein - Barr	a. Vaginal/vulval cancer, penile cancer
2. Cytomegalovirus	b. Primary liver cancer
3. Human papillomavirus	c. Leukemia and lymphomas
4. Hepatitis B virus	d. Burkitt's lymphoma, nasopharyngeal cancer
5. Human T-cell lymphotrophic/leukemia virus	e. Kaposi's sarcoma

LABELING

Label the following figure.

Fig. 25.1 Viral replication.

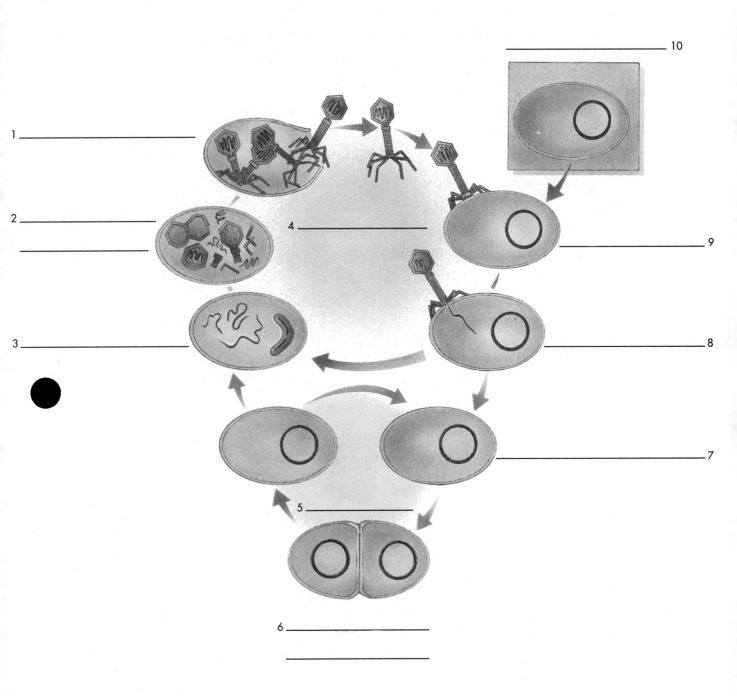

1 _____

2 _____

   _____

3 _____

4 _____

5 _____

6 _____

   _____

7 _____

8 _____

9 _____

10 _____

MULTIPLE CHOICE QUESTIONS

_____ 1. What is the first step in the lytic cycle?
   a. Release of viral particles from an infected cell
   b. Viral particles replicated by cellular machinery
   c. Viral particle attaches to a receptor site
   d. DNA or RNA injected into a cell
   e. Viral parts are assembled into a complet virus.

_____ 2. Which of the following type of cells would be most resistant to bacterial infections?
   a. Plants
   b. Animals
   c. Neither plants nor animals
   d. Both plants and animals

_____ 3. Which of the following type of cells would be most resistant to viral infections?
   a. Plants
   b. Animals
   c. Bacteria
   d. Both plants and animals
   e. Bacteria and animals

_____ 4. Herpes simplex is most likely to affect which of the following tissues?
   a. Muscle
   b. Genital
   c. Lung
   d. Skin
   e. Nerve

_____ 5. The chicken pox virus can go into a latent period, causing which of the following diseases later in life?

    a.   Gout

    b.   Cancer

    c.   Goiter

    d.   Shingles

    e.   Measles

_____ 6. Which of the following diseases would NOT be caused by a virus?

    a.   Cholera

    b.   Influenza

    c.   Smallpox

    d.   Rubella

    e.   Hepatitis

_____ 7. In which of the following kingdoms do bacteria belong?

    a.   Protista

    b.   Animalis

    c.   Monera

    d.   Fungi

    e.   Plantae

_____ 8. In which of the following kingdoms do viruses belong?

    a.   Protista

    b.   Animalis

    c.   Monera

    d.   Fungi

    e.   none of the above

_____ 9. In the classification schema of bacteria, which of the following would follow the kingdom?

    a.   Species

    b.   Sections

    c.   Divisions

    d.   Categories

    e.   Families

_____ 10. Certain bacteria are able to live in the gut of cattle and in land fills. Which of the following products would be associated with these bacteria?

    a. Alcohol

    b. Methane

    c. Carbon dioxide

    d. Ozone

    e. Hydrogen sulfide

_____ 11. Which of the following bacteria would NOT be considered a photoautotroph?

    a. Blue-green

    b. Purple

    c. Green

    d. Red

    e. None of the above

_____ 12. Chemoautotrophs derive their energy from which of the following?

    a. Glucose

    b. Methane

    c. Hydrogen sulfide

    d. A and B

    e. B and C

_____ 13. Nitrogen fixing archaebacteria and eubacteria are associated with which of the following?

    a. Photoautotrophs

    b. Heterotrophs

    c. Chemoautotrophs

    d. Virotrophs

    e. Organotrophs

_____ 14. Bacteria that fix nitrogen would be found on roots of which of the following plants?
- a. Corn
- b. Tomatoes
- c. Clover
- d. Peas
- e. C and D

_____ 15. Which of the following cycles would NOT be associated with bacteria?
- a. Nitrogen cycle
- b. Water cycle
- c. Carbon cycle
- d. Sulfur cycle
- e. All of the above

SHORT ANSWER QUESTIONS

1. How do viruses cause latent infections?

2. How are viruses classified?

3. What features distinguish the prokaryotes from the eucaryotes?

4. Why are archaebacteria classified differently from other bacteria?

5. What two indicators were first used to rule out bacteria as the causitive agents for tobacco mosaic disease?

6. What are the criteria for classifying bacteria?

CRITICAL QUESTIONS

1. Offer a sound rationale for the argument that viruses are nonliving.

2. What type of evidence might you gather to determine that bacteria, not viruses, were the infectious agents for a disease?

3. Why do you think some viral diseases are so difficult to control or eradicate?

TELLING THE STORY

Develop a story that demonstrates your understanding of viruses and bacteria. If you have difficulty with any of the associated areas or topics as you write, go back and review that information in Chapter 25.

CHAPTER REVIEW ANSWERS

MATCHING

1. F	5. J	9. H	13. G
2. I	6. N	10. E	14. A
3. M	7. B	11. C	
4. D	8. K	12. L	

TUMOR VIRUSES

1. D   2. E   3. A   4. B   5. C

LABELING

1. Lysis of cell
2. Assembly of new viruses
3. Replication of vegetative virus
4. Lytic cycle
5. Lysogenic cycle
6. Reproduction of lysogenic bacteria
7. Viral DNA integrated into bacterial chromosome
8. Viral DNA injected into cell
9. Virus attaching to cell wall
10. Uninfected cell

MULTIPLE CHOICE QUESTIONS

1. B	5. D	9. C	13. C
2. A	6. A	10. B	14. E
3. A	7. C	11. D	15. B
4. D	8. E	12. E	

SHORT ANSWER

1. If the viral nucleic acid is incorporated or integrated into the host cell's nucleic acid, it will be passed on to daughter cells when the host cell divides. The virus will not enter the lytic cycle until an appropriate stimulus triggers that action. The virus can remain inactive for long periods of time.

2. They are first classified by the kind of organism they infect. Following that, their morphology, type of nucleic acid, and method of replication can be used to classify them.

3. The most significant distinguishing characteristic is the lack of membrane-bound organelles and nucleus in prokaryotes.

4. The chemical composition of the cell walls and cell membranes of archaebacteria are different enough to set them apart from other bacteria. Additionally, because these bacteria have such unusual chemical processes, they live in very unusual places, like landfills, the intestinal tract of cattle, salt marshes, and deep-sea vents.

5. The viruses were not retained by the fine-pored filters and the culture yielded no organisms.

6. Shape of cells; arrangements of groups of cells; presence of flagella; staining characteristics; nutritional characteristics; temperature, pH, and oxygen requirements; biochemical nature of cellular components; genetic characteristics

CRITICAL QUESTIONS

1. Answers will vary, but you should mention that viruses are not able to reproduce on their own. Without the assistance of a living cell, viruses would not have a means to multiply.

2. A culture from the infected organism would grow on a specialized medium if the causitive agent were bacterial. Antibiotics are generally effective against such agents, but have no direct effect on viruses.

3. Viruses can lie dormant or latent for long periods of time and can be difficult to combat with conventional therapy. Additionally, the structural arrangement of the protein coats of viruses can change making it difficult for them to be identified.

CHAPTER 26

PROTISTS AND FUNGI

KEY CONCEPTS OVERVIEW

1.   Protists

Protists are a very diverse group of eukaryotic organisms. Those that are more animal-like, that is, they are **heterotrophic**, are referred to a **protozoans**. The plant-like protists, those that are photosynthetic autotrophs, are called **algae**. The fungus-like protists consist of the **slime molds** and the **water molds**.

Protozoans

There are four phyla of protozoans: the amoebas, flagellates, ciliates, and sporozoans. The **amoebas** are shapeless masses of protoplasm that develop cellular projections called **pseudopods** formed by protoplasmic streaming. Amoebas surround and engulf food, forming a **food vacuole** into which they secrete digestive enzymes. The process is similar to phagocytosis by the amoeba-like white blood cells. The **contractile vacuole** of the amoeba controls the water content within the cell by pumping out excess water. These organisms reproduce by binary fission.

Some of the interesting types of amoebas include *Entamoeba histolytica*, which produces amebic dysentery, a disease associated with poor sanitary conditions. There are also groups of amoebas that secrete shells that cover and protect their cells. These shelled amoeba are exemplified by the *radiolarians* and *foraminifera*.

The **flagellates** are also a very diverse group of protozoans. All of them have a relatively simple structure. They move by a whip-like motion produced by at least one flagellum. One of the most interesting representatives of the flagellates is the group called **euglenoids**. These flagellates have chloroplasts that allow them to produce their own food. Each of these organisms has two flagella. They reproduce by transverse fission. The Euglena, an example of euglenoids, exhibits an interesting process called **positive phototaxis.** As light is filtered through pigments in a structure called an **eyespot,** a receptor senses the direction and intensity of the light source and assists the movement of the organism.

The **ciliates** are cells that totally or in part are covered by short, hair-like extensions called **cilia.** These structures beat in unison, causing the environment around the cell to move. Food particles are moved toward the **gullet** of the cell. An example of ciliates are the **paramecia,** which can reproduce by transverse fission or by a sexual form of reproduction called **conjugation.** Most of the ciliates live in fresh water.

The **sporazoans** are non-motile, spore-forming parasites of vertebrates. These protozoans have complex life cycles that involve both asexual and sexual phases. An example of a sporazoan is the *Plasmodium vivax,* which is the causative agent for malaria.

Algae

There are three phyla of multicellular algae: the brown algae, the green algae, and the red algae. The phyla of unicellular algae includes the dinoflagellates and the golden algae. Placing algae in the Protista kingdom suggests that there is a close evolutionary relationship among the members of these various phyla.

The unicellular **dinoflagellates** have outer coverings of cellulose plates that make them very unique in appearance. Their flagella are attached in such a way that causes the organism to move in a spinning pattern, which is how their name was derived. These unicellular organisms, which live primarily in the sea, can produce different colored pigments, carry out photosynthesis, and reproduce by longitudinal division and, in some cases, sexual reproduction. The best known dinoflagellates are the ones that produce red tide. The organisms undergo such an explosive population growth that they cause a discoloration in the sea water. This "bloom" produces and releases very poisonous toxins that kill fish, birds, and marine mammals.

The phylum Chrysophyta includes the **yellow-green algae, golden-brown algae,** and **diatoms.** The most abundant of these are the diatoms, which are found in both fresh and salt water. Diatoms have top and bottom shells that fit together with one overlaping the other. They reproduce asexually by separating at the paired shells and regenerating the missing half. When diatoms die, the shells sink to the sea floor forming a thick deposit, which is used commercially as water filters, additives to paint, and as abrasive material. This material is referred to as **diatomaceous earth.**

The **brown algae** are found on the rocky, northern shores of the world. This group of algae is characterized by **sargasso weed**, a type of rockweed for which the Sargasso Sea is named, and the giant **sea weeds**, such as kelp, which provide an important food source for fish, invertebrates, marine mammals, and birds. These large, multicellular organims have a life cycle similar to that of plants.

The **green algae** includes both unicellular and multicellular forms, which are both aquatic and semiterrestrial. There are more than 7,000 different species, exemplified by the **Chlamydomonas**, which are very simple, unicellular, haploid organisms and the **Spirogyra**, which are multicellular, filamentous algae with spiral chloroplasts.

The **red algae**, or **Rhodophyta**, are multicellular and primarily marine. They play a major role in the formation of coral reefs and produce substances that are both economically and ecologically significant. **Agar** and **carregeenan** are made from the glue-like, gelatinous extracts of the red algae and used for a variety of purposes.

Slime Molds

The fungus-like, heterotrophic protists are subdived into two classes: the **cellular slime molds** and the **plasmodial (acellular) slime molds**. The cellular slime molds resemble amoebas, and the plasmodial slime molds are multinucleated masses of cytoplasm. The most notable feature of the plasmodial slime molds is the streaming of the protoplasm in a mass called a plasmodium. The plasmodia engulf and digest food as they move. If conditions are favorable, the plasmodia will reproduce asexually. However, if conditions warrant, the slime mold may move to another area and form spores. The spores will remain dormant until conditions are again favorable and then germinate into flagellated, haploid cells called **swarm cells**. These cells divide and produce more swarm cells, or they can act as gametes, fuse, and form new plasmodia.

Water molds are somewhat of an enigma to taxonomists. Since water molds have flagellated spores, they are not characteristic of fungi. These molds live in fresh and salt water, in moist soil, and on plants. During sexual reproduction, the water molds have very large egg cells, which explains their other name: **egg fungi**. These organisms are parasitic, causing such diseases as late blight and downy mildew.

2.   Fungi

Most of the eukaryotic organisms in this kingdom are multicellular, colorless **saprophytes**. These types of saprophytes feed on dead or decaying organic material by secreting enzymes that break down the food. The fungi then absorb the products. Fungi can also be parasitic, meaning they feed off of living organisms. Their plant bodies (**mycelia**) are composed of thin filament structures called **hyphae**. Fungi can reproduce both sexually and asexually and these differences serve as a basis for classification. Those organisms that have a distinct sexual phase are the zygote-forming fungi, the sac fungi, and the club fungi. The imperfect fungi are a division that have never been observed having a sexual stage of reproduction.

The **zygote-forming fungi** are represented by common bread mold or *Rhizopus*. These types of fungi form **zygospores,** which will undergo meiosis to produce haploid spores capable of germinating and producing new hyphae.

The **sac fungi**, which are the largest division of fungi, have special sexual reproductive structures called **asci** or sacs that contain spores. Their life cycles are similar to that of the club fungi. This group of fungi includes the yeasts, cup fungi, and truffles, and some pathogenic examples, such as the causitive agent for Dutch Elm disease.

Club fungi, probably the best known of the fungi, include the mushrooms, toadstools, puffballs, jelly fungi, shelf fungi, rusts, and smuts. The mushrooms include a wide variety of types from those used for food to those that are very dangerous plant pathogens. The mycelia of the club fungi appear similar to those of the sac fungi until they develop their large, above-ground, fruiting structures. Underneath the **cap** on the **gills**, there are sexually reproducing, club-shaped **basidia,** which produce **basidospores.**

The **imperfect fungi** are so named because they have no sexual stage of reproduction. They asexually reproduce by forming spores which have varied spore cases allowing them to be dispersed in different ways. Some members of this division are sources of antibiotics (*Penicillium*) and others are used in foods (*Aspergillus*). Most of the fungi that cause disease in humans are imperfect fungi.

3. Lichens

Lichen are mutualistic associations of fungi and algae. The fungi supply water and minerals while the plants supply food. Lichens are able to grow very slowly in very harsh environments and tend to be the first organisms to establish new environments for growth that would then be suitable for other organisms. Scientists believe that they may be some of the oldest organisms on Earth, living for thousands of years.

CHAPTER REVIEW ACTIVITIES

CONCEPT MAPPING

Develop a concept map that demonstrates your understanding of protists and fungi. Use as many of the key terms as possible, but make sure you really understand how they fit into the theme of the map.

MATCHING

Select the term from the list provided that best relates to the following statements. If you are unsure of the correct answer or chose an incorrect response, review the related concept in Chapter 26.

_____ 1. These nonmotile protozoans are parasites of vertebrates, including humans.

_____ 2. These plant-like protists are eukaryotic photosynthetic autotrophs.

_____ 3. These protozoans are characterized by fine, short, hair-like extensions.

_____ 4. These sexual spores are formed by cerain kinds of fungi, which are therefore termed zygote-forming fungi.

_____ 5. These slime molds have an amoeboid stage to their life cycle.

_____ 6. These structures enclose the sexual spores of sac fungi.

_____ 7. This type of flagellated, unicellular algae is characterized by stiff, outer coverings and a spinning motion.

_____ 8. This term describes an organism that feeds on dead or decaying organic matter.

_____ 9. These protists are fungus-like in one phase of their life cycle and amoeba-like in another phase.

_____ 10. These members of a class of flagellates have chloroplasts and make their own food by photosynthesis.

_____ 11. This phyla of unicellular algae has a gold-green pigment and stores food as oil.

_____ 12. These animal-like protists are heterotrophs.

_____ 13. These are club-shaped structures in club fungi from which unenclosed sexual spores are produced.

_____ 14. These protozoans are characterized by fine, long, hair-like cellular extensions.

_____ 15. This term describes the movement of an organism toward a source of light.

_____ 16. These rather simple organisms are characterized by unicellular, and some multicellular, forms.

TERMS LIST

a. Protists	f. Cilitates	k. Positive phototaxis
b. Slime molds	g. Dinoflagellates	l. Saprophytic
c. Zygospores	h. Golden algae	m. Algae
d. Basidia	i. Sporozoans	n. Flagellates
e. Asci	j. Euglenoids	o. Protozoans
		p. Plasmodial slime molds

LABELING

The pictures below are examples of certain protists and fungi. Match the pictures with the correct name and type of organism.

1. Fig. 26.1   _____

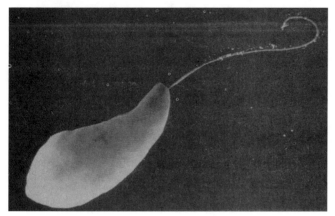

2. Fig. 26.2   _____

3. Fig. 26.3   _____

4. Fig. 26.4 _____

NAMES	TYPES
a. Aspergillis	aa. Ciliate
b. Paramecium	bb. Golden algae
c. Spirogyra	cc. Brown algae
d. Fructicose lichen	dd. Sac fungus
e. Euglena	ee. Imperfect fungus
f. Rockweed	ff. Lichen
g. Diatoms	gg. Flagellate
h. Cup fungus	hh. Green algae

MULTIPLE CHOICE QUESTIONS

_____ 1. Which of the following protists is NOT motile?

    a. Amoebas

    b. Flagellates

    c. Sporozoans

    d. Ciliates

    e. None of the above

_____ 2. Which of the following includes only unicellular algae?

    a. Red algae

    b. Brown algae

    c. Green algae

    d. Golden algae

    e. Blue-green algae

_____ 3. The organisms which cause red tide belong to which of the following?
   a. Dinoflagellates
   b. Red algae
   c. Ciliates
   d. Brown algae
   e. Sporozoans

_____ 4. The Chloroophyta include which of the following algae?
   a. Blue-green
   b. Green
   c. Red
   d. Golden-brown
   e. Brown

_____ 5. The microscopic "pill boxes" are which of the following?
   a. Diatoms
   b. Dinoflagellates
   c. Sporozoans
   d. Ciliates
   e. Radiolarians

_____ 6. Which of the following algae consists of sheets of tissue only two cells thick?
   a. Chlamydomonas
   b. Spirogyra
   c. Sea lettuce
   d. Bossiella
   e. Volvox

_____ 7. In which of the following kingdoms do lichen belong?
   a. Protista
   b. Monera
   c. Fungi
   d. A and B
   e. C and A and/or B

8. The large marine algae are members of which of the following?
   a. Blue-green
   b. Green
   c. Red
   d. Golden-brown
   e. Brown

9. Which of the following types of algae are thought to give rise to the higher plants?
   a. Blue-green
   b. Green
   c. Red
   d. Golden-brown
   e. Brown

10. What color is the pigment that is adapted to absorb light that penetrates to the deeper levels of the ocean?
    a. Blue
    b. Green
    c. Violet
    d. Yellow
    e. Red

11. Which of the following types of algae contribute to the hard material of coral reefs?
    a. Blue-green
    b. Brown
    c. Green
    d. Red
    e. Golden-brown

12. Which group of fungi has no sexual reproduction?
    a. Sac fungi
    b. Club fungi
    c. Imperfect fungi
    d. Zygote-forming fungi
    e. All fungi reproduce only by asexual reproduction

13. The fungus responsible for the Dutch Elm disease is carried from tree to tree by which of the following?
    a. Ants
    b. Elm beetle
    c. Bark beetle
    d. Woodpeckers
    e. The wind

14. Yeasts reproduce by which of the following methods?
    a. Gametes
    b. Binary fission
    c. Conjugation
    d. A and B
    e. A and C

15. Which of the following is NOT a member of the club fungi?
    a. Cup fungi
    b. Mushrooms
    c. Puffballs
    d. Shelf fungi
    e. Rusts and smuts

16. Which of the following statements is true of gills?
    a. Structures that produce basidia underneath the cap of the mushroom.
    b. Gills are cavities produced in substrates, resulting from the growth of hyphae.
    c. Gills are underground connections between mushrooms.
    d. Gills are divisions that separate spores in the ascus and basidium.
    e. None of the above

SHORT ANSWER QUESTIONS

1. What are the classes of protists?

2. What are the two kinds of slime molds and how do they differ?

3.   What are the different types of lichens?

4.   What role do lichens play in succession?

5.   Compare the different ways protists obtain food.

CRITICAL QUESTIONS

1.   What would be the significance of a drastic extinction in the fungi kingdom?

2.   Coral reefs are often destroyed by natural disasters such as hurricanes. Why does it take so long for a reef to reestablish and mature?

3.   Some fungi that cause disease in humans are referred to as opportunistic. Why is this term used?

TELLING THE STORY

Develop a story that demonstrates your understanding of protists and fungi. If you have difficulty with any of the associated areas or topics as you write, go back and review that information in Chapter 26.

CHAPTER REVIEW ANSWERS

MATCHING

1. I	5. P	9. B	13. D
2. M	6. E	10. J	14. N
3. F	7. G	11. H	15. K
4. C	8. L	12. O	16. A

LABELING

Number	Name	Type
1	E	GG
2	H	DD
3	G	BB
4	A	EE

MULTIPLE CHOICE QUESTIONS

1. C	5. A	9. B	13. C
2. D	6. C	10. E	14. D
3. A	7. E	11. D	15. A
4. D	8. E	12. C	16. A

SHORT ANSWER QUESTIONS

1. Flagellates, ciliates, amoebas, and sporozoans

2. Cellular slime molds look and behave like amoebas, forming a large moving mass called a slug to obtain food. This slug eventually forms fruiting bodies which contain cyst-like forms called spores, which become new cells if they land on a suitable habitat. The plasmodial slime molds stream along as a plasmodium, which is a nonwalled, multinucleated mass of cytoplasm. This mass resembles a moving slime body. This type of slime mold may reproduce asexually. It also forms spores that germinate into haploid cells, called swarm cells, which can act like gametes.

3. The three types of lichens are: crustose (encrusting), fruticose (shrubby), and the foliose (leafy).

4. In the primary succession, the lichens may be the first invaders of a newly exposed habitat. They slowly contribute to the chemical and physical degradation of exposed rock surface, providing organic material and a suitable environment for other colonists.

5.
Amoebas	They extend pseudopods to surround food, engulf it, and incorporate it within a food vacuole.
Euglenoids	They absorb food or photosynthesize and make their own food.
Ciliates	They take food into their gullet and form a food vacuole.
Algae	Some have chloroplasts and produce their own food; sporozoans are parasitic and eat off of their host.

CRITICAL QUESTIONS

1. Since fungi are the primary decomposers, disruption in many of the important cycles would occur. Natural breakdown of dead or decaying material would be slowed and undesirable opportunistic organisms would thrive.

2. The photosynthetic process of food production, and therefore the fueling of cellular functions, requires light and carbon dioxide. If other organisms have left the area and the water is clouded by debris, the process is slowed.

3. Some fungi only cause disease when a person is already compromised by another disease or condition. Then pathogenic fungi are able to establish themselves and cause infection in this susceptible individual.

CHAPTER 27

PLANTS: REPRODUCTIVE PATTERNS AND DIVERSITY

KEY CONCEPTS OVERVIEW

1.   Plant Characteristics

The kingdom Plantae was designated to classify the multicellular, eukaryotic, photosynthetic autotrophs that live on land. Since classification is a tool used to help us understand the evolution and diversity of organisms, there are those who would have other opinions about the accuracy of the above description. Taxonomists do agree, however, that plants can be placed into two distinct groups: vascular and nonvascular plants. More than 80% of living plants are **vascular**, meaning they have specialized tissues that function in fluid transport. Within the vascular plant category, there are three subgroups: the seedless plants, plants with naked seeds, and plants with protected seeds. **Seeds** are structures that protect the embryo from dehydration or damage; they provide nutrients for the embryo, and they give rise to new sporophyte plants. There are also plants that do not have specialized transport tissues. These are called **nonvascular** plants. The **bryophytes** , or true mosses, are the only major group of nonvascular plants.

The plant kingdom has ten distinct **divisions** (an equivalent classification level to phyla). These divisions and example organisms are listed in Table 27.1.

Table 27.1 Divisions of plants.

Group	Division	Examples
Nonvascular plants	Bryophyta	"True"mosses, liverworts, hornworts
Seedless vascular plants	Psilophyta	Whisk ferns
	Lycophyta	Club mosses
	Sphenophyta	Horsetails
	Pterophyta	Ferns
Vascular plants with naked seeds	Coniferophyta	Conifers
	Cycadophyta	Cycads
	Ginkgophyta	Gingos
	Gnetophyta	Gnetae
Vascular plants with protected seeds	Anthophyta	Flowering plants
	Class Monocotyledons	Grasses, irises
	Class Dicotyledons	Flowering treees, shrubs, roses

2.   Patterns of Reproduction

The events that occur during the life span of an organism are referred to as a life cycle. The life cycle of plants is characterized by **alternations of generations**. This means that plants have alternating generations that are multicellular haploid phases (**gametophytes**) and multicellular diploid phases (**sporophytes**). These phases are not equal throughout the life span of plants, nor are they equal throughout the plant divisions. The gametophyte generation dominates the life cycles of the nonvascular plants, while the sporophyte generations dominate the life cycle of the vascular plants.

A close look at the general characteristics of the vascular and nonvascular plants reveals that each group has its own variation on the alternation of generations. These differences point to the ways plants have adapted as they evolved. One characteristic that is noteworthy is the way plants produce gametes and spores in specialized structures. In the bryophytes and some vascular plants, the female gametes (eggs) are formed in **archegonia** and the male gametes (sperm) are formed in structures called **antheridia.** During the course of evolution, these structures have decreased in size to accommodate the dominance of the sporophyte generation in the life cycle.

3.   Nonvascular Plants

Bryophytes are small, low-growing plants that include the mosses, liverworts, and hornworts, which have their life cycle dominated by the gametophyte generation. The mosses are the largest class of bryophytes. The top of some of the leafy gametophytes contain the antheridium, which produces many sperm. The tops of others contain the archegonia, each producing only one egg. The sperm are splashed onto the tip of the female, where they swim to the egg and fertilize it. A diploid zygote forms, becoming the first of the sporophyte generation, then grows out of the archegonia and differentiates into a slender stalk, which gets its nutrients from the gametophyte. On the tips of these stalks is a spore capsule, called a **sporangium,** which contains haploid spores that have been produced by meiosis. The spores are carried by the wind to new sites where they germinate and grow into gametophytes.

4.   Vascular Plants

There are some plants (four divisions) within the vascular group that do not form seeds. An example of a **seedless vascular plant** is the fern. Mature ferns, the sporophyte generation, produce spores by meiosis and bear these haploid structures on the undersurface of their leaves or **fronds.** These spores are dispersed, settle on suitable soil, and grow into small, heart-shaped gametophytes called **prothallus.** These are anchored to the ground by filaments of cells called **rhizoids.** Sperm from the antheridia swim through available moisture and fertilize the egg within the archegonia. The zygote begins developing in this protected environment until it is able to grow on its own.

There are also four divisions within the vascular plants that have **naked seeds.** This group, called the **gymnosperm,** include the conifers, cycads, ginkgo, and gnetophytes. The most familiar of these are the conifers, or **cone-bearing** trees, and include the pines, spruces, firs, redwoods, and cedars. The sporotype generation (tree) produces male and female cones that develop spores, which undergo meiosis. The resulting structures are male and female gametophytes. The male gametophytes contain prodigious amounts of **pollen grains,** which contain sperm. The female gametophytes produce two or three eggs, which develop within ovules inside the scales of the female pine cones. Male cones release pollen that is carried by the wind to the female cone. Contact with the outer portion of the ovule causes the germination and the formation of a **pollen tube,** which provides a passageway to the egg. This process takes about 15 months to result in fertilization. The zygote develops within the ovule and, eventually the seed falls to the ground and begins growing into a new tree.

Those vascular plants with protected seed are called **angiosperm** or **flowering plants.** The **flower,** which is actually a modified leaf, is structured to promote sexual reproduction. Its consists of four whorls of parts designed to attract insects for pollination. The outermost whorl is composed of **sepals** that protect the flower in the bud. Within the sepals are a whorl of petals that contain the male parts of the flower called the **stamen.** Each stamen consists of an **anther** where the haploid pollen grains are produced. The innermost whorl is the female organ, the pistil, and it consists of three parts. The sticky tip, called the **stigma,** is the site where the pollen can adhere. The stigma is supported by the style, which gives it the best advantage for fertilization. The ovary sits at the base of the pistil and protects the ovules. The cells within the ovules meitotically divide to produce four cells, one of which becomes the female gametophyte. When mature, this structure is called the **embryo sac.** Upon fertilization, the ovary becomes a seed.

For fertilization to occur, the pollen grains must attach to the stigma. **Pollination** occurs when the pollen grains are spread by such means as insects, animals, the wind, etc. Sometimes the pollen never spreads to another plant, but falls directly onto the stimga of the same plant. This process is called **self-pollination.** Once the pollen reaches the stimga, these grains are directed to the ovule by a long **pollen tube,** which develops down the style to the ovary. One male gamete fuses with the egg to form the diploid zygote that begins the sporophyte generation. The other male gamete fuses with two other nuclei of the embryo sac to form the triploid **endosperm** that will form the food for the developing embryo. The ovary develops into the **fruit** while the ovules become the seeds.

Seed **germination** depends on many environmental factors, water being the main one. Additionally, germination and growth require the mobilization of food reserves, and the location of these nutrients is the basis for differentiation between angiosperm. The **monocots,** such as irises, lilies, grasses, and orchids are characterized by one-seed leaf or **cotyledon.** The **dicots,** such as roses, legumes, sunflowers, trees, and shrubs are characterized by two-seed leaves.

Once the embryo is ready to emerge, the first portion to be exposed is the **radicle,** or root. This provides a means for the developing plant to anchor to the soil and absorb water and nutrients. The radicle is followed by the **shoot** and then the first true leaves develop.

Some plants reproduce asexaully by a method known as **vegetative propagation.** This process involves the development of a new plant from a portion of the parent plant. These new plants can emerge from underground stems called **rhizomes,** or horizontal stems called **runners** or **stolons.** New plants can also develop from specialized underground storage stems called **tubers.**

New technology has allowed scientists to remove cells from parent plants and grow them on media to produce new plants. This process is called **cell culture technique.** This allows botanists to produce genetically identical stock and propagate plants that are slow to multiply on their own.

5.   Plant Hormones

Plant growth is regulated by many environmental factors but is also controlled by chemicals called hormones. **Hormones** are chemical messengers produced in very small quantities in one place in the organism and transported to another to cause a physiological response. The **auxins** and **gibberillins** promote growth by cell elongation. The auxins are also responsible for the growth response of plants to directional stimuli. The **cytokinins** stimulate cell division. **Abscisic acid** is a growth inhibitor that induces and maintains dormancy. **Ethylene** is a gaseous hormone given off by ripening fruits to trigger other fruits to develop and ripen faster. The pigment **phytochrome** controls the plant's flowering response to day length.

CHAPTER REVIEW ACTIVITIES

CONCEPT MAPPING

Develop a concept map that demonstrates your understanding of plant reproduction and diversity. Use as many of the key terms as possible. Develop the map as thoroughly as possible, but make sure the linkages between concepts are accurate.

MATCHING

Select the term from the list provided that best relates to the following statements. If you are unsure of the correct answer or chose an incorrect response, review the related concept in Chapter 27.

_____ 1. These structures in vascular plants with protected seeds and flowers act as sexual reproduction organs.

_____ 2. These modified leaves house the sexual reproductive organs in the vascular plants.

_____ 3. These structures, from which new sporophyte plants grow, protect the embryonic plant from dehydration and damage.

_____ 4. These specialized structures are where eggs are formed in bryophytes and several divisions of vascular plants.

_____ 5. This is the term for a seed leaf.

_____ 6. This group of angiosperm stores most of their extra food in extraembryonic tissue called endosperm.

_____ 7. The haploid phase of a plant life cycle tends to dominate the life cycle of nonvascular plants.

_____ 8. These plants reproduce by means of spores, rather than seeds.

_____ 9. These structures are where sperm are produced in several divisions of vascular plants.

_____ 10. This structure is formed when a sperm nucleus fuses with two nuclei in the embryo sac.

_____ 11. This term describes the sprouting of a seed, which begins when it receives water and appropriate environmental cues.

_____ 12. This series of events takes place from one stage during the life span of an organism, through a reproductive phase, until a stage similar to the original is reached in the next generation.

_____ 13. These vascular plants have naked seeds.

_____ 14. These chemical substances are produced in small quantities that bring about physiological responses.

_____ 15. This group of angiosperms stores food in their cotyledons.

_____ 16. This diploid generation tends to dominate the life cycles of vascular plants.

_____ 17. These major taxonomic categories correspond approximately to phyla and are used primarily in botany.

_____ 18. These plants lack specialized transport tissue.

_____ 19. This type of life cycle has both a multicellular haploid phase and a multicellular diploid phase.

_____ 20. These plants have systems specialized for transporting fluids.

TERMS LIST

a. Divisons	f. Seedless vascular plants	k. Flowers	p. Monocots
b. Antheridia	g. Life cycle	l. Vascular plants	q. Archegonia
c. Seeds	h. Angiosperm	m. Dicots	r. Cotyledon
d. Gymnosperm	i. Sporophyte generation	n. Endosperm	s. Nonvascular plants
e. Gametophyte generation	j. Hormones	o. Alternation of generation	t. Germination

PLANT COMPARISON

Fill the spaces with the correct information regarding the various plants.

PLANTS	DIVISION	SPORE OR SEEDS	VASCULAR OR NONVASCULAR	DOMINANT GENERATION
Liverworts				
Roses				
Firs				
Fern				
Maple tree				
Mosses				

LABELING

Label the following life cycles. Review any of the material associated with these diagrams in the text if you are unsure of the correct answer.

Fig. 27.2

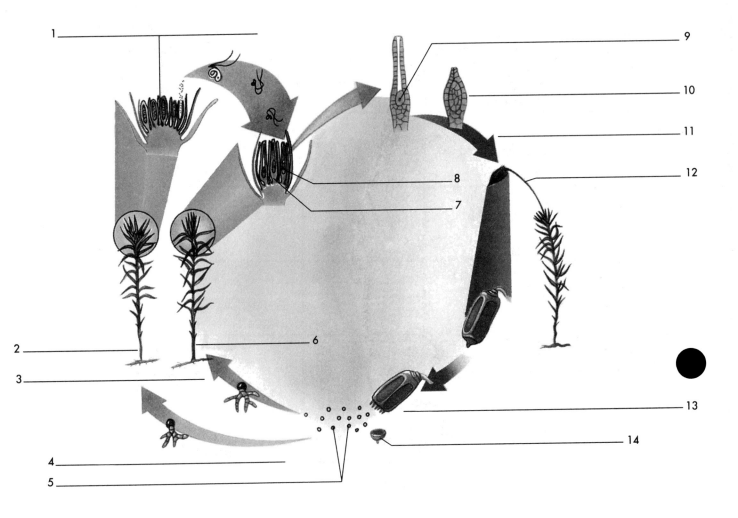

Fig. 27.3

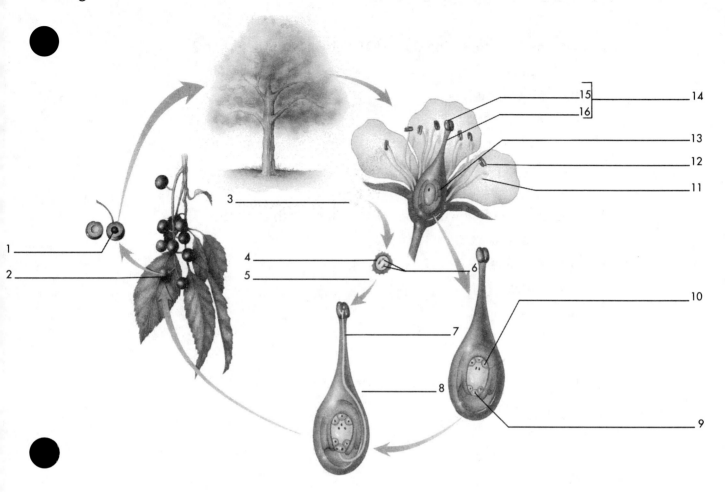

Fig. 27.4

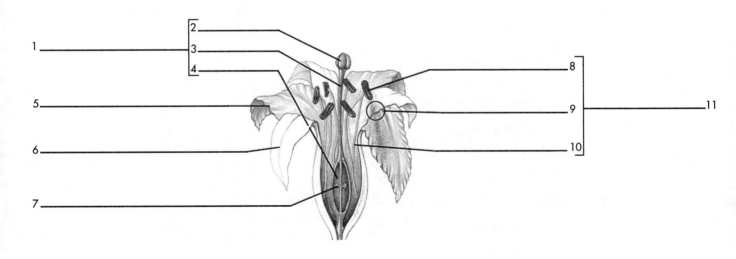

MULTIPLE CHOICE QUESTIONS

_____ 1. Which of the following does NOT belong with the others?
  a. Club mosses
  b. Liverworts
  c. Bryophytes
  d. Hornworts
  e. True mosses

_____ 2. Which of the following does NOT belong with the others?
  a. Grasses
  b. Conifers
  c. Roses
  d. Irises
  e. Oak trees

_____ 3. Which of the following are plants with naked seeds?
  a. Mosses
  b. Ferns
  c. Conifers
  d. Dicots
  e. Monocots

_____ 4. To which of the following divisions do the most highly evolved plants belong?
  a. Cycadophyta
  b. Gnetophyta
  c. Coniferophyta
  d. Anthrophyta
  e. Pterophyta

_____ 5. Spanish moss is a relative of which of the following?
  a. Bryophytes
  b. Pineapples
  c. Roses
  d. Lilies
  e. Grasses

6. The antheridia and archegonia of Marchantia, a liverwort, are shaped like which of the following?
   a. Paddles
   b. Umbrellas
   c. Boxes
   d. Hearts
   e. Wings

7. Pollinaton takes place on which of the following structures?
   a. Stigma
   b. Style
   c. Stamen
   d. Pistil
   e. Ovule

8. Fertilization takes place on which of the following structures?
   a. Stigma
   b. Ovary
   c. Style
   d. Anther
   e. Pistil

9. Which of the following is the male gametophyte generation?
   a. Stigma
   b. Stamen
   c. Pollen grain
   d. Endosperm
   e. Cotyledon

10. Where are the sproangia of a fern found?
    a. At the tips of rhizomes
    b. Extending beyond the stolon
    c. On the underside of the fronds
    d. At the petiole
    e. On the underside of the prothallus

_____ 11. During the life cycle of seed plants, which of the following best relates to the process of meiosis?
   a.   It does not take place.
   b.   It results in the production of spores.
   c.   It results in female gametes only.
   d.   It results in male gametes only.
   e.   It results in female and male gametes.

_____ 12. Which of the following is TRUE regarding moss?
   a.   Moss lacks vascular tissue.
   b.   Sporophyte generation is dependent on gametophyte generation.
   c.   Moss has two completely separate generations.
   d.   A and B
   e.   A and C

SHORT ANSWER QUESTIONS

1.   What are the three common members of the bryophytes?

2.   What is the function of seeds?

3.   What is vegetative propagation?

CRITICAL QUESTIONS

1.   Explain why vegetables are considered to be fruits.

2.   Forest fires often result in numerous new conifers growing in the damaged area. Why is new growth so plentiful?

3.   Explain how it can be possible for plants to be pollinated without being fertilized.

TELLING THE STORY

Develop a story that demonstrates your understanding of plant reproduction and diversity. If you have difficulty with any of the associated areas or topics as you write, go back and review that information in Chapter 27.

CHAPTER REVIEW ANSWERS

MATCHING

1. H	5. R	9. B	13. D	17. A
2. K	6. P	10. N	14. J	18. S
3. C	7. E	11. T	15. M	19. O
4. Q	8. F	12. G	16. I	20. L

PLANT COMPARISON

PLANTS	DIVISION	SPORE OR SEED	VASCULAR OR NONVASCULAR	DOMINANT GENERATION
Liverworts	Bryophyte	Spore	Nonvascular	Gametophyte
Roses	Tracheophyte	Seed	Vascular	Sporophyte
Firs	Tracheophyte	Seed	Vascular	Sporophyte
Fern	Tracheophyte	Seed	Vascular	Sporophyte
Maple tree	Tracheophyte	Seed	Vascular	Sporophyte
Mosses	Bryophyte	Spore	Nonvascular	Gametophyte

LABELING

Fig.27.2

1. Spores	6. Female	11. Mature sporophyte
2. Germinating spores	7. Archegonium	12. Sporophyte stalk
3. Gametophytes	8. Egg	13. Spore capsule
4. Male	9. Zygote	14. Spores released
5. Antheridium	10. Developing sporophyte in archegonium	15. Top of spore capsule

Fig.27.3

1. Seed	5. Style	9. Pollen grains (male)	13. Embryo sac with 8 nuclei
2. Sporophyte	6. Ovary	10. Nuclei	14. Egg
3. Pistil (female)	7. Anther	11. Gametophytes	15. Sexual fusion
4. Stigma	8. Filament	12. Pollen tube	16. Fruit

Fig. 27.4

1. Ovule	5. Stigma	9. Anther
2. Sepal	6. Style	10. Pollen grain
3. Petal	7. Ovary	11. Filament
4. Pistil	8. Stamen	

MULTIPLE CHOICE QUESTIONS

1. A	5. B	9. C
2. B	6. B	10. C
3. C	7. A	11. B
4. D	8. B	12. D

SHORT ANSWER QUESTIONS

1. Mosses, liverworts, and hornworts

2. Seeds are the protective structures for embryonic sporophytes. They are agents of dispersal, so that plants may find favorable environments to grow. They also allow the plant to escape extreme environments by being in a dormant condition until suitable conditions prevail. They also provide a nutrition storage area for the embryo to survive on until it is capable of acquiring water and nutrients on its own.

3. Vegetative propagation is an asexual reproductive process in which new plants develop from a portion of the parent plant. These new plants can develop from rhizomes, stolons, or tubers. They can form on the leaves of maternity plants, also.

CRITICAL QUESTIONS

1. Vegetables produce seeds within the fibrous, edible material and therefore are called fruits.

2. New growth results from so many seeds being dispersed during the fire. The extreme heat forces the cones to open and release the seeds on the ground. Rain or water from the fire fighting efforts provide the necessary water for these seed to begin growing. The forest reestablishes a large population of new trees, but it takes a long time to reestablish a mature forest.

3. The process of pollination involves the dispersal of pollen from the anther of the angiosperm to the stigma where the pollen grains stick. This process does not insure fertilization, but enhances the opportunity for this process to occur.

CHAPTER 28

PLANTS:
PATTERNS OF STRUCTURE
AND FUNCTION

KEY CONCEPTS OVERVIEW

1. Vascular Plant Organization

    For living organisms to grow, reproduce, move, respire, metabolize, etc., they must be equipped to do so. And they must be organized structurally to accomplish these processes with relative efficiency. The relationship between structure and function is fundamental to studying biology.

    Vascular plants are organized along a vertical axis, and their structures are divided into the shoot and the root systems. The **root** is usually below ground and absorbs water and nutrients. It also functions to anchor the plant for stability. The shoot system is comprised of the **stem** and **leaves**. The stem is a framework for the leaves, which are essential for gathering light energy and carrying out the process of photosynthesis.

2. Tissues of Vascular Plants

    Plants have organs made up of different types of tissues. Remember that **tissues** are a group of cells that work together to carry out a specific function. There are three basic types of tissues within the root and the shoot systems: vascular tissue, ground tissue, and dermal tissue. **Ground tissue** stores plant-produced carbohydrates and basically surrounds the transport system made up of **vascular tissue**. The **dermal tissue** protects the plant and helps to conserve water.

    (See Figure 28-1, P. 616 in your text)

Another type of tissue, called the **meristematic tissue (meristem)**, contains cells that divide to facilitate plant growth. These undifferentiated cells eventually develop into one of the three types of tissues mentioned above.

Vascular tissue can be divided into two types: xylem and phloem. Only after they die do these tissues conduct water and dissolved minerals. The **xylem** conducts water and minerals specificially within the transport vessels. It is made up of structures called **tracheids**, which are stacks of cells with tapering ends connected by **pits**, and **vessel elements**. These are stacks of cells that are connected by structures called **perforations**. Pores allow water to move laterally in both of these types of cells.

The **phloem** conducts carbohydrates and other nutrients needed by the plant and is made up of two types of cells called sieve tube member (conducting cells) and companion cells. **Conducting cells** lack nuclei and are linked end to end by pits. The **companion cells** are support cells for the conducting cells, and they secrete substances into and remove substances from the conducting cells.

Within the ground tissue there are **parenchyma cells**, which function during photosynthesis and food storage. It is this tissue that makes up the fleshy portion of the roots, fruits, and vegetables we eat. There are also **sclerenchyma cells**, which offer support and protection for the ground tissue.

The dermal tissue is made up of **epidermal cells**, which offer protection against water loss. They are covered by a waterproof structure called a **cuticle**. In addition to the epidermal cells, the dermal tissue contains **guard cells**, which surround openings on leaves for gas exchange and **trichomes**, which can increase surface area, reflect sunlight, and offer a first line of defense against potential dangers.

The location of meristematic tissue dictates where plants can grow. **Apical meristem** is located at the tips of stems and roots and is responsible for the increase in their length. This type of growth is called **primary** growth. Lateral meristem is a cylinder of tissue found in the stems and roots of plants. It allows a plant to increase in width, a **secondary** growth that occurs in all woody trees and shrubs. You are able to observe that not all plants have the same type or location of meristematic tissue.

3.   Vascular Plant Organs

The roots, stems, and leaves are the organs of vascular plants. Root and shoots of vascular plants tend to share the same basic structure, but differ in the relative distribution of vascular and ground tissue systems.

The roots of **dicots** have a central column of xylem with laterally radiating arms. Strands of phloem can be found between these lateral branches. This grouping is surrounded by a cylinder of **pericycle,** which is made up of parenchymal cells. The pericycle is, in turn, surrounded by a single row of cells, the endodermis, which makes up the inner layer of the **cortex.** The root is covered by epidermal cells that have hair-like extensions on them. These **root hairs** increase the surface area. This greatly enhances water and nutrient absorption. **Monocots** have a similar structural arrangement of root tissues, with the exception of having a central **pith** composed of parenchymal cells. A **root cap** covers and protects the growing tip. Some roots have a single **tap root**, while others, such as grasses, have a **fibrous** or diffused root system. Roots that arise from the stem are called **adventitious roots.** Corn plants would exhibit this type of root structure.

The shoot system, consisting of stems and leaves, is that part of the plant that grows above ground. Stems develop leaves at locations called **nodes.** The function of stems is to display leaves to the light and to transport material between the leaves and the roots. The stem area between nodes is referred to as **internodes.** In monocots, the veins are scattered and diffused throughout, but dicots have their veins arranged in a circular form around the central pith. In dicots, those plants with woody stems develop a **vascular cambium,** which is actually a lateral meristem. This cambium produces a **secondary xylem** and a **secondary phloem.** The secondary xylem is actually the wood of the dicot tree or shrub. This material can vary in density. Hardwoods are more dense than softwoods. Differential growth rates produce a visible pattern each year of the tree's life. Each year's growth is marked by an **annual ring** that can be used to determine the age of the stem. As the plant ages, a second lateral meristem, called the **cork cambium,** develops and produces **bark.**

The leaf is composed of a flattened structure called the **blade** and a stalk called the **petiole.** Veins can be observed in parallel patterns on monocot leaves and in net-like patterns in dicots. Conifers have needles that are modified leaves and are more suitable for the climates and conditions in which they live. The leaf parenchyma , called **mesophyll,** is found in the middle layer. This mesophyll, in most dicots, is made up of two layers: the **palisade** layer and the **spongy** layer. The cells of these layers are filled with chloroplasts and photosynthesis occurs here. The palisade layer has many air spaces between cells. These spaces are connected to openings on the surface of the leaves called **stomata.** The size of these openings is controlled by turgor pressure in the two guard cells surrounding the pore opening. Water is lost through the stomata in a vapor form through a process called **transpiration.**

4. Transport Within Plants

Water is moved through a plant by a process known as **transpiration**. Transpiration occurs when water vapor passes out of a plant through the stomata. Water is pulled toward the leaves from the xylem. The evaporation of water from the leaves creates a concentration gradient, which causes tension or pulling on the column of water.

Sap is transported in plants by the phloem. Sap contains about 10% to 25% sucrose and other substances that are dissolved in the water matrix of this fluid. This fluid moves very fast by a pressure-flow or mass-flow system. This system is based on the pressures created by different concentrations of solutes.

5. Nonvascular Plant Organization

Bryophytes have no true roots, stems, or leaves, but they do have structures that function much like these organs. While there are no roots, for example, there are small structures called rhizoids that anchor the plants. For additional comparative information on the bryophytes, you may want to review the material in Chapter 27.

CHAPTER REVIEW ACTIVITIES

CONCEPT MAPPING

Develop a concept map that demonstrates your understanding of plant structure and function. Incorporate as many of the key terms as possible. Where appropriate, try utilizing terms you learned from the previous chapters.

MATCHING

Select the term from the list provided that best relates to the following statements. If you are unsure of the correct answer or chose an incorrect response, review the related concept in Chapter 28.

_____ 1. This pair of cells brackets a stoma and regulates its opening and closing.

_____ 2. This plant tissue stores the carbohydrates produced by the plant itself.

_____ 3. This part of a plant's structure serves as a framework for the positioning of the leaves.

_____ 4. This innermost layer of cells makes up the cortex.

_____ 5. This group of cells works together to perform specific functions.

_____ 6. This type of plant growth occurs in all woody trees and shrubs and causes plants to increase in width.

_____ 7. These openings in the epidermis of a leaf allow the carbon dioxide needed for photosynthesis to enter and the oxygen produced during photosynthesis to escape.

_____ 8. These parts of a plant are where most photosynthesis takes place.

_____ 9. This type of vascular tissue in plants conducts water and dissolved inorganic nutrients.

_____ 10. This type of root develops from the stems.

_____ 11. This term describes the stem area between the leaf attachment sites.

_____ 12. This is an undifferentiated type of tissue in a vascular plant in which cell division occurs during growth.

_____ 13. This slender, root-like projection anchors bryophytes to a substrate.

_____ 14. This process enables water to be moved from the xylem to the leaves for evaporation.

_____ 15. This is the flattened portion of a leaf.

_____ 16. This outer protective covering of most plants protects them from water loss and injury.

_____ 17. This part of the vascular plant exists above the ground.

_____ 18. These locations on the stem of plants are where leaves form.

_____ 19. This type of modified leaf is found on conifers.

_____ 20. This type of plant growth occurs mainly at the tips of the roots and the shoots.

_____ 21. This type of tissue conducts water and nutrients up the plant and carries the products of photosynthesis throughout the plant.

_____ 22. This type of vascular tissue conducts carbohydrates and other substances used by the plant.

_____ 23. This is the slender stalk of a leaf.

_____ 24. This is another term for the secondary xylem.

TERMS LIST

a. Rhizoids	f. Stomata	k. Endodermis	p. Needles	u. Xylem
b. Guard cells	g. Dermal tissue	l. Adventi- tious roots	q. Vascular tissue	v. Meriste- matic tissue
c. Internodes	h. Stem	m. Wood	r. Shoot	w. Secondary growth
d. Leaves	i. Trans- piration	n. Primary growth	s. Phloem	x. Ground tissue
e. Tissue	j. Node	o. Blade	t. Petiole	

LABELING

Label the following diagrams. If you have any difficulty with the appropriate terms, review the material in the text.

Fig. 28.1 Root tip.

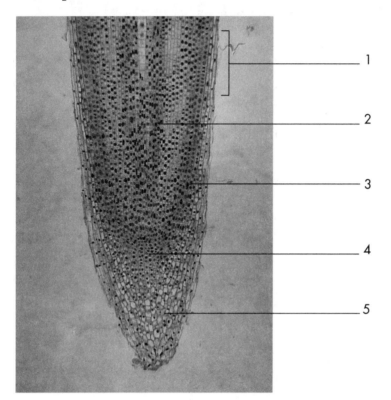

Fig. 28.2 Plant vascular tissue.

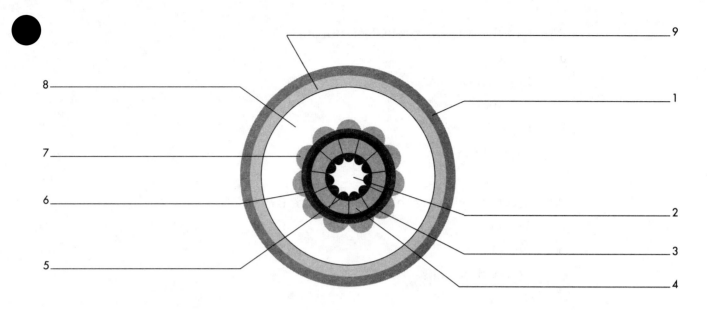

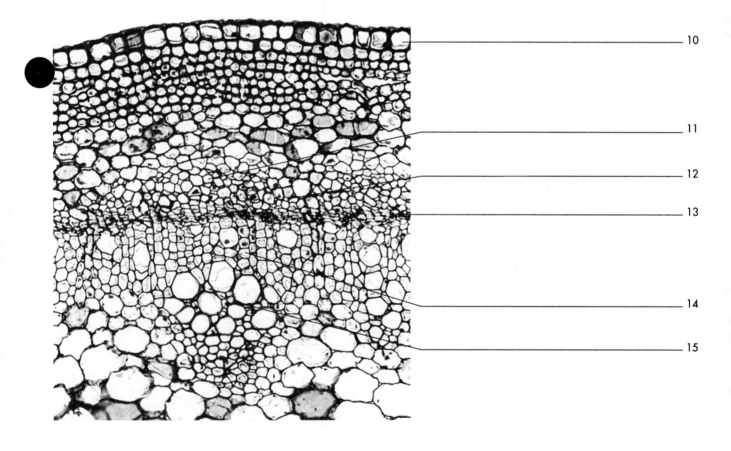

MULTIPLE CHOICE QUESTIONS

_____ 1. Which of the following cells lack nuclei?
  a. Epidermal
  b. Companion
  c. Conducting
  d. Guard
  e. Parenchyma

_____ 2. Which of the following cells are xylem cells?
  a. Tracheids
  b. Vessels
  c. Companion cells
  d. Sieve tubes
  e. Both A and B

_____ 3. Which of the following cells are closely associated with sieve tubes?
  a. Tracheids
  b. Vessels
  c. Companion cells
  d. Parenchyma
  e. Schlernchyma

_____ 4. Which of the following would be the most common type of cell found in plants?
  a. Tracheids
  b. Vessels
  c. Companion cells
  d. Parenchyma
  e. Schlerenchyma

5. Which of the following tissues are growth centers characterized by cell division?
   a. Ground
   b. Vascular
   c. Meristem
   d. Dermal
   e. Pith

6. For what purpose is the water absorbed by roots primarily used?
   a. Photosynthesis
   b. Replacement of the water lost by transpiration
   c. Turgor pressure
   d. Digestion of complex compounds
   e. The production of nitrogen compounds deposited in the soil

7. Which of the following are cells found in the center of dicot roots?
   a. Pericycle
   b. Cortex
   c. Xylem
   d. Phloem
   e. Endodermis

8. Which of the following cells of a root are specialized for storage?
   a. Pericycle
   b. Cortex
   c. Xylem
   d. Phloem
   e. Endodermis

9. In which of the following would the mesophyll be found?
   a. Leaves
   b. Fruit
   c. Root
   d. Stem
   e. Pits

_____ 10. Which of the following would best relate to the cork cambium?
    a. Apical root
    b. Palisade layer
    c. Spongy layer
    d. Bark
    e. Pith

_____ 11. Which of the following would best relate to leaves?
    a. Blade
    b. Petiole
    c. Palisade layer
    d. Spongy layer
    e. All of the above

_____ 12. In which area of a plant would you most likely find apical meristem?
    a. Leaf edges
    b. Nodes
    c. Root tips
    d. Flowers
    e. Stems

_____ 13. Which of the following plants would be the best example of adventitious roots?
    a. Corn
    b. Ivy
    c. Roses
    d. A and C
    e. A and B

_____ 14. The trichomes within the dermal tissue perform which of the following functions?
    a. Regulate water movement into and out of the plant
    b. Increase surface area
    c. Promote cell division
    d. Attract insects for pollination
    e. None of the above

SHORT ANSWER QUESTIONS

1. What are the major systems in vascular plants?

2. What are the three types of differentiated tissues found in vascular plants?

3. What is the cause of annual rings?

4. How do lateral branches form in vascular plants?

5. What are fibrous roots? Give an example of a plant that has this characteristic.

6. In mass flow, what is a source and a sink?

CRITICAL QUESTIONS

1. Would you expect to find annual rings in all woody plants? Explain.

2. The leaves on your house plant have started to turn yellow. What deficiency might be present and how would you correct the problem?

3. What is the advantage of having xylem cells adjacent to phloem cells in vascular plants?

TELLING THE STORY

Develop a story that demonstrates your understanding of plant structure and function. If you have difficulty with any of the associated areas or topics as you write, go back and review that information in Chapter 28.

CHAPTER REVIEW ANSWERS

MATCHING

1. B	5. E	9. U	13. A	17. R	21. Q
2. X	6. W	10. L	14. I	18. J	22. S
3. H	7. F	11. C	15. O	19. P	23. T
4. K	8. D	12. V	16. G	20. N	24. M

LABELING

Fig. 28.1

1. Epidermis
2. Vascular tissues
3. Cortex
4. Apical meristem
5. Root cap

Fig. 28.2

1. Cork	6. Vascular cambium	11. Primary phloem
2. Pith	7. Primary phloem	12. Secondary phloem
3. Secondary phloem	8. Cortex	13. Vascular cambium
4. Secondary xylem	9. Cork cambium	14. Secondary xylem
5. Primary xylem	10. Cpidermis	15. Primary xylem

MULTIPLE CHOICE QUESTIONS

1. C	5. C	9. A	13. E
2. E	6. B	10. D	14. B
3. C	7. C	11. E	
4. D	8. B	12. C	

SHORT ANSWER

1.   The shoot system includes the stems, leaves, and flowers. The root system, made up of root structures, is the portion found primarily underground.

2.   Ground, dermal, and vascular tissues

3.   In the early spring, the environmental conditions favor rapid growth and large, relatively thin-walled cells are formed. During the rest of the year, the

cell division process is much slower, and the cells produced are smaller and have thicker walls. This pattern forms what is known as annual rings.

4.   At the node in the axis between the petiole and stem, there are lateral buds. If the terminal bud is removed by pruning, the lateral buds are no longer suppressed by hormones in the terminal bud cells. The bud will then begin to develop and produce a new stem from which leaves will eventually emerge.

5.   Fibrous roots are root systems with no predominant root. Most monocots have these types of roots. An example of a plant with this type of root system would be grasses.

6.   A source is the location of the production of sucrose, and a sink is a site that uses the sucrose. A source might be found in a leaf, and a sink might be a stem or root.

CRITICAL QUESTIONS

1.   No. The plants that grow in uniform conditions, such as the tropical rainforest, will not have differentiated growth rates. Therefore, such patterns would not be expressed.

2.   Deficiencies in macro- or micronutrients can produce diagnostic symptoms in plants. Yellowing leaves indicate a magnesium deficiency and could therefore be treated with additional magnesium in the soil. Magnesium deficiencies produce an inability to synthesize chlorophyll and therefore the leaves of the plant yellow.

3.   The advantage of having xylem and phloem adjacent to each other is that this arrangement allows the plant to utilize the differences in water potential to accelerate the movement of carbohydrates and other plant nutrients. When the hydrostatic pressure in the phloem is greater at the source than the sink, this pressure causes the contents of the phloem to move toward the sink.

CHAPTER 29

INVERTEBRATE ANIMALS: PATTERNS OF STRUCTURE, FUNCTION, AND REPRODUCTION

KEY CONCEPTS OVERVIEW

1.   Animal Characteristics

Since animals are heterotrophs, they must be able to move, gather food, respire, metabolize, digest, etc. These functions necessitate having specialized structures that will facilitate the activities. Since there is enormous diversity among animals, you can well imagine that these structures will range from the simplest form to very complex systems.

Animals are divided into two major groups: the invertebrates and **vertebrates**. Most scientists agree that animals have common evolutionary ancestry.

As you can see from the ancestral tree, there is great variation in the appearance of animals. The word **symmetry** is used to describe the distribution of the parts of organisms. Other than the sponges, which lack symmetry, animals have either **radial** or **bilateral** symmetry. The cnidarians and echinoderms exemplify radial symmetry. This structural arrangement enables them to react to their aquatic enviroment most effectively. The remainder of the animals exhibit bilateral symmetry, meaning that the right and left sides of an organism are mirror images of each other. These animals have a dorsal and ventral side, and a cephalic and caudal end.

The body cavity, or coelom, is another way that animals can be distinguished from one another. This fluid-filled enclosure is lined with connective tissue and functions to transport materials and dispose of waste. Organisms having a coelom are referred to as **coelomates,** and those lacking a coelom are **acoelomates**.

(See Figure 29.1, P. 636 in your text)

The coelomate animals represent two distinct evolutionary lines: the protostomes and the deuterostomes. During development, the morula undergoes spiral cleavage in the **protostomes,** and the first indentation becomes the mouth. The **deuterostomes** exhibit radial cleavage and develop two openings, which become the anus and the mouth.

2.   Invertebrate Symmetry and Coelom

This section of the review will look at the diversity in symmetry and the coelom among the various types of animals.

Asymmetry

The simplest animals belong to the phylum Porifera. The sponges are asymmetrical with a central cavity. There are openings, or pores, that allow the movement of water into the central cavity and create a way for the sponge to acquire food. **Collar cells** create a current to facilitate this water movement. Sponges reproduce both sexually and asexually. The asexual process is accomplished by **fragmentation.** Fragmentation occurs when a group of cells separate from the sponge body and begin to develop into new individuals. The sexual process occurs, because the sponge is **hermaphroditic,** meaning one organism has both male and female sex cells.

Radial Symmetry

Only two phyla of animals are classified as radially symmetrical: the Cnidaria (jellyfish, hydra, sea anemones, and corals) and the Ctenophora (comb jellies). These organisms have an outer epidermis and an inner gastrodermis with a jelly-like mesoglea between them. There are two basic body plans in the life cycle: the polyp and the medusa. The polyp, shaped like an immobile hydra, forms a large mass and reproduces by asexual budding. The polyps release free-swimming medusae that reproduce sexually. The tentacles of cnadaria, which bear stinging cells called cnidocytes, help them capture prey.

Bilateral Symmetry

All the remaining animal phyla are bilaterally symmetrical. They also all developed from three layers of tissue: the endoderm, the mesoderm, and the ectoderm. The simplest of the organisms possessing this type of symmetry are the flatworms or **platyhelminthes.** They are the first phyla to have exhibited the three layer development. There are three classes of flatworms: tubullarians, flukes, and tapeworms. All the flatworms are free-living. They can live in freshwater, saltwater, or moist soil. The flukes and tapeworms are parasitic. As a group, flatworms are the most primitive bilaterally symmetrical acoelomates. The turbellarians swim around and ingest food through a protruding pharynx and digest this material in a sac-like digestive system with only one opening. Their nervous system consists of sensory pits, eyespots, a small cephalic mass of tissue, and a ladder-like paired nerve cord. Flatworms are hermaphroditic, with better developed reproductive organs than the sponges. Each partner deposits sperm in the other's copulatory sac. These sperm fertilize the eggs, which are then laid in cocoons. Some parasitic flatworms have a larval stage. Still others are capable of asexual reproduction by regeneration.

The roundworms, or nematodes, have a pseudocoelomate body plan. Many members of the group of seven phyla are parasitic, and about 50 species are parasitic to humans. Roundworms are covered by a tough outer cuticle, which they shed as they grow. Their longitudinal muscular architecture provides locomotion in muddy water and distribution of food and oxygen within the pseudocoelom. Their digestive system has two openings, and food moves through a tract where it is broken down and absorbed. They contain very primitive nervous and excretory systems.

The remaining invertebrates are coelomates and belong to the mollusk, annelid, and arthropod phyla. These organisms are all protostomes.

Mollusks have a soft body surrounded by a mantle, a muscular foot, and most are covered with a hard shell. Their habitat is primarily marine and fresh waters, but some live on land. They exhibit four body plans with very distinct differences: gastropod, cephalopod, bivalve, and chiton. Mollusks have an open and closed circulatory system. They have a heart in each system to pump blood. In the closed system, there are vessels through which blood is transported. In the open system, blood flows in vessels leading to and from the heart but through **blood sinuses** throughout the rest of the body. Waste is removed from mollusks by tubular structures called **nephridia.** Mollusks that are aquatic use gills for respiration. Terrestrial mollusks have adaptations for breathing.

Annelids and arthropods have segmented bodies. The annelids are the segmented worms and include three classes: the sea worms, the earthworms, and the parasitic leeches. Annelids have a digestive system that consists of a tube within a tube with openings at both ends. Segmented worms differ from nematodes in that they have circular muscles in addition to longitudinal muscles. Earthworms contain a well-developed cerebral ganglion, which is a mass of cell bodies representing a brain. There are several muscular blood vessels that act as pumps within a closed circulatory system. These organisms also have structures called **setae,** or bristles, which they use to anchor themselves in the soil so they can stretch out and pull forward during locomotion. Reproduction differs among the classes of annelids. Some have separate sexes and some are hermaphroditic.

Arthropods have a rigid external skeleton called an **exoskeleton.** This covering provides a place for muscle attachment, protection, and conservation of water. All arthropods are grouped as those with or without **mandibles,** or jaws. Many arthropods have a **compound eye** that contains many independent visual units. These organisms have a tubular gut that extends from the mouth to the anus. They also have an open circulatory system. The respiratory system in terrestrial arthropods consists of small air ducts called **tracheae.** Aquatic arthropods breathe by means of gills. The excretory system is associated with the open circulatory system. The principle structures are the malpighian tubules.

The phylum Echinodermata has five living classes. These organisms differ from other invertebrates in that they are deuterostomes. Echinoderms have no well organized circulatory system. Respiration takes place by means of skin gills, which also account for water removal. Most reproduction is sexual and external, although some exhibit asexual regeneration.

CHAPTER REVIEW ACTIVITIES

CONCEPT MAPPING

Develop a concept map that demonstrates your understanding of invertebrate structure, function, and reproduction. Incorporate as many of the key terms as possible. Where appropriate, try utilizing terms you learned from the previous chapters.

MATCHING

Select the term from the list provided that best relates to the following statements. If you are unsure of the correct answer or chose an incorrect response, review the related concept in Chapter 29.

_____ 1. This animal or plant has both male and female reproductive organs.

_____ 2. This rigid skeleton is found on the outside of arthropods.

_____ 3. This term describes the way certain animals' body parts emerge from a central point.

_____ 4. This outer layer of cells is formed during the early development of the embryos of all bilaterally symmetrical animals.

_____ 5. These members of a branch of coelomate aminals have gastrulas with two indentions, which become the anus and mouth.

_____ 6. These are free-floating and generally umbrella shaped aquatic animals.

_____ 7. This term describes the sessile hydra-like body plan of cnidarians.

_____ 8. These organisms have no body cavity.

_____ 9. This inner layer of cells forms during the early development of the embryos of all bilaterally symmetrical animals.

_____ 10. This body cavity is found within most bilaterally symmetrical organisms and the echinoderms.

_____ 11. These mollusks develop feet on their "head" regions.

_____ 12. This is the principle excretory structure of arthropods.

_____ 13. These members of a branch of coelomate animals have gastrulas with only one indentation, which becomes the mouth of the organisms.

_____ 14. This material is found between the epidermis and the gastrodermis in jellyfish.

_____ 15. These branched air ducts make up the respiratory systems of arthropods.

_____ 16. These organisms have a partially lined coelom that houses the organs of an organism.

TERMS LIST

a. Cephalopods	e. Ectoderm	i. Exoskeleton	m. Polyps
b. Protosomes	f. Tracheae	j. Coelom	n. Medusae
c. Endoderm	g. Acoelomates	k. Mesoglea	o. Hermaphrodites
d. Pseudocoelomate	h. Radial symmetry	l. Malpighian tubules	p. Deuterostomes

LABELING

Label the following diagrams completely. If you have difficulty , go back and review the related material in the text.

Fig. 29.1 Segmented worm.

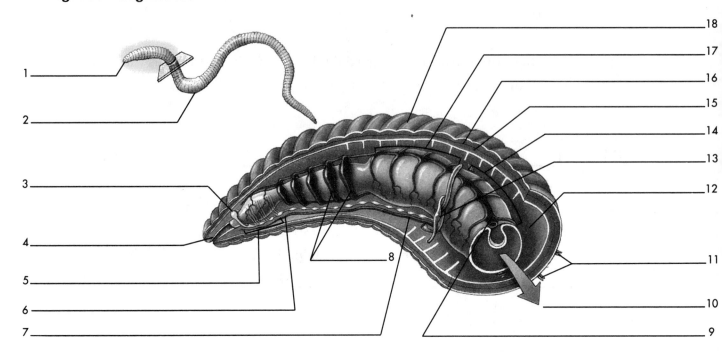

Figure 29.2 Grasshopper

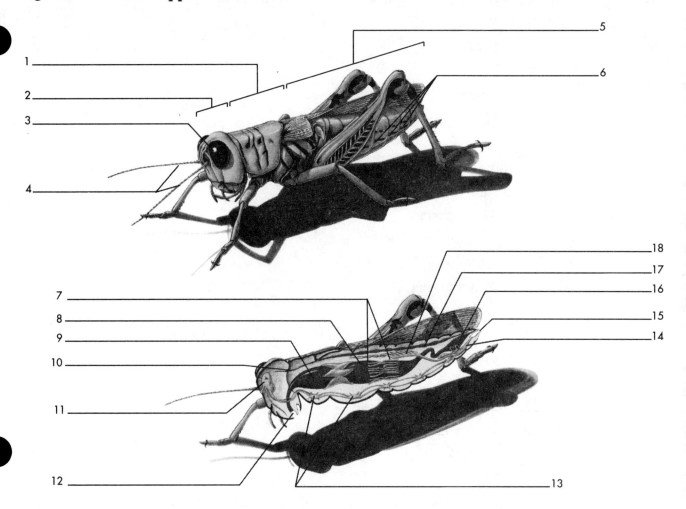

MULTIPLE CHOICE QUESTIONS

_____ 1. To which phylum do corals belong?

    a. Cnadaria

    b. Annelida

    c. Arthropoda

    d. Porifera

    e. Mollusca

_____ 2. Which of the following is NOT an invertebrate?

   a.   Worm

   b.   Snail

   c.   Jellyfish

   d.   Spider

   e.   Fish

_____ 3. Members of which phylum have radial symmetry/

   a.   Annelida

   b.   Arthropoda

   c.   Mollusca

   d.   Porifera

   e.   Echinodermata

_____ 4. Which of the following is an animal that has a false coelom?

   a.   Cnararian

   b.   Flatworm

   c.   Round worm

   d.   Crustacean

   e.   Mollusk

_____ 5. The protostome animals include all but which of the following?

   a.   Annelids

   b.   Echinoderms

   c.   Crustaceans

   d.   Mollusks

   e.   Arthropods

_____ 6. Which of the following is NOT characteristic of the sponges?

   a.   Flagellated collar cells

   b.   Needle-like crystals of calcium carbonate or silica

   c.   Pores or canals into a central cavity

   d.   Stinging cells called cnidocytes

   e.   Ameboid cells that wander around the matrix

_____ 7. Cnidarians include all but which of the following?

   a. Jellyfish
   b. Hydra
   c. Sea anemone
   d. Sea cucumber
   e. Corals

_____ 8. Which of the following phyla was the first to develop all three germ layers?

   a. Annelida
   b. Mollusca
   c. Platyhelminthes
   d. Porifera
   e. Cnidaria

_____ 9. Which of the following is a class of free-living animals?

   a. Turbellaria
   b. Hirundinae
   c. Cestoda
   d. Trematoda
   e. None of the above

_____ 10. The first group of animals to have members with eyespots able to detect light belong to which of the following?

   a. Porifera
   b. Platyhelminthes
   c. Nematodes
   d. Cnidarians
   e. Annelids

_____ 11. The first group to develop a digestive system with two openings are members of which of the following?

   a. Porifera
   b. Platyhelminthes
   c. Nematodes
   d. Cnidarians
   e. Annelids

_____ 12. A mantle is characteristic of which of the following?
   a.   Mollusca
   b.   Arthropoda
   c.   Annelida
   d.   Echinodermata
   e.   Nematoda

_____ 13. Which of the following is the simplest phylum to develop segmentation?
   a.   Mollusca
   b.   Arthropoda
   c.   Annelida
   d.   Echinodermata
   e.   Nematoda

_____ 14. To which of the following phyla do animals with jointed appendages belong?
   a.   Mollusca
   b.   Arthropoda
   c.   Annelida
   d.   Echinodermata
   e.   Nematoda

_____ 15. Which of the following arthropods lack mandibles?
   a.   Insects
   b.   Millipedes
   c.   Crustaceans
   d.   Spiders
   e.   Centipedes

_____ 16. Which of the following is NOT an echinoderm?
   a.   Sea stars
   b.   Sea anemoneS
   c.   Sea cucumberS
   d.   Sea lilies
   e.   Sand dollars

_____ 17. Which of the following is associated with roundworms?
    a. Schistosomiasis
    b. Pinworms
    c. Trichinosis
    d. Elephantiasis
    e. Roundworms are not parasitic.

_____ 18. In which of the following would you find the digestive and reproductive systems of mollusks?
    a. Mantle cavity
    b. Visceral mass
    c. Gastrovascular cavity
    d. Coelom
    e. Cuticle

_____ 19. Protosomes and deuterostomes can be distinguished by which of the following?
    a. Nervous system
    b. Digestive system
    c. Circulatory system
    d. Embryonic development
    e. A and D

SHORT ANSWER QUESTIONS

1. List some of the general characteristics of the animal kingdom.

2. Describe the function of an exoskeleton in arthropods.

3. Describe the alternations in body shapes in the life cycle of cnidarians.

4. What are the Ctenophora?

5. Why are sponges considered to be the most primitive form of animal life?

CRITICAL QUESTIONS

1. Why do you think the number of parasitic infections in humans is less in the United States than in Central America?

2. Starfish are natural predators of clams. New England fishermen would cut up any starfish that were caught in an attempt to reduce the predator population. Explain what was wrong with this method of controlling the predator population.

TELLING THE STORY

Develop a story that demonstrates your understanding of invertebrate structure, function, and reproduction. If you have difficulty with any of the associated areas or topics as you write, go back and review that information in Chapter 29.

CHAPTER REVIEW ANSWERS

MATCHING

1. O	5. P	9. C	13. B
2. I	6. N	10. J	14. K
3. H	7. M	11. A	15. F
4. E	8. G	12. L	16. D

LABELING

Fig. 29.1

1. Mouth	10. Intestinal cavity
2. Clitellum	11. Setae
3. Cerebral ganglion	12. Coelom
4. MouthCrop	13. Nephridium
5. Pharynx	14. Gizzard
6. Blood vessels	15. Longitudinal muscle
7. Ventral nerve cord	16. Circular muscle
8. Hearts	17. Crop
9. Dorsal blood vessel	18. Epidermis

Fig. 29.2

1. Thorax	10. Aorta
2. Head	11. Brain
3. Compound eye	12. Mouth
4. Antennae	13. Nerve ganglia
5. Abdomen	14. Vagina
6. Spiracles	15. Ovipositor
7. Malpighian tubules	16. Rectum
8. Stomach	17. Heart
9. Crop	18. Ovary

MULTIPLE CHOICE QUESTIONS

1. A	6. D	11. C	16. B
2. E	7. D	12. A	17. D
3. E	8. C	13. C	18. B
4. C	9. A	14. B	19. B
5. B	10. B	15. D	

SHORT ANSWER QUESTIONS

1. Animals are multicellular and heterotrophic. Their cells have nuclei and membrane-bound organelles. Most eat food they have to gather and have a one or two opening digestive cavity. Most have a stage that is able to move in pursuit of food, reproduction, or escape from predators.

2. The exoskeleton prevents desiccation, provides protection from predation, and serves as a point of attachment for muscles.

3. Sessile polyps grow to form a massive colony. Individual polyps reproduce asexaully. Some buds produce free-floating medusae that produce ovaries and testes when they are mature. The eggs and sperm from the medusae fuse to reproduce sexually to form a zygote, which then produces free-swimming larvae. The larvae settle on a substrate and begin the formation of a new colony of sessile polyps.

4. Ctenophores are animals called comb jellies that belong to a small primitive phyla somewhat related to the cnadarians.

5. Sponges lack tissues, organs, systems, and a coelom. Lacking these characteristics makes these animals very simple and relatively primitive.

CRITICAL QUESTIONS

1. There are fewer parasitic infections in the United States as compared to Central America because of the climate, the amount of developed land, health and sanitation regulations, food and drug regulations, meat inspection, technology, and water purification. These reduce the potential for infection and increase the diagnostic and treatment capabilities.

2. When fishermen cut up the starfish, they did not realize that this would increase the opportunity for population growth of starfish. These organisms are capable of regeneration and therefore this method of control was actually working against them, increasing the population of starfish and increasing the predators on the clams.

CHAPTER 30

CHORDATES AND VERTEBRATE ANIMALS: PATTERNS OF STRUCTURE, FUNCTION, AND REPRODUCTION

KEY CONCEPTS OVERVIEW

1. Unity in Symmetry and Coelom

Members of the phylum Chordata are the most highly evolved members of the animal kingdom. These organisms are classified into three subphyla: the **urochordates or tunicates**, the **cephalochordates or lancelets**, and the **vertebrates.** They are characterized by three distinct structures that appear in all chordates in the embryo stage of development: a single hollow dorsal nerve cord, a rod-shaped **notochord**, and **pharyngeal (gill) arches.** Pharyngeal arches develop into gills in fishes and into the ears, jaws, and throats of terrestrial vertebrates. The presence of the gill arches is also an evolutionary clue as to the aquatic ancestry of vertebrates.

In addition to these traits, chordates have a true coelom and bilateral symmetry. Chordate embryos exhibit segments of tissue called somites, which develop into skeletal muscles in humans. Most chordates have an endoskeleton that serves a number of functions (muscle attachment, movement, support, and protection). The exception is the larval tunicates and adult lancelets, which lack an internal skeleton. Their muscles attach to their notochords.

2. Tunicates

Urochordata means "tail notochord." This group, made up of organisms called tunicates, is comprised of about 2,500 species of marine animals. The larvae are free-swimming for a few days, until they settle onto a surface and become a sessile filter feeder. Tunicates pull water through an incurrent siphon into the mucus lined pharynx and force water out the excurrent siphon. This ability has given them the name **sea squirts**. Colonial tunicates reproduce asexually by budding. Individual tunicates are hermaphroditic.

3. Lancelets

Lancelets are fish-like marine chordates that are only a few centimeters in length and pointed at both ends. Their name is derived from their shape, which resembles a lance or lancet. These scaleless organisms have a segmented appearance, caused by their muscle tissue arrangement. Lancelets have no real head, eyes, ears, or nose, but they are characterized by their notochord, which spans the entire length of the dorsal nerve cord. Males and females are separate, however, and there are no notable external differences between them.

4. Vertebrates

Among the vertebrates, the notochord is found in the larvae and in the adult it is replaced by a vertebral column that protects the spinal cord. The jawless fish are an exception to this standard. Vertebrates have brains that are encased in a skull and a closed circulatory system with a pumping heart. Most of these organisms have other organs (liver, kidneys, etc.) that contribute to their success in their environments.

The **Agnatha,** or jawless fish, are the most primitive of the vertebrates. They lack paired jaws and paired fins or scales. These tube-like organisms generally live in the sea or brackish waters. Because of their primitive structure, the agnatha are not well adapted as predators and therefore spend much of their time as bottom dwellers. The lampreys are parasitic, feeding on the fish to which they attach themselves with their suction cup-like mouths and spine-covered tongues. Both male and female jawless fish spawn, deposit sperm and eggs respectively, directly into the water and fertilization occurs outside their bodies. This process is known as **external fertilization**. Fertilized eggs develop into larvae that feed on plankton, then mature and metamorphosize into parasitic adults. Other jawless fish hatch directly from fertilized eggs.

The Chondrichthyes are fish with cartilaginous skeletons. They include the sharks, skates, and rays. They possess an external skin made of teeth-like scales called **dentacles**. The sharks are predators to many marine animals. Their mouths are filled with triangular shaped teeth that rip pieces of flesh when the shark is actively feeding. Other sharks are plankton feeders. The skates and rays have a different structure from the sharks. Their bodies are flat, with enlarged pectoral fins. They do not use their tails as a major structure for locomotion. The mouth of these organisms is on their undersurface and their main source of food is invertebrates located on or near the bottom of the water. All Chondrichthyes have a very sophisticated sensory system. Sharks have a **lateral line system,** containing mechanoreceptors, that is sensistive to vibrations, movement, and pressure. Addtionally, these organisms have electroreceptors used for navigation and detection of electrical fields emitted by other fish and chemoreceptors that enhance their ability to smell.

Sharks lack the hydrostatic organ called the **swim bladder** that is characteristic of most fish. The buoyancy in sharks is partially provided by oils, but the shark continually swims to maintain depth and to provide a fresh flow of water over the gills. Male sharks have a **pelvic clasper,** which is inserted into the female **cloaca** and is used to deliver sperm to the female during copulation. This denotes an internal fertilization process. Additionally, fish are **ovoviparous,** meaning that they retain fertilized eggs within their oviducts until the young hatch and receive nutrition from the yolk of eggs , not from the female. This is in contrast to **oviparous** organims that lay eggs, and the young hatch outside the female and also to **viviparous** organisms that bear young alive. Both types of developing embryos derive nourishment directly from the female.

The largest number of known species of fish belong to the class **Osteichthyes or bony fish.** There are two subclasses of bony fish: the ray-finned fishes and the lobe-finned fishes. They reside in fresh and salt water. These organisms have an internal skeleton and skin covered with scales. They have a flat extention called an **operculum** that covers the gills and increases water flow for maximizing oxygen availability. In addition to these characteristics, the bony fish are oviparous, have tube-like hearts with four chambers, and regulate water balance by action of the kidneys.

The Amphibians ("two lives") depend on water during the early stages of their development and can then move to land as they mature. They return to water to lay eggs and undergo larval development. To avoid desiccation as adults, they live in moist places to lessen the loss of water through their skin. A few amphibians are ovoviviparous, and a few are viviparous. They have lungs rather than gills and a three-chambered heart.

There are three major orders of reptiles: the alligators and crocodiles, the turtles and tortoises, and the lizards and snakes. Reptiles have dry skin covered with scales, which aid in the prevention of water loss. They have a more efficient heart with a **septum** between the ventricles. Although it closely resembles a four-chambered heart, it has only three. An amnionic egg, which contains a yolk and albumin, protects the embryo from dessication, provides nutrition for the developing embryo, and allows it to develop outside the water. Reptiles are considered ectothermic, which means that they regulate body temperature by taking in heat from the environment.

Birds are equipped with beaks without teeth, feathers which are light, flexible and waterproof, amniotic eggs, and hollow bones that are adapted for flight. Birds also have highly efficient lungs that provide oxygen to accommodate muscle contraction during flight. They have a four-chambered heart and regulate body temperature internally (**endothermic**).

The 4,500 species of mammals have a number of associated characteristics. They are endothermic vertebrates that have hair. Females of this group develop and secrete milk from mammary glands to feed their young. They have a four-chambered heart that circulates blood to the lungs and the rest of the body. They have developed specialized groups of teeth including the incisors, canines, premolars, and molars.

There are three subclasses of mammals: the monotremes (egg-laying platypus and spiny anteaters), the marsupials (pouched animals), and the placentals (animals that develop to maturity in the mother). There are 14 orders of placental mammals and the most highly evolved of these is the primates, which include the apes and humans.

CHAPTER REVIEW ACTIVITIES

CONCEPT MAPPING

Develop a concept map that demonstrates your understanding of vertebrate structure, function, and reproduction. Incorporate as many of the key terms as possible. Develop the map with as much detail as you can.

MATCHING

Select the term from the list provided that best relates to the following statements. If you are unsure of the correct answer or chose an incorrect response, review the related concept in Chapter 30.

_____ 1. These mammals nourish their developing young within the body of the mother by means of a placenta.

_____ 2. These animals are capable of living on land and in the water.

_____ 3. This term describes thee control of water movement in and out of an organism's body.

_____ 4. This group of marine animals looks like sacs attached to the floor of the ocean.

_____ 5. This rod-shaped structure forms between the nerve cord and the gut during the development of all vertebrates.

_____ 6. This is the method of reproduction in which the female retains the fertilized egg within her oviducts until the young hatch.

_____ 7. This is the process by which most fish reproduce.

_____ 8. This extention flap covers the gills of bony fish and increases the water flow for oxygen availability.

_____ 9. This subclass of mammals gives birth to immature young that are carried in a pouch.

_____ 10. This is the terminal part of the gut of an aquatic animal into which ducts from the kidneys and reproductive systems open.

_____ 11. These mammals lay eggs with leathery shells similar to those of reptiles.

_____ 12. This is the method of reproduction in which the female lays her eggs, and the young hatch outside of the mother.

_____ 13. This complex system of mechanoreceptors, possessed by all fish and amphibians, detects mechanical stimuli such as sound, pressure, and movement.

_____ 14. These are the sharp, teeth-like scales on the skin of sharks.

_____ 15. This is the method of reproduction in which the young are born alive and the developing embryos derive nourishment from the mother not the egg.

_____ 16. This type of fertilization takes place outside the body of the female.

_____ 17. This ova, which protects the embryo from dessication, provides nourishment, and enables it to develop outside of water.

_____ 18. This term refers to the ability to regulate body temperature by taking in heat from the environment.

_____ 19. This hollow cord along the back is a principal feature of chordates.

_____ 20. This term refers to the ability of an organism to regulate body temperature internally.

_____ 21. This principal feature of chordate embryos develops into gills in fish and the ear, jaw, nose, and throat of terrestrial vertebrates.

TERMS LIST

a. Oviporous	h. Spawning	o. Viviporous
b. Osmoregulation	i. Endothermic	p. Marsupial
c, Pharyngeal arches	j. Notochords	q. Lateral line system
d. Amniotic egg	k. Cloaca	r. Ovoviviporous
e. Placental mammals	l. Denticles	s. External fertilization
f. Nerve cord	m. Tunicates	t. Operculum
g. Amphibians	n. Monotremes	u. Ectothermic

LABELING

Label the following structure. If you encounter any difficulty, review the related material in the text.

Fig. 30.1 Structure of an amniotic egg.

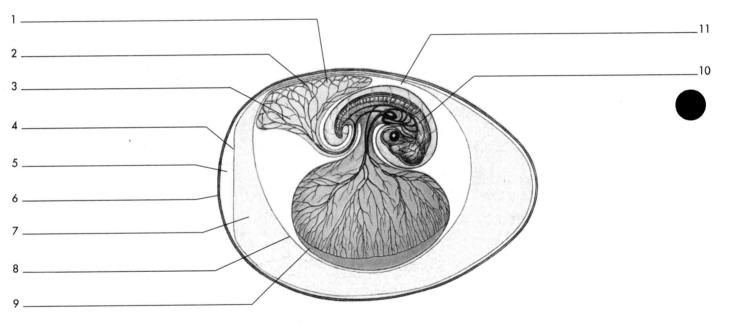

MULTIPLE CHOICE QUESTIONS

_____ 1. Which of the following part(s) would be associated with the development of the pharyngeal arches in adult chordates?

a. Jaw

b. Ear

c. Nose

d. Throat

e. All of the above

_____ 2. The tunicates differ from other animals in that they possess which of the following structures?

a. Siphons

b. Swimming larvae, sessile adults

c. Tunic made of cellulose

d. Notochords

e. A nerve net

_____ 3. The lancelets possess which of the following characteristics?

a. Filter feeders

b. Parasites

c. Predators

d. Herbivores

e. Scavengers

_____ 4. The lancelets appear segmented because of which of the following?

a. Muscle arrangement

b. Nerve bundles

c. Scales

d. Ciliated exterior

e. Brown pigments under the skin

_____ 5. The lancelet has which of the following?

    a.   Real head

    b.   Eyes

    c.   Nose

    d.   Anus

    e.   Feet

_____ 6. The class of vertebrates in which the vertebral column is made out of cartilage, not bones is which of the following?

    a.   Agnatha

    b.   Chondrichthyes

    c.   Reptilia

    d.   Amphibia

    e.   Osteichthyes

_____ 7. Which of the following is NOT classified as a tetrapod?

    a.   Amphibia

    b.   Aves

    c.   Mammalia

    d.   Reptilia

    e.   Osteichthyes

_____ 8. The lamprey are considered which of the following?

    a.   Filter feeders

    b.   Parasites

    c.   Scavengers

    d.   Herbivores

    e.   Photosynthetic

_____ 9. Agnatha are fish that do NOT have which of the following?

    a.   Scales

    b.   Paired jaws

    c.   Bony vertebral column

    d.   Paired fins

    e.   All of the above

10. The cartilaginous fish differ from the bony fish by their lack of which of the following?
    a. Swim bladder
    b. Pectoral and pelvic fins
    c. Lateral line system
    d. Dermal scales
    e. Liver

11. Which of the following classes of organism has a loop of Henle within the kidneys?
    a. Osteichthyes
    b. Aves
    c. Reptiles
    d. Chondrichthyes
    e. Amphibia

12. Which part of the amnionic egg controls the nitrogenous waste produced by the embryo?
    a. Amnion
    b. Albumin
    c. Allantois
    d. Yolk sac
    e. Chorion

13. The gas exchange in the amnionic egg is associated with which of the following?
    a. Amnion
    b. Albumin
    c. Allantois
    d. Yolk sac
    e. Chorion

_____ 14. Which of the following have some type of scales?
  a. Fishes
  b. Reptiles
  c. Birds
  d. A and B
  e. All of the above

_____ 15. Which would be the best reason that amphibians have not totally adapted to land?
  a. Dependency on water environments for external fertilization
  b. Young are not born alive
  c. The skin plays a greater role than the lungs in gas exchange
  d. Poor locomotion ability
  e. No well developed excretory system

_____ 16. Which of these characteristics would NOT be found in fishes?
  a. Endoskeletons
  b. Scales
  c. Endothermic
  d. Dorsal hollow nerve cord
  e. Operculum

SHORT ANSWER QUESTIONS

1. What are the three principal features shared by the chordates and vertebrates?

2. Where are the notochords located in vertebrates?

3. What is the difference between the processes of spawning and copulation?

4. What is metamorphosis?

5. What are the three subphyla in the phylum Chordata? Give an example of each.

CRITICAL QUESTIONS

1. It is not always easy to classify organisms phylogenically. The tunicates presented a problem for taxonomists. What feature gave enough information to allow for their classification?

2. What are the distinguishing characteristics that separate the tunicates and lancelets from the vertebrates?

3. Which class of vertebrate was the first to adapt to land and why?

TELLING THE STORY

Develop a story that demonstrates your understanding of vertebrate structure, function, and reproduction. If you have difficulty with any of the associated areas or topics as you write, go back and review that information in Chapter 30.

CHAPTER REVIEW ANSWERS

MATCHING

1. E	8. T	15. O
2. G	9. P	16. S
3. B	10. K	17. D
4. M	11. N	18. U
5. J	12. A	19. F
6. R	13. Q	20. I
7. H	14. L	21. C

LABELING
Fig. 30.1

1. Oxygen	7. Albumin
2. Carbon dioxide	8. Chorion
3. Allantois	9. Yolk sac
4. Egg membrane	10. Embryo
5. Air space	11. Amnion
6. Egg shell	

MULTIPLE CHOICE QUESTIONS

1. E	5. D	9. C	13. E
2. C	6. B	10. A	14. E
3. A	7. E	11. B	15. A
4. A	8. C	12. C	16. C

SHORT ANSWER QUESTIONS

1.   All have a dorsal nerve cord, pharyngeal arches, and a notochord.

2.   The Urochrodates have a notochord in the tail region. The Cephalochordates have a notochord in the head region.

3.   In spawning, fertilization occurs externally after the gametes are released into the water. In copulation, fertilization occurs internally within the female.

4.   Metamorphosis is a change in body form in the life cycle. For example, the tadpole is a larval stage that undergoes considerable change in becoming an adult frog.

5.   The three subphyla of the Chordates are as follows:

Tunicates	Sea Perch
Lancelets	Cephalochordate
Vertebrates	Jawless fishes, Cartilaginous fish, Bony fish, Amphibians, Reptiles, Birds, Mammals

CRITICAL QUESTIONS

1.   Taxonomists were working with the sessile adult forms. Once they found the immature larval stage, they determined how to properly classify these organisms. The larvae possessed a notochord, placing them in the phylum Chordata.

2.   Vertebrates have a segmented backbone (vertebral column), a head and brain, two pairs of appendages, a jointed endoskeleton, and a chambered, ventral heart.

3.   The reptiles were the first to adapt completely to land. The main requirement for this change was the formation of an egg that would survive without drying out. The development of the three embryonic membranes (amnion, allantois, and chorion) made the egg structurally able to protect itself from desiccation.

•Chapter 31

BIOTECHNOLOGY
AND
GENETIC ENGINEERING

KEY CONCEPTS OVERVIEW

1.   Classical and Molecular Biotechnology

Classical biotechnology did not originate in a science laboratory setting. In fact, some of the decisions made by humans very early on were the first trials in this developing process.

The notion that certain species of plants and animals are selected by humans for their functional and esthetic characteristics is not a new one. For at least 10,000 years, humans have been removing specific organisms from the wild and breeding them for continuation. Gregor Mendel was one of the frontrunners in gathering evidence regarding the variability of characteristics among living things. Breeders then were able to cross organisms of the same species having desired characteristics to produce **hybrid** offspring. Mendel determined that organisms had specific hereditary factors that produced the characteristics associated with that species and with that particular member of the species. Scientists soon began developing breeding processes involving the selection and recombination of those hereditary factors, now referred to as **genes.** These carefully designed and controlled breeding programs could produce stronger and improved hybrid offspring. These improvements from the parent to the offspring generation are called **hybrid vigor.**

When the agriculture industry began incorporating the use of selection and recombination in crops, a **"green revolution"** occurred. Production of food crops increased dramatically. Plants are now stronger, more resistant to drought, harsh climates, and poor soil conditions, and yield substantially more food.

During the time work on hybridization was being done by Mendel and peers, interesting discoveries were occurring in the area of microbiology. Although the fermentation had been utilized for many years prior, the process was still not well understood. This changed when it was determined that yeasts were actually living organisms. About 30 years later, Louis Pasteur determined that some types of bacteria and yeast were capable of forming by-product molecules, such as lactic acid, acetic acid, alcohol, and carbon dioxide. From this discovery, he determined that by killing microorganisms responsible for spoiling products like wine and milk, negative fermentative changes could be avoided. The process is referred to as **pasteurization.** Fermentation has since been used for a number of purposes, such as the production of glycerol, acetone, butanol, and citric acid.

Classical biotechnology was also employed to combat diseases. Robert Koch, developer of the germ theory, discovered the bacterium responsible for anthrax in cattle and humans and the bacterium that causes tuberculosis. During his work with these bacteria, he also determined how to produce pure cultures, so that specific organisms could be isolated and studied without contamination by other organisms.

In 1797, Edward Jenner developed the first successful vaccine against smallpox. He hypothesized that if individuals were injected with a virus containing solution, they would develop a resistance to the disease that is caused by the active form of the virus. Although he did not understand how this process worked, his success set the stage for the likes of Pasteur and Alexander Fleming some 100 years later. The dawn of antibiotic therapy came in 1927, when Fleming discovered penicillin. The first purified form of this substance was used in 1940 on soldiers in World War II.

The discovery of the molecular structure of DNA was the dawn of molecular biology. It was determined that the molecular "packets" of information responsible for each inherited trait were stored within the nucleus on long, coiled strands called chromosomes, which are made up of DNA and proteins. In 1952, Alfred Hershey and Martha Chase identified DNA as the hereditary material and not the protein. Rosalind Franklin and Maurice Wilkins determined that DNA was shaped as a helix, and then James Watson and Francis Crick developed the structural details for the DNA molecule.

The challenge since the structure of DNA was determined has been to locate and manipulate genes to enhance the natural process of genetic recombination. Multiple genes are often responsible for coding for one specific trait. Because these genes are not always sequenced on the chromosome one after the other, locating and identifying them is challenging.

2. Natural Gene Transfer

Bacteria naturally transfer genetic material from one cell to another by three methods: transformation, transduction, and conjugation. During the process of **transformation**, a bacterial cell lyses and releases genetic material into its surrounding matrix. Certain bacterial cells, referred to as a competent cells, incorporate this genetic material into their own genome and become transformed. A **genome** can be described as a full complement of genetic material. The ability to do this is itself an inherited characteristic.

In the process of **transduction**, DNA from donor bacteria are transferred to recipient bacteria by viruses. The viruses must first combine their genetic material with that of donor bacteria, and only certain types of viruses (lysogenic viruses) are capable of doing so. Additionally, not all bacteria are vulnerable to the invasion of these viruses.

Conjugation involves donor and recipient bacteria connecting, so that the DNA from the donors can be injected into the recipients. Only bacteria having extrachromosomal pieces of DNA, called **plasmids**, are capable of this transaction. Plasmids, which contain codes for the ability to transfer chromosomes, are DNA fragments capable of replicating independently of the main chromosome and make up about 5% of the DNA in many bacteria. **F plasmids** contain specialized genes, called the **fertility factor**, that promote the transfer of plasmids to other cells. A tube-like structure called a **pilus** forms on the surface of a bacterial cell having the F plasmid. When this cell encounters another lacking a pilus, the replication and transfer of the plasmid occurs.

3. Human Engineered Gene Transfer

The combination of DNA among different species of organisms and the manipulation of genes in eukaryotic cells did not begin until the mid 1970s. Techniques involved in these processes are referred to as **genetic engineering** or **recombinant DNA technology**. One of the first uses of recombinant DNA technology was the insertion of human genes into bacterial cells to produce human insulin and human growth hormone.

One method for developing human products in bacterial and yeast cells, called **shotgun cloning**, involves cutting the entire DNA genome into pieces, isolating and purifying the DNA, and then inserting these fragments into the microbe for cloning. When the transformed organism reproduces, it makes copies of the entire genome in fragments, which can then be screened for specific genes. Most screening techniques involve DNA hybridization and the use of **molecular probes**.

Once a gene has been isolated, multiple copies are necessary to make enough recombinant cells to culture significant amounts of the desired product. The **polymerase chain reaction (PCR)** process, developed in 1983 by Kary Mullis, is one effective method for doing this. Because of the rapid advancements in biotechnology, this method can now be done mechanically in the laboratory.

In addition to human insulin and growth hormone, genetic engineering has been used to produce a variety of human proteins. These proteins have been used to treat a variety of disorders or diseases such as certain types of cancer, heart attacks, and certain deficiencies of the human immune system.

Until very recently, this technology had not been used to correct genetic defects. However, **gene therapy** has now been successfully attempted in several situations. In order for this type of treatment to be more available, there are three types of human DNA "maps" that must be completed: (1) a genetic linkage map that shows the distances between genetic markers on a chromosome, (2) a physical map that shows the number of nucleotides between markers, and (3) an ultimate map that shows the sequence of nucleotides in a chromosome and describes the gene and the proteins they make. In 1990, the Human Genome Project was initiated in an effort to accomplish this task.

The food industry has also used biotechnology is many ways. Agricultural crops are genetically altered to produce high yield and resist pathogens and harsh growing conditions. These types of plants are called **transgenic plants.** Transgenic animals are also produced, but as of yet have not been approved for commercial use.

CHAPTER REVIEW ACTIVITIES

CONCEPT MAPPING

Develop a concept map that demonstrates your understanding of biotechnology and genetic engineering. Incorporate as many of the key terms as possible. Develop the map with as much detail as you can.

MATCHING

Select the term from the list provided that best relates to the following statements. If you are unsure of the correct answer or chose an incorrect response, review the related concept in Chapter 31.

_____ 1. This molecule binds to a specific gene or nucleotide sequence.

_____ 2. These specific proteins recognize certain nucleotide (base) sequences in a DNA strand and break the bonds between the nucleotides at those points.

_____ 3. This term is used to describe plants or animals that are genetically altered in the hope of making them more appealing to humans or resistant to pathogens.

_____ 4. This is an organism's total complement of genetic material.

_____ 5. This is a collection of clones of DNA fragments, which together represent the entire genome of an organism.

_____ 6. This sealing enzyme helps re-form the bonds between broken pieces of DNA, including the fragments of human DNA and bacterial DNA formed during the cloning process.

_____ 7. These extrachromosomal pieces of DNA replicate independently of the main chromosome.

_____ 8. This is a process for cloning genes that can be performed if the nucleotide sequence of the gene is known and the genetic engineer is able to synthesize that gene.

_____ 9. These are smaller pieces of DNA that scientists can produce using enzymes that recognize certain nucleotide sequences.

_____ 10. These linkage points in DNA molecules are where restriction enzymes identify certain nucleotide (base) sequences and then break the links between the nucleotides.

_____ 11. This process of exchanging genetic material occurs when DNA from a donor bacterium is transferred to a recipient bacterium by a virus.

_____ 12. This is the treatment of a genetic disorder by the insertion of a "normal" gene into the cells of a patient.

_____ 13. This is a method of genetic recombination in bacteria during which a donor and a recipient bacterium make contact and the DNA from the donor is injected into the recipient cell.

_____ 14. This is the use of scientific and engineering principles to manipulate organisms.

_____ 15. These techniques of molecular biology involve the manipulation of genes themselves, rather than the organism.

_____ 16. These are copies of DNA fragments.

_____ 17. This process of genetic transfer occurs when a donor bacterial cell lyses and releases its genetic material into the surrounding matrix, and another type of bacterial cell incorporates this material into its genome.

_____ 18. This is the process that relies on traditional techniques of selection, mutation, and hybridization.

_____ 19. This is a worldwide project to decipher the DNA code of all the 46 human chromosomes.

_____ 20. This process for replicating genes can be performed if the mRNA transcribed from a gene is available.

TERMS LIST

a. Gene therapy	k. Conjugation
b. Gene synthesis cloning	l. Restriction enzymes
c. Complimentary DNA cloning	m. Classical biotechnology
d. Probe	n. Clones
e. Recognition sites	o. Ligase
f. Transduction	p. Genome
g. Human Genome Project	q. Transformation
h. Plasmid	r. Genetic engineering
i. Biotechnology	s. Transgenic
j. Restriction fragments	t. Gene library

MULTIPLE CHOICE QUESTIONS

_____ 1. In which of the following could recombinant DNA be used ?

    a. To clone plant cell DNA for crop enhancement

    b. To clone animal cell DNA for heavy, lean muscle mass

    c. To mass produce human proteins

    d. To produce nonpathogenic strains of normally pathogenic organisms

    e. All of the above

_____ 2. Which of the following would NOT be needed to develop recombinant DNA molecules?

    a. Donor DNA

    b. DNA ligase

    c. Recipient DNA

    d. DNA polymerase

    e. Restriction enzymes

_____ 3. Which of the following acts as a seal when there are breaks in the DNA molecule?
   a. DNA ligase
   b. DNA polymerase
   c. Restriction enzymes
   d. DNA cyclase
   e. None of the above

_____ 4. To which of the following molecules would DNA probes bind?
   a. RNA
   b. Plasmids
   c. DNA
   d. Restriction enzymes
   e. Viruses

_____ 5. Which of the following individuals is credited with the production of the polymerase chain reaction process?
   a. Edward Jenner
   b. Maurice Wilkins
   c. Martha Chase
   d. Kary Mullis
   e. Rosalind Franklin

_____ 6. Which of the following is a tube-like structure developed on certain bacteria to transfer genetic information?
   a. Plasmid
   b. Pilus
   c. Inoculum
   d. Clone
   e. Probe

_____ 7. Which of the following is characteristic of molecular probes?
   a. They can be molecules of mRNA.
   b. They can be complimentary sequences of DNA or RNA.
   c. They can be antibodies specific for certain proteins.
   d. A, B, and C
   e. B and C only

_____   8. Which of the following is NOT true of the vaccines?
        a.   They are presently effective against HIV.
        b.   They stimulate the immune response.
        c.   They are produced by culturing pathogenic agents and then
             killing them.
        d.   They can be made of attenuated pathogens.
        e.   They can be produced by genetic engineering.

_____   9. Which of the following is NOT true of the genetically engineered
            product Humalin?
        a.   It is less expensive than animal preparations.
        b.   It is used to treat tuberculosis.
        c.   It is used to treat diabetes.
        d.   More than 50% of new cases of diabetes in the US. are treated with
             it.
        e.   Fewer allergic reactions result from using this product than the
             animal preparation types.

_____ 10. Which term best describes the process of binding between a probe and
            the complementary DNA sequences?
        a.   Autoradiography
        b.   Polymerase chain reaction
        c.   Hybridization
        d.   Cloning
        e.   Conjugation

_____ 11. Which of the following describes the physical maps being produced
            through the Human Genome Project?
        a.   Distances between genetic markers on the chromosomes
        b.   Sequence of nucleotides in a chromosome
        c.   The gene names and the proteins they produce
        d.   Number of nucleotides between the markers on the chromosomes
        e.   Number of chromosomes

_____ 12. Which of the following individuals is credited with the discovery of antibiotics?
   a. Alexander Fleming
   b. Louis Pasteur
   c. Edward Jenner
   d. Gregor Mendel
   e. Robert Koch

_____ 13. Which of the individuals below determined that DNA was the hereditary material in the nucleus of cells and not the proteins?
   a. Martha Chase and Alfred Hershey
   b. James Watson and Francis Crick
   c. Kary Mullis
   d. Rosalind Franklin and Maurice Wilkins
   e. Robert Koch

_____ 14. Why does wine that is not pasteurized turn to vinegar?
   a. The yeast continues to ferment and produce too much alcohol.
   b. Microorganisms produce carbon dioxide that cause the wine to be too acidic.
   c. Bacteria produce acetic acid that lowers the pH and turns the wine sour.
   d. The lack of oxygen causes the vinegar to form.
   e. None of the above

_____ 15. Which term best relates to viruses that infect a cell but do not immediately replicate?
   a. Plasmidic
   b. Lysogenic
   c. Transformed
   d. Conjugated
   e. Recombined

SHORT ANSWER QUESTIONS

1. What is a plasmid?

2. What is the major difference between classical biotechnology and molecular biotechnology?

3. Define the role of a restrictive enzyme in recombinant DNA technology.

4. What is shotgun cloning, and why is this name used for this process?

5. Which types of organisms are most widely used to produce genetically engineered products and why?

6. What are the three types of natural gene transfer methods among bacteria?

CRITICAL QUESTIONS

1. What is the significance of having a gene library available in the recombinant DNA technology process known as shotgun cloning?

2. Edward Jenner's successful use of a vaccine containing the smallpox virus is considered a benchmark in immunology. Why would his techniques not be well accepted or approved today?

TELLING THE STORY

Develop a story that demonstrates your understanding of genetic engineering. If you have difficulty with any of the associated areas or topics as you write, go back and review that information in Chapter 31.

CHAPTER REVIEW ANSWERS

MATCHING

1. D	5. T	9. J	13. K	17. Q
2. L	6. O	10. E	14. I	18. M
3. S	7. H	11. F	15. R	19. G
4. P	8. B	12. A	16. N	20. C

MULTIPLE CHOICE QUESTIONS

1. E	4. C	7. D	10. C	13. A
2. D	5. D	8. A	11. D	14. C
3. A	6. B	9. B	12. A	15. B

SHORT ANSWER QUESTIONS

1. Plasmids are fragments of DNA separate from the main circular DNA molecule that occurs in bacteria.

2. Classical biotechnology invovles using techniques such as selection, mutation, and hybridization on organisms to effect the genetic characteristics of their offspring. Molecular biotechnology can effect both the genes of the organism and the organism itself, through the use of genetic engineering processes.

3. Restrictive enzymes are used to cut long strands of DNA, into fragments that contain one or more genes.

4. Shotgun cloning involves cutting the DNA of the entire genome into pieces, isolating and purifying the genomic DNA and then inserting these pieces or fragments into bacteria. The bacteria incorporate the entire genome, reproduce, and make identical copies. The term shotgun implies that no one gene is targeted for cloning, but rather the entire genome is cloned into DNA fragments that can be screened for more detailed identification of specific genes.

CRITICAL QUESTIONS

1. The gene library allows researchers to study fragments of the DNA rather than entire strands. These pieces are significantly shorter and contain fewer genes to screen. When a specific gene or gene sequence is desired, this arrangement makes a very complex process a bit more managable.

2. The Food and Drug Administration has very rigid guidelines for experimenting with drug therapies. Many laboratory experiments must be completed and more definitive information determined about the short and long term effects of the drug on the subjects. Animal models are used prior to human subjects, and the clinical trials usually involve a double blind design to avoid biasing the results. Jenner's work was not regulated by anyone. His volunteers relied on his reputation and the desire to put an end to the smallpox epidemic.

CHAPTER 32

INNATE BEHAVIOR AND LEARNING IN ANIMALS

KEY CONCEPTS OVERVIEW

1.   Ethology

The diversity among animals regarding their structure and their various habitats and needs have long been associated with the biological sciences. However, there are many very astute biologists whose interests in animals focuses on patterns of behavior. The study of animal behavior in natural environments is known as **ethology**. The behaviors of significance include the patterns of movement, vocalization (sound), postures, coloration changes, and scents released at certain times. Behaviors can be either simple or complex. Simple behaviors are automatic responses to some stimuli. The complex behaviors involve receiving a stimulus, processing the stimulus input, and responding. This type of behavior is limited to multicellular animals with a neural network.

Before the late 1950s, ethological research was based on a physiological perspective to determine patterns within animals. Austrian scientist Konrad Lorenz is probably the most famous of ethologists. His work is based on the premise that behaviors in animals are evolutionary adaptations. He, along with one of his countrymen Karl von Frisch, and Dutch ethologist Nikolaas Tinbergen, won the Nobel prize for their work in the study of animal behavior.

There was another group of animal researchers who gained recognition about the same time as the aforementioned ethologists. These psychologists, called **behaviorists**, studied behavior in the laboratory, without focusing on the cognitive processes that occur during behaviors exhibited by animals. By the early 1960s, ethology and behavioral psychology blended to form the discipline of animal behavior. As is with most science, this field expanded, and in the late 1970s the science of behavioral ecology emerged.

2. Link Between Genetics and Behavior

Behavior cannot be studied without understanding the physiological mechanisms that allow animals to function. Since these functional characteristics are genetically endowed, an understanding of genetics is also required. Genetically determined neural programs either develop at an appropriate point in maturation or are part of the nervous system at birth. These programs allow an animal to perform an **innate behavior** with relative completeness the first time it is exhibited. Innate behaviors provide a certain level of adaptive advantages that protect and sustain animals.

The influence of environment and learning on innate behaviors has been well debated. Today there seems to be less division between innate and learned behavior, because evidence suggests that there are many aspects of behavior that are influenced by both genetics and experience. Innate behaviors are now viewed as those without *obvious* influence from the environment.

3. Coordination and Orientation

Animals would not survive if they were unable to respond to their environment. This implies an innate ability to recognize a stimulus and respond in some way. In addition, this response must "make a positive difference" or it would be of no value. Reflex, an example of a **coordination behavior**, is the simplest type of innate response to a stimulus. **Kineses**, which involve random movements, and **taxes**, which are directed movements, are simple types of **orientation behaviors**.

Reflexes are automatic responses to neural stimulation. They play an important role in survival, but they do not require cognitive processing to function.

4. Fixed Action Patterns

Fixed action patterns are a sequence of innate behaviors. This type of innate behavior seems to be more complex than simple reflex responses. Fixed action patterns occur because the resulting behaviors follow an unalterable sequence of muscle movements. Once a releaser initiates a behavioral response, the activity is carried out to completion.

5. Learning

Behaviors based on the experiences of an animal are learned. The ability to learn helps animals alter behavior to adapt to changes in their environment. They can then become better suited to a particular environment. Such behaviors can be grouped into five categories: imprinting, habituation, trial-and-error learning, classical conditioning, and insight.

Imprinting is a learning pattern that occurs early in the life of some animals. A typical example occurs in young birds when a new hatchling will adopt the first large moving object it sees as its mother. The work with the graylag geese performed by Lorenz is a classic study in imprinting. Other studies have demonstrated that birds must hear the songs of their specific species if they are to be able to repeat them as adult birds. Additionally, migrations are possible because of imprinting. Migrating animals learn to recognize their birthplace by locality imprinting. They also have the ability to orient themselves in relation to an environmental cue.

Habituation involves an organism learning that a particular stimulus is nonthreatening or not important and allows the animal to ignore it. This "getting used to" certain types of stimuli prevents stimulus overload and allows the animal to be keen toward those that influence the safety and survival of the species.

Trial-and-error learning, or **operant conditioning**, is a form of learning whereby animals associate some action with reward or punishment. American psychologist B.F. Skinner studied such conditioning by placing rats in a box with levers and other experimental devices. Rewards and punishments were soon associated with certain levers and the animals learned to use only the levers that provided rewards. It has been determined that operant conditioning works only for stimuli and responses that have meaning for animals in nature.

Classical conditioning involves the animal learning a new stimulus along with a natural stimulus that the animal would deem as meaningful. The experimenter presents the stimulus and the experimental animal learns to respond to the stimulus. The best known example of classical conditioning was developed by Ivan Pavlov, a Russian psychologist. The natural stimulus-response connection in an animal is an inborn reflex, or **unconditioned response**. These responses are important as protective mechanisms. Pavlov's work demonstrated that behavior depends on innate neural systems, but that their programs can be modified by learning.

Insight, or reasoning, is the most complex form of learning. Animals with this capability are able to recognize a problem and solve it mentally prior to trying a possible solution. Wolfgang Kohler, a German psychologist, is credited with the first description of learning by insight.

CHAPTER REVIEW ACTIVITIES

CONCEPT MAPPING

Develop a concept map that demonstrates your understanding of innate behavior and learning in animals. Incorporate as many of the key terms as possible and choose the linking terms that will best correlate one term to another. Develop the map with as much detail as you can.

MATCHING

Select the term from the list provided that best relates to the following statements. If you are unsure of the correct answer or chose an incorrect response, review the related concept in Chapter 32.

_____   1.   These are sequences of innate behaviors in which the actions follow an unchanging order of muscular movements.

_____   2.   This is the prominent view in psychology today, which suggests that individuals acquire and then store information in memory, and that new information merges with the old, leading to change in behavior.

_____   3.   This study of animal behavior includes patterns of movement, sounds, and body position in the natural environment.

_____   4.   This is a directed movement by an animal toward or away from a stimulus, such as light, chemicals, or heat.

_____   5.   This is the most complex form of learning, where an animal is capable of recognizing a problem and mentally solving it prior to trying a possible solution.

_____   6.   This is an automatic response to nerve stimulation.

_____   7.   This term describes the ability of animals to become used to certain types of stimuli.

_____   8.   This school of thought believes that learning takes place in a stimulus/response fashion, possibly reinforced by some type of reward that may or may not be readily available.

_____   9.   This is the change in the speed of the random, nondirected movements of an animal, with respect to changes in certain environmental stimuli.

_____   10.   This rapid and irreversible type of learning takes place during an early developmental stage of some animals.

_____ 11. This scientific discipline was formed when the fields of ethology and behavioral psychology merged.

_____ 12. These are the ranges of the pattern of movement, vocalization, and posture exhibited by an animal.

_____ 13. This term describes an alteration in behavior based on experience.

_____ 14. These behaviors result from genetically determined neural programs that are part of the nervous system at the time of birth or develop at an appropriate point in maturation.

_____ 15. These long-range, two-way movements by animals, often occur yearly with the change of seasons.

_____ 16. In this form of learning, an animal is taught to associate a new stimulus with a natural stimulus that normally evokes a response in the animal.

_____ 17. This scientist is often referred to as the father of modern ethology.

_____ 18. This American psychologist studied the trial-and-error learning process using rats in enclosures equipped with levers that elicited rewards and punishments.

_____ 19. This person was the first to describe the process of learning by insight.

_____ 20. This is the form of learning in which an animal associates something that it does with a reward or punishment.

_____ 21. This is the inborn reflex in an animal.

_____ 22. This psychologist is associated with studying classical conditioning in dogs.

TERMS LIST

a. Ethology	i. Ivan Pavlov	q. Migrations
b. Reflex	j. Insight	r. B. F. Skinner
c. Wolfgang Kohler	k. Unconditional reflex	s. Cognitivism
d. Behaviors	l. Classical conditioning	t. Behaviorism
e. Imprinting	m. Taxis	u. Konrad Lorenz
f. Operant conditioning	n. Innate behaviors	v. Animal behavior
g. Fixed action patterns	o. Habituation	
h. Kinesis	p. Learning	

MULTIPLE CHOICE QUESTIONS

_____ 1. Birds such as cardinals are stimulated to feed their young when signaled by which of the following stimuli?
   a.   Peeping of the young
   b.   The standing and outstretching of the young
   c.   Gaping mouths
   d.   Using the beak to peck at the mother
   e.   Set time intervals or biological clock mechanisms

_____ 2. From which of the following perspectives would ethologists analyze animal behavior?
   a.   Teleological
   b.   Hormonal
   c.   Structural
   d.   Physiological
   e.   Psychological

_____ 3. Which of the following individuals was (were) NOT among the Nobel prize recipients for work in animal behavior?
   a.   Lorenz
   b.   Kohler
   c.   von Frisch
   d.   Tinbergen
   e.   B and C

_____ 4. Which of the following is an example of the change in speed of random movement in response to environmental stimuli?
   a. Taxis
   b. Reflex
   c. Kinesis
   d. Fixed action pattern
   e. Learned behavior

_____ 5. Which of the following would NOT be a type of orientation behavior?
   a. Reflex
   b. Kinesis
   c. Taxis
   d. B and C
   e. A and C

_____ 6. When the graylag geese began recognizing Lorenz as the parent figure, which type of learning were they exhibiting?
   a. Insight
   b. Imprinting
   c. Migration
   d. Classical conditioning
   e. Operant conditioning

_____ 7. Which of the following is an example of the movement of an organism toward or away from a stimulus?
   a. Chemotaxis
   b. Phototaxis
   c. Gravotaxis
   d. All of the above
   e. None of the above

_____ 8. Mosquitoes find their prey by which of the following mechanisms?
   a. Taxis
   b. Kinesis
   c. Fixed action pattern
   d. Learned behavior
   e. Totally by accident

9. What type of behavior is exemplified by jerking the hand back after being burned?
   a. Taxis
   b. Reflex
   c. Fixed action pattern
   d. Learned behavior
   e. Kinesis

10. Which of the following is NOT a fixed action pattern?
    a. Egg retrieval movement in birds
    b. Nest building
    c. A dog scratching flea bites
    d. A cat washing its face
    e. Courtship behavior

11. What term do we use to denote the time-dependent forms of learning?
    a. Trial-and-error
    b. Classical conditioning
    c. Imprinting
    d. Habituation
    e. Insight

12. Salmon are able to locate their birthplace by which of the following processes?
    a. Sense of smell
    b. Visual cues
    c. Auditory environmental stimuli
    d. Inborn navigation patterns
    e. Temperature changes

13. Which of the following would be used by migrating animals to assist with directional navigation?
    a. Magnetic fields of the earth
    b. Stars
    c. Sun position
    d. Biological clock
    e. All of the above

_____ 14. Which of the following describes the process of learning to not respond to a nonthreatening stimulus?
   a.   Habituation
   b.   Classical conditioning
   c.   Imprinting
   d.   Operant conditioning
   e.   Insight

_____ 15. When a bird learns how to sing, which of the following describes the learning process involved?
   a.   It must practice for a long time to learn the song
   b.   It is born with the ability to sing the song
   c.   At maturity it just develops the skill to perform the song
   d.   Motoric imprinting occurs listening to the father sing
   e.   The song is coded for in the genes of the bird

_____ 16. If a toad is stung while feeding on a bee, it will spit it out and learn not to eat bees. What type of learning behavior is this?
   a.   Operant conditioning
   b.   Insight
   c.   Imprinting
   d.   Classical conditioning
   e.   Habituation

_____ 17. Which of the following animals is (are) capable of the most complex form of learning?
   a.   Dogs
   b.   Porpoises
   c.   Parrots
   d.   Chimpanzee
   e.   Elephants

SHORT ANSWER QUESTIONS

1.   What is the distinction between classical and operant conditioning?

2.   Give examples of reflex actions, other than the knee jerk response to patellar tendon stimulation.

3.   How can innate behavior be a detriment to an animal?

4.   What are some behavior patterns that would be different between adult male birds raised in isolation and adult male birds raised in their natural environment?

5.   How long does it take for behavior patterns to be expressed?

CRITICAL QUESTIONS

1.   This chapter has reviewed material relating to animal behavior. Do plants exhibit behavior? Explain.

2.   How might one go about investigating the premise that behavior is genetically determined?

3.   Which of the behaviors presented in the chapter would be most associated with the term "instinct" and why?

TELLING THE STORY

Develop a story that demonstrates your understanding of innate behavior and learning in animals. Write as if you were teaching the information to someone. If you have difficulty with any of the associated areas or topics as you write, go back and review that information in Chapter 32.

CHAPTER REVIEW ANSWERS

MATCHING

1. G	9. H	17. U
2. S	10. E	18. R
3. A	11. V	19. C
4. M	12. D	20. F
5. J	13. P	21. K
6. B	14. N	22. I
7. O	15. Q	
8. T	16. L	

MULTIPLE CHOICE QUESTIONS

1. C	6. B	11. C	16. A
2. D	7. D	12. A	17. D
3. B	8. D	13. E	
4. C	9. B	14. D	
5. A	10. C	15. D	

SHORT ANSWER QUESTIONS

1. In operant conditioning, the experimental animal controls the stimulus that leads to the reward or punishment. In classical conditioning, the experimenter controls the stimulus.

2. Answers will vary. Here are some examples of reflex responses: blinking, sneezing, coughing, vomiting, moving arms in front of you when falling, "goose bumps" when you get cold, shivering, pupil dilation and constriction to light differences, ducking when something falls toward you

3. An innate behavior is consistent and when this pattern does not lend itself to the present conditions, the situation could be detrimental to survival. Animal coloring other than those that would blend into an Arctic environment would be an example.

4. There would be several differences, but two are most evident. The birds raised in isolation would be unable to give the male territorial song for that

particular species and there might be some defects resulting in failure to imprint on their natural mother.

5.   The length of time it takes for a behavior pattern to be expressed varies depending on the type of behavioral pattern. The suckling response is quickly exhibited, while courtship behavior would not be expressed in sexually immature forms.

CRITICAL QUESTIONS

1.   Plants do exhibit behavior patterns. The nastic or "sleep" patterns often seen in the leaves and flowers of plants would be an example. Other plants respond to being touched by closing their leaves, such as the Venus Fly trap.

2.   Answers will vary. Here is an example response. An experiment might be set up to deprive an animal reared in isolation so that they receive no input from other members of their particular species. If they still exhibit the species-specific behavior, it can be linked to genetic determination rather than experiences.

3.   Instinct would be most closely associated with innate behavior made up of fixed action patterns. This process is an all-or-none mechanism triggered by a sign stimulus. It is consistent and is always executed in the exact same way. Once it has started, it will continue to completion.

CHAPTER 33

SOCIAL BEHAVIOR IN ANIMALS

KEY CONCEPTS OVERVIEW

1.   Social Behaviors

The biology of social behavior, or **sociobiology,** applies the knowledge of evolution to social behavior. This modern concept attempts to explain animal behavior on the basis of genetic and evolutionary foundations. Since social behaviors tend to be very complex and interactive patterns of behavior, discussion about the origin of human behavior is usually lively, to say the least. One thing that is agreed upon is that social behavior allows members of the same species to communicate and interact with one another.

Competitive behavior occurs when resources needed by two or more individuals are scarce. Acquisition of these resources creates a competition among organisms. Intimidation or **threat displays** are usually the first form of aggressive behavior. Such behavior varies among species, but can be exemplified by showing fangs or claws, growling, postural alterations, and coloration changes. These displays are important social signals and communicate the intent to fight. These same signals can serve to attract members of the opposite sex.

Submissive behavior is employed by animals to avoid fighting. This type of behavior may include making the body appear smaller, retracting fangs and claws, and turning away or facing away from an aggressor. Rarely do members of the same species fight to the death. This is probably due to natural selection which led to the development of a set of behavior patterns that reduced or eliminated this type of behavior. Scientists believe that aggressive behavior within a species is meant to result in chasing off an intruder, not death.

Territorial behavior can be seen in situations where animals mark off boundaries and defend the marked area as their own. Usually aggression results when same sex members of a species intrude. Territoriality is a common behavior in all vertebrates and in some invertebrates. Marking techniques will vary, but can include movements, body posture, sound, and odors. Territoriality can result in negative effects on a species by limiting nesting sites, food, water, and reproduction. Only those that have secured a territory with adequate resources will reproduce.

Reproductive behavior that promotes sexual reproduction involves patterns of male-female interactions. The term "**mating**" refers to these behaviors. Prior to mating however, animals work at attracting appropriate mates. This process of **courtship** is generally the task of the male members of the species, which use various means to attract females. Marking and defending a territory, displaying body colors, sound production, dances, and odors are used by different animals as means of courting. Additionally, courtship behaviors usually consist of a series of fixed action patterns of movement, each triggered by some action of the partner. This process assists with the formation of a pair relationship. Courtship not only eliminates the chance that different species would attempt to mate, but also insures that the male and female gametes will be released simultaneously.

After hatching or birth, some animals exhibit **parenting behavior** that increases the chance of the offspring survival. This behavior is found in almost all animal groups. Female mammals produce milk requiring their offspring to stay near and because of this assume the parenting role. In other animal forms, the male is the sole provider for the offspring. These types of behaviors are referred to as **altruistic behaviors.**

Some animals form groups, such as herds, flocks, schools, or prides. This behavior is designed to provide greater safety in numbers. The groups can be temporary or permanent, but the social group members all work together for common purposes. Many species of insects are organized into social groups that are so structured that they have a division of labor among the members. These are called **insect societies.**

One of the better known societies is exhibited by honeybees. In this society there are different **castes**, which provide for the division of labor. There are three castes: the male **drone**, the female **workers**, and the **queen**. The queen is the only fertile female in the hive, and she is able to maintain the dominant role by producing a pheromone that makes the other females sterile and suppresses the workers from making the compartments in which fertilized eggs can develop into queen bees. Workers communicate with each other regarding the location of food by body language called the waggle dance. The angle of the dance indicates the direction she wants other workers to fly. The drones are only responsible for mating with the queen. They congregate near larvae to keep them warm and they help to distribute food. At the time for mating, the drones follow the pheromone trail of the queen and compete to mate with her in flight. Following mating, the drone dies and those that do not mate with the queen are eventually killed by the workers in the hive.

Within groups there is a social hierarchy or **rank order**. In birds this social structure is called peck order. This graded series of dominance is exhibited differently through the species. In some species this order is linear, with one form dominant to all, a second dominant to all but the first, and so forth. In other animals, the dominance sequence is mixed so that individual A is dominant to individual B, and B is dominant to C, but C may be dominant to A. Such systems reduce the aggression and confrontations.

2.   Human Behavior

Needless to say, humans are very unique animals. They may exhibit some of the characteristics described in this chapter such as territoriality or rank order. However, many scientists resist the notion that the principles of sociobiology apply to human behavior. Because of many studies that have been undertaken (twin studies for example), most biologists hold that the behavior of humans is genetically determined, but that the neural circuits specified by genes can be shaped and molded to some degree by learning.

CHAPTER REVIEW ACTIVITIES

CONCEPT MAPPING

Develop a concept map that demonstrates your understanding of social behavior in animals. Incorporate as many of the key terms as possible and choose the linking terms that will best articulate one term to another. Develop the map with as much detail as you can.

MATCHING

Select the term from the list provided that best relates to the following statements. If you are unsure of the correct answer or chose an incorrect response, review the related concept in Chapter 33.

_____ 1. This social hierarchy exists in many groups of fishes, reptiles, birds, and mammals.

_____ 2. This is behavior that animals often exhibit in response to a threat display in order to avoid fighting.

_____ 3. These are the behavior patterns that lead to mating.

_____ 4. This science applies the knowledge of evolutionary biology to the study of animal behavior.

_____ 5. This term is used to describe the rank order in birds.

_____ 6. This type of behavior occurs when two or more individuals strive to obtain the same needed resource.

_____ 7. This type of behavior benefits one at the cost of another.

_____ 8. This chemical produced by one individual alters the physiology or behavior of the individuals of the same species.

_____ 9. Animals display this form of aggressive behavior during a competitive situation.

_____ 10. These behaviors help members of the same species communicate and interact with one another, each responding to stimuli from others.

_____ 11. These are behaviors an animal may exhibit that involve marking off an area as its own and defending it against the same sex members of its species.

_____ 12. These male-female behaviors result in fertilization, regardless of whether copulation occurs.

_____ 13. These social groups are formed by many species of insects, particularly bees and ants, and are characterized by a strict division of labor.

_____ 14. This ritualistic movement by worker bees helps direct co-workers to a particular location.

_____ 15. These activities are designed to lead to mating.

TERMS LIST

a. Waggle dance	f. Rank order	k. Pheromone
b. Submissive behavior	g. Mating	l. Altruistic behavior
c. Courtship	h. Peck order	m. Social behaviors
d. Threat display	i. Courtship ritual	n. Competitive behavior
e. Insect societies	j. Sociobiology	o. Territorial behavior

MULTIPLE CHOICE QUESTIONS

_____ 1. Various types of social behaviors are exhibited by animals to accomplish which of the following?
   a. Caring for and feeding offspring
   b. Courting potential mates
   c. Defending territory
   d. Seeking food
   e. All of the above

_____ 2. Which behavior would be exhibited if the cichlid fish turned dark all over with light spots on the side?
   a. On the alert while caring for young
   b. Full aggression
   c. Rising aggression
   d. Neutral mood
   e. Frightened but exposed

_____ 3. Which of the following animals would NOT mark their territory by odor chemicals?
   a. Dogs
   b. Monkeys
   c. Rodents
   d. Wolves
   e. Rhinoceroses

_____ 4. Female elephants attract males by which of the following means?
   a. Odors
   b. Visual cues
   c. Infrasound mating songs
   d. Touch
   e. Pawing in the dirt

5. Which of the following would describe courtship behaviors?

    a. Fixed action patterns

    b. Operant conditional response

    c. Reflexes

    d. Classical conditioning

    e. Taxis

6. Which of the following animals would group together and encircle their young when threatened?

    a. Geese

    b. Llamas

    c. Musk oxen

    d. Cattle

    e. Porpoises

7. Which of the following social "positions" would determine the character of a particular insect society?

    a. Drones

    b. Female workers

    c. Sterile females

    d. Queen

    e. Larvae

8. Which of the following is the first task of worker bees?

    a. Guarding the hive

    b. Foraging for food

    c. Feeding larvae

    d. Producing wax

    e. Building compartments

9. Which of the following activities would be restricted to the older worker bees (> 20 days old)?

    a. Guarding the hive

    b. Foraging for food

    c. Feeding larvae

    d. Producing wax

    e. Building compartments

10. The queen substance in honeybees produces which of the following effects?
    a. Makes the workers sterile
    b. Inhibits the formation of queen cells in the hive
    c. Controls the production of new queens
    d. Prevents bees from leaving the hive
    e. All of the above

11. Why do leaf cutter ants gather leaves as part of their behavior?
    a. To construct nests
    b. To use as food
    c. To prepare a site for mating
    d. To use as a substrate for growing fungi
    e. To camouflage themselves on the floor of the tropics

12. Which of the following types of animals would NOT exhibit rank order?
    a. Insects
    b. Birds
    c. Fish
    d. Mammals
    e. Reptiles

SHORT ANSWER QUESTIONS

1. What prevents animals from the same species from fighting to the death when defending their territories?

2. Why is asexual reproduction inferior to sexual reproduction and therefore not seen in higher order animals?

3. Why do some birds remove broken shell fragments from their nests immediately after the young have hatched?

4. What is the difference between courtship and mating?

5.   What is submissive behavior, and what purpose does it serve as a behavior in animals?

CRITICAL QUESTIONS

1.   Your text suggests that behavior that promotes successful sexual reproduction is highly adaptive behavior. Explain this statement.

2.   Is competition greater between members of a species or between members of different species, and what types of behavior would be exhibited in such competition?

3.   Offer some explanation as to why human behavior differs from other mammals.

TELLING THE STORY

Develop a story that demonstrates your understanding of social behavior in animals. Write as if you were teaching the information to someone. If you have difficulty with any of the associated areas or topics as you write, go back and review that information in Chapter 33.

CHAPTER REVIEW ANSWERS

MATCHING

1. F	6. N	11. O
2. B	7. L	12. G
3. C	8. K	13. E
4. J	9. D	14. A
5. H	10. M	15. I

MULTIPLE CHOICE QUESTIONS

1. E	4. C	7. D	10. E
2. B	5. A	8. C	11. D
3. B	6. C	9. B	12. A

SHORT ANSWER QUESTIONS

1. Fighting to the death does not usually occur in animals of the same species. It is an adaptive behavior to scare off competitors of this type rather than wound or kill them. Fighting tends to disperse the population among the available resources rather than decrease the population size and compromise the species numbers. Confrontations usually end in one competitor displaying submissive behavior and backing away from the situation.

2. Asexual reproduction does not lend itself to genetic variability. If environment changes occur, little adaptation would be possible among the members of the species and the population might die out. Additionally, the types of animals that reproduce sexually usually have means to defend their offspring and assure greater success for them.

3. It has been determined that the egg shell fragments attract predators by producing an attractive scent. Removal prevents predators from coming near the nesting sites and offers protection for the other eggs which have not hatched as well as the young hatchling.

4. Courtship behaviors are activities that occur between male and female members of a species prior to mating. Mating is the action that results in fertilization. This action may also be called spawning in animals that exhibit external fertilization and copulation in those that exhibit internal fertilization.

5. Submissive behavior may be the antithesis of aggressive behavior. This type of behavior serves to end aggression in the same species and prevents injury or death, which could deplete the species population.

CRITICAL QUESTIONS

1. Adaptations are genetically controlled traits that are passed on to the next generations. Any behavior that would increase the chance that these genes would be passed on to the next generation would be viewed as adaptive. Such genes would be incorporated into the genome and result in an increased possibility that they would appear in larger numbers of species members as time progressed.

2. The competition among members of the same species is greater because they are more likely to have the same needs and demands on the environment. In the same species, such behavior patterns as territoriality and peck order reduce direct competition between members. Confrontations are usually more ritualistic and do not result in severe injury or death. Such results would be detrimental to the success of the species.

3. Answers to this question may vary. Human behavior differs from other mammals based on the fact that insight is the principal means that governs behavior. Because humans have the ability to communicate through speech, writing, and nonverbal means and are capable of devising and using symbolism and computation, sets them apart from other mammals. Additionally, humans plan ahead and reason from new and past knowledge and experiences. There is ongoing debate among humans as to whether human behavior is more than evolved animal behavior.

CHAPTER 34

POPULATION ECOLOGY

KEY CONCEPTS OVERVIEW

1.   Population Growth

The study of ecological interactions is usually divided into four areas: populations, communities, ecosystems, and the biosphere. This chapter is devoted to looking at the concept of populations.

Populations are groups of individuals of the same species that live in the same general area at the same time. The members of a population can increase and decrease in number over time. There are factors that influence the growth or decline in a population, and these factors do not impact all populations in the same way or to the same degree. The size of a population is the difference between the additions to the population from new births within the population and **immigration** and the deletions from the population caused by death or **emigration**.

Population Change = (births + immigration) - (death + emigration)

Population changes are usually presented as a rate or the number of individuals per thousand per year. It can be assumed that generally conditions are more favorable for a population to be successful and grow at a given rate. While the rate of increase in population size may not increase, the actual number of individuals within the population can rise. When the rate increase remains constant, but the actual increase in number of individuals accelerates rapidly as the population grows, we call this process **exponential growth**. Exponential growth cannot continue indefinitely for obvious reasons. An increase in population numbers can only occur when conditions are ideal, such as abundant resources, excellent climate, and appropriate predator/prey relationships.

The population level that can be supported by a given environment is limited. This stabilization in population number is called the **carrying capacity** of a particular location. Carrying capacity is the number of individuals in a population that can be supported within a particular environment for an indefinite period. The population will grow at an exponential rate until limiting factors begin to slow the growth near the carrying capacity of that particular location. A graphic representation of the relationship between the time interval and individual numbers would result in an S-shaped or **sigmoid curve**.

Fig. 34.1 Sigmoid growth curve.

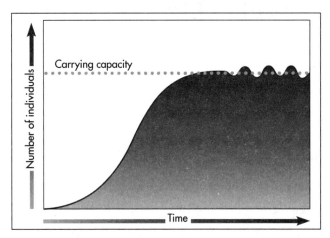

2. Population Size

It would seem logical to assume that larger populations would have a better change to survive than smaller ones. An unexpected negative event could drastically effect a smaller population and cause its demise. Smaller populations offer less opportunity for variation in the genetic pool. Inbreeding among the members of a smaller population can result in the expression of many recessive traits that are usually masked by dominant ones. Any loss of genetic diversity can increase the probability of extinction for that species.

3. Population Density and Dispersion

Populations usually increase in density as they grow. Population **density** reflects the number of organisms per unit of area. One of the critical factors related to the density of a population is the dispersion or the way in which the individuals of a population are arranged. Populations in nature exhibit three different types of dispersion: uniformed or evenly spaced, random, and clumped. Clumped distribution is the most frequently observed pattern in nature.

4. Population Regulation

A growing population will begin to experience competition among its members. Since the population relies on the same resources, individuals are responsible for securing these resources to be successful. If two species are competing with one another for the same limited resource, the species able to use that resource more efficiently will eventually eliminate the other. This process, called the principle of **competitive exclusion**, is a form of **density-dependent limiting**. Other factors of this type that limit the population size include increased numbers of predators, increased numbers of parasites, and increased environmental pollution.

There are also **density-independent limiting factors**. These factors operate regardless of the population density and include weather conditions, soil nutrient availability, physical disruptions of an area caused by such things as earthquakes, volcano eruptions, erosion, etc.

5. Mortality and Survivorship

Population growth rate also depends on the ability of the individuals within the population to survive and compete effectively for necessary resources. As organisms age, their competitive edge is compromised. Scientists express the mortality characteristics of a population by means of a survivorship curve. **Mortality** is the death rate in a population and **survivorship** is the proportion of an original population that lives to a certain age. Within animal populations, the survivorship curves are closely linked to parental care for offspring. Organisms demonstrate three different types of survivorship curves based on mortality rates. Organisms such as oysters and insects, which have a very high rate of death when they first hatch, exhibit Type III survivorship. Type II survivorship is seen when the organisms have equal chance to die at any age. Humans and other animals that provide consistent care for their young exhibit Type I survivorship, characterized by few deaths when young and most deaths toward the end of the normal life span.

6. Demography

Demography is the statistical study of human populations. One of the techniques used to accomplish this type of study involves plotting the percentage of the population by age and gender, to create a population pyramid. This pyramid is actually a bar graph that shows the composition of a population by age and gender. In stable populations, the pyramid is of approximately even dimensions throughout the age groups. In rapidly growing populations, there are proportionally fewer older individuals and dramatically more younger ones.

7.   Human Population Explosion

Your text asks the question: "How did the human population reach its present-day size?" Over time there have been significant social and technological advances that have promoted a major increase in the world population. Of these advances, agriculture, cities and urbanization, the development of the germ theory and the discovery of antibiotics, the development of public health education and policies, and the Industrial Revolution had enormous impacts. Despite all these contributing factors, the size of the human population is still limited and controlled by the environment and the carrying capacity of Earth.

So, are there changes in the human population? In developed countries, the birth and death rates have declined and many countries are approaching stability or even a decrease in population. Some countries, such as those in Africa, are demonstrating the most rapid growth rate. The impact of additional rapid growth will not go unnoticed. As we add more and more people to the population, the quality of life will decline.

CHAPTER REVIEW ACTIVITIES

CONCEPT MAPPING

Develop a concept map that demonstrates your understanding of population ecology. Incorporate as many of the key terms as possible and choose the linking terms that will best articulate one term to another. Develop the map with as much detail as you can.

MATCHING

Select the term from the list provided that best relates to the following statements. If you are unsure of the correct answer or chose an incorrect response, review the related concept in Chapter 34.

_____ 1. This bar graph shows the composition of a population by age and gender.

_____ 2. This term describes the movement of organisms out of a population.

_____ 3. This principle states that when two species are competing for the same resource in a specific location, the species most efficient in using that resource will eventually win out over the other.

_____ 4. This term describes the number of individuals added to a population during a given time period.

_____ 5. This term describes the death rate of a population.

_____ 6. These environmental factors result from the growth of a population but act to limit its subsequent growth.

_____ 7. This term describes a population whose size remains the same through time.

_____ 8. This is an interaction in which an organism of one species lives in or on another at their expense.

_____ 9. This is the statistical study of human populations.

_____ 10. This term describes the individuals of a given species that occur together at one place and at one time.

_____ 11. This is the movement of organisms into a population.

_____ 12. This term describes the number of individuals within a population that can be supported within a particular environment for an indefinite period.

_____ 13. These environmental factors operate to limit a population's growth, regardless of its density.

_____ 14. This is the proportion of an original population that lives to a certain age.

_____ 15. This is an interaction among populations in which an organism of one species kills and eats an organism of another species.

_____ 16. These are scientists who study how populations grow and interact.

_____ 17. In this type of growth pattern a population increases as its number is multiplied by a constant factor.

_____ 18. This is the proportion of individuals in the different age categories of a population.

TERMS LIST

a. Immigration	g. Population ecologists	m. Mortality
b. Age distribution	h. Population	n. Stable population
c. Population pyramid	i. Carrying capacity	o. Emigration
d. Predation	j. Survivorship	p. Demography
e. Growth rate	k. Parasitism	q. Exponential growth
f. Density-independent limiting factors	l. Competitive exclusion	r. Density-dependent limiting factors

MULTIPLE CHOICE QUESTIONS

_____ 1. Which number below best describes the net increase in population in the United States from births and deaths in 1989?
a. 1.2 million
b. 1.8 million
c. 2.6 million
d. 3.3 million
e. 4.1 million

_____ 2. The prickly pear grew out of control when it was first introduced into which continent?
a. Africa
b. Asia
c. Australia
d. North America
e. South America

_____ 3. Which of the following control agents was used to curb the excessive growth of the cactus?
a. Moth
b. Grass
c. Wasp
d. Rabbit
e. Locust

_____ 4. Which of the following distribution patterns is characteristic of a seed bearing plant population?
a. Clumped
b. Random
c. Uniform
d. Plants don't have a distribution pattern.
e. A combination of A, B, and C

_____ 5. Which term describes the secretion of toxic chemicals from one plant that harms other plants and results in uniform distribution?
   a. Toxicity
   b. Allelopathy
   c. Insipidity
   d. Parasitism
   e. Predation

_____ 6. Which of the following would be an example of a density-independent limiting factor?
   a. Shelter
   b. Mating sites
   c. Water
   d. Food
   e. Weather

_____ 7. In what year was or will 50% of the world's population exhibit urbanization?
   a. 1800
   b. 1900
   c. 1950
   d. 2000
   e. 2100

_____ 8. Which of the following presently has the largest city population in the world?
   a. Calcutta
   b. Tokyo
   c. Mexico City
   d. London
   e. Buenos Aries

_____ 9. In which of the following survivorship curves would there be an equal chance for organisms to die at any age?

  a. Type I
  b. Type II
  c. Type III
  d. At no time is the death rate constant

_____ 10. Which organism exhibits Type III survivorship curves?

  a. Oysters
  b. Humans
  c. Cats
  d. Hydra
  e. Elephants

_____ 11. Which of the following countries is exhibiting negative growth?

  a. Austria
  b. Canada
  c. Brazil
  d. Mexico
  e. India

_____ 12. In which country would the age structure diagram have a broader base than the others?

  a. Ireland
  b. Japan
  c. Kenya
  d. United States
  e. France

_____ 13. What is the present growth rate percentage in the United States?

  a. 0.4%
  b. 0.7%
  c. 1.0%
  d. 1.4%
  e. 2.1%

_____ 14. What is the present growth rate percentage in the world?
      a.   1.2%
      b.   1.8%
      c.   2.1%
      d.   2.7%
      e.   5.0%

_____ 15. How many millions of humans are added to the population each year?
      a.   45 million
      b.   75 million
      c.   95 million
      d.   125 million
      e.   200 million

SHORT ANSWER QUESTIONS

1.   Why would smaller populations be less likely to survive than larger ones?

2.   Explain the concept of exponential growth.

3.   What are the three patterns of population distribution?

4.   Name some of the density-dependent limiting factors that limit the size of a population.

5.   Name some of the density-independent limiting factors that limit the size of a population.

6.   What are some of the factors that can account for the dramatic increase in human population?

7.   What are the factors that control growth rate in a population?

CRITICAL QUESTIONS

1. Suppose you have a bottle that contained the necessary ingredients for a certain organism to be successful. You introduce the organism into the bottle and its numbers double each minute. After 110 minutes, the bottle's capacity is 1% full. Answer the following questions about the population in the bottle.

   a.) When would the bottle be approximately half full?

   b) Approximately when would the bottle be completely full?

   c) If someone gave you an additional 100 bottles (which would increase the available space 100 times) how many minutes would it take to completely fill all the bottles?

TELLING THE STORY

Develop a story that demonstrates your understanding of population ecology. Write as if you were teaching the information to someone. If you have difficulty with any of the associated areas or topics as you write, go back and review that information in Chapter 34.

CHAPTER REVIEW ANSWERS

MATCHING

1. C	4. E	7. N	10. H	13. F	16. G
2. O	5. M	8. K	11. A	14. J	17. Q
3. L	6. R	9. P	12. I	15. D	18. B

MULTIPLE CHOICE QUESTIONS

1. B	4. B	7. D	10. A	13. B
2. C	5. B	8. C	11. A	14. B
3. A	6. E	9. B	12. C	15. C

SHORT ANSWER QUESTIONS

1. In smaller populations there would be less genetic variability and as a population they would be unable to cope with environmental changes. Many members of the population would perish, and the population would be compromised.

2. Any constant growth rate would represent exponential growth. For example, if the population was 100 members and a 1% growth rate was experienced, this would mean that one new member would be added to the population in the first year, and the population would double every 70 years.

3. The three distribution patterns are random, uniform, and clumped.

4. Density-dependent factors adversely affect a population as it grows. One of the limiting factors is competition for any needed resource. Predators, parasites, environmental changes, etc. are additional factors that begin to cause a decline in population growth.

5. The density-dependent factors that can limit populations would include abiotic factors such as soil nutrients, weather related events such as floods, droughts, and severe freezes, and earthquakes, and volcano eruptions.

6. Human population has grown because of major developments in technology and industry. Agricultural development, urbanization, the Industrial

Revolution, improved public health and sanitation, understanding germ theory of disease, development of antibiotics, and public education have all contributed to improving the human conditions for success population growth.

7.   Growth rate is controlled by birth rate and immigration rate and by the death rate and emigration rate. Growth rate can be determined mathematically using the following equation:

Population growth rate = (birth rate + immigration rate) - (death rate + emigration rate).

CRITICAL QUESTIONS

1.   Part A

Time (minutes)	% Filled
111 min	2% filled
112 min	4% filled
113 min	8% filled
114 min	16% filled
115 min	32% filled
116 min	64% filled

Part B
By 117 minutes the bottle would be totally filled.

Part C

Time (minutes)	# Filled
118 min	2 bottles
119 min	4 bottles
120 min	8 bottles
121 min	16 bottles
122 min	32 bottles
123 min	64 bottles
124 min	128 bottles

It would take less than 7 minutes to complete fill the 100 extra bottles.

CHAPTER 35

INTERACTIONS WITHIN COMMUNITIES OF ORGANISMS

KEY CONCEPTS OVERVIEW

1.   Ecosystems

The interaction between the **abiotic** factors and the biotic factors within an environment make up what is known as an ecosystem. The abiotic factors would include all the nonliving components and the biotic factors include any living components. The organisms within an ecosystem reside in areas called **habitats.** You can think of a habitat as a space suitable for living. Any organism that resides in an ecosystem has a role to play and an impact on the environment. The term ecological **niche** is used to describe an organism's use of biotic and abiotic resources within its environment. The living portion of this environment is composed of populations of organisms that interact with each other to form a **community.** To be thorough in describing or studying an organism's niche, it would be important to include the organism's behavior and the behavioral changes that occur at different intervals of time.

2.   Communities

Populations of organisms interact with each other, forming communities. A **community** is defined as a grouping of populations of different species living together in a particular area at a particular time. An extension of a community would be called an ecosystem. An **ecosystem** would be the interaction between a community of organisms and all the biotic and abiotic factors with which the community interacts.

3.   Interactions within Communities

There are five basic types of reactions that occur within a community: competition, predation, commensalism, mutualism, and parasitism. These reactions occur in conjunction with other interactions between and among organisms and their environments. The study of these complex interactions and their impact on environments is called **ecology**.

When different organisms with the same needs live in close proximity to one another, they exhibit **competition** for specific limited resources. The organism best adapted and most efficient will be able to outcompete other organisms for the resource.

The **predation** reaction is exemplified when an organism of one species kills and eats an organism of another species. The intricate interactions between predators and their prey often impact the populations of other communities. Predation can result in preventing or reducing competitive exclusion by limiting the population of one of the competing species. Any predator/prey relationship will be a key factor in population balance within a natural community.

Commensalism is the relationship between organisms where one species benefits from its interactions with another organism that will neither benefit or be harmed in the process. This rather one-sided relationship can be accomplished between organisms by being physically attached to each other or by having a free-moving relationship.

Sometimes there is mutual benefit from organismal relationships. Such interactions would be viewed as examples of **mutualism**. These types of interactions may have evolved from those that were originally parasitic. You might wonder if this type of relationship results when two-way evolutionary adaptations occur.

There are situations where an organism of one species lives in or on another. When this interaction has negative results for the harboring organism or host, the relationship demonstrates **parasitism**. This type of parasite/host interaction is closely associated with predation but does not usually result in the death of the host. If death were to occur for the host, the parasite would die as well. Coevolution generally results in a stabilized relationship between the parasite and the host. Because of this, you might predict that parasitism is usually host specific.

If you think about the complex interactions that have to occur in a community among plants, animals, protists, fungi, and bacteria, it should make sense that these organisms have had to adjust to one another. Of course, these changes and adaptations have occurred over millions of years and are continuing to do so. **Coevolution** is the process that involves such long term, mutual evolutionary adjustments in members of a biological community in relation to one another. Take, for example, the interaction between plants and herbivores in a community. Plants could be considered to be prey for herbivore populations and have evolved various defense mechanisms to reduce predation. Structural and chemical modifications include the presence of spines and thorns, the development of hard, abrasive and impalatable parts, and even the development of toxins.

Some groups of animals develop protective coloration. One example, **camouflage**, allows an organism to escape predation by resembling its background. Other animals, quite capable of defending themselves against natural predators, exhibit bright or bold colored patterns that advertise their presence. This mechanism is referred to as **warning coloration** and predators learn to associate such display with danger. Some organisms have come to resemble these distasteful species as a way to protect themselves. These otherwise defenseless organisms exhibit warning coloration. Such patterns of resemblance are called **Batesian mimicry**, after British naturalist H.W. Bates. In another type of coloration mechanism, the protective colorations of different animal species come to resemble one another, but each possesses similar defense mechanisms. This is known as **Mullerian** mimicry, after German biologist Fritz Muller.

There are many ways to study the competition among organisms. Some scientists use laboratory settings to better control the conditions in which interactions and resulting changes occur. Another way to research the competitive relationships among or between organisms is to observe environments and gather data from natural settings. There are limitations and benefits to both types of research methods.

4.   Succession

It makes sense that, if the organisms within communities change and the interactions within the community are affected by these changes, the community itself would change over time. This dynamic process of change is called **succession**. As new individuals, or **pioneer communities**, enter an area, they modify the environment and create conditions that favor a new population. This leads to a predictable sequence of communities replacing one another and ending with a stable climax community composed of organisms adapted to the environment. In **primary succession**, the organisms invade a new environment where no organisms existed before. **Secondary succession** occurs in places that have been disturbed and are in the process of recovery, such as areas that have experienced natural disasters.

CHAPTER REVIEW ACTIVITIES

CONCEPT MAPPING

Develop a concept map that demonstrates your understanding of communities of organisms. Incorporate as many of the key terms as possible and choose the linking terms that will best articulate one term to another. Develop the map with as much detail as you can.

MATCHING

Select the term from the list provided that best relates to the following statements. If you are unsure of the correct answer or chose an incorrect response, review the related concept in Chapter 35.

_____ 1. This is the process of new communities replacing old ones in areas that have been disturbed and that were originally occupied by organisms.

_____ 2. This principle states that when two species are competing for the same resource in a specific location, the species most efficient in using that resource will eventually win out over the other.

_____ 3. This is the role each organism plays within an ecosystem.

_____ 4. These are living factors within the environment.

_____ 5. This is the study of the interactions among organisms and between organisms and their environment.

_____ 6. This is the place where an organism lives or grows.

_____ 7. This is a community in which the mix of plants and animals becomes stable.

_____ 8. This is an interaction in which an organism of one species lives in or on another at their expense but does not result in death to the host species.

_____ 9. This is the role an organism actually plays in the ecosystem.

_____ 10. This term describes two organisms using the same limited resource for survival.

_____ 11. This group of organisms consists of individuals of the same species.

_____ 12. This term describes the dynamic process of change, during which a sequence of communities replaces one another in an orderly and predictable way.

_____ 13. This is an organism of one species that kills and eats organisms of another.

_____ 14. This form describes all the biotic and abiotic factors within a certain area.

_____ 15. This is an interaction between organisms where one organism benefits while the other is neither harmed or benefited.

_____ 16. This term describes a grouping of populations of different species living together in a particular area at a particular time.

_____ 17. This is the role that an organism might play in an ecosystem if competitors were not present.

_____ 18. This is an organism that is killed and eaten by a member of another species.

_____ 19. This type of community replacement takes place in areas not previously supporting organisms.

_____ 20. These are nonliving factors within the environment.

TERMS LIST

a. Biotic	k. Succession
b. Population	l. Niche
c. Competition	m. Secondary succession
d. Parasitism	n. Habitat
e, Abiotic	o. Climax community
f. Ecosystem	p. Prey
g. Competitive exclusion	q. Ecology
h. Primary succession	r. Realized niche
i. Predator	s. Commensalism
j. Community	t. Fundamental niche

MULTIPLE CHOICE QUESTIONS

_____ 1. Based on the role a rabbit plays in a community, which of the following terms best describes it characteristically?

a. Producer

b. Decomposer

c. Carnivore

d. Herbivore

e. Autotroph

_____ 2. Which of the following examples would NOT be a characteristic of a population niche?

a. Fed on by herbivores

b. Usually reproduces in the early spring

c. Stores food for the winter

d. Prefers bright lighted areas

e. None of the above

3. Mutualism between two species would occur in which of the following situations?
   a. Only one species benefits
   b. One species is harmed, the other benefits
   c. Both species benefit
   d. Both species are harmed
   e. None of the above

4. Barnacles feed by which of the following methods?
   a. Drilling into their substrate
   b. Using appendages like a net to obtain food
   c. Filtering water through their bodies
   d. Absorbing food from their surroundings through their foot
   e. Sending out tentacles and capturing food

5. Organisms that interact and mutually adjust to one another over time exhibit which of the following evolutionary processes?
   a. Coevolution
   b. Parallel evolution
   c. Regressive evolution
   d. Divergent evolution
   e. Convergent evolution

6. Which of the following refers to the interaction among organisms of different species that live near one another and strive to obtain the same limited resources?
   a. Mutualism
   b. Commensalism
   c. Parasitism
   d. Competition
   e. Predation

7. Which of the following individuals is credited with the formulation of the principle of competitive exclusion?
   a. Fritz Muller
   b. John Harper
   c. G.F. Gause
   d. H.W. Bates
   e. J.H. Connell

8. Which of the following terms is used to refer to a display of bold coloration to draw attention?
   a. Batesian mimicry
   b. Warning coloration
   c. Camouflage
   d. Mullerian mimicry
   e. None of the above

9. Which of the following describes the interaction between barnacles and the whales to which they attach?
   a. Mutualism
   b. Commensalism
   c. Parasitism
   d. Competition
   e. Predation

10. Which of the following describes the interaction that occurs when a population of *Didinia* are placed in an environment with a population of *Paramecia*?
    a. Mutualism
    b. Commensalism
    c. Parasitism
    d. Competition
    e. edation

_____ 11. By which of the following means are clownfish protected from predation while living in the sea anemones?
   a.   Camouflage
   b.   Poisonous spines located on the fish
   c.   Swelling and appearing larger than it really is
   d.   The stinging tentacles of the anemone
   e.   A toxin produced by the clownfish

_____ 12. In which of the following did the killer bees originate?
   a.   Africa
   b.   Brazil
   c.   Chile
   d.   Mexico
   e.   France

SHORT ANSWER QUESTIONS

1.   List the five types of interactions that occur between or among organisms that live in the same community.
2.   What is the advantage of using genetic engineering to grow plants to repel predators?

3.   Give some examples of different types of protective coloration.

4.   How does parasitism differ from predation?

5.   What happens if the niches of two organisms overlap?

6.   Differentiate betweeen a niche and a habitat.

7.   Differentiate between Batesian and Mullerian mimicry.

CRITICAL QUESTIONS

1. Offer an explanation as to how mimicry evolves.

2. Succession can occur in any environment. Given the fact that pasteurized milk contains nonpathogenic bacteria, use a carton of milk to illustrate the processes that occur during succession. Which type of succession would this be?

TELLING THE STORY

Develop a story that demonstrates your understanding of organismal communities. Write as if you were teaching the information to someone. If you have difficulty with any of the associated areas or topics as you write, go back and review that information in Chapter 35.

CHAPTER REVIEW ANSWERS

MATCHING

1. M	5. Q	9. R	13. I	17. T
2. G	6. N	10. C	14. F	18. P
3. L	7. O	11. B	15. S	19. H
4. A	8. D	12. K	16. J	20. E

MULTIPLE CHOICE QUESTIONS

1. A	4. B	7. C	10. E
2. E	5. A	8. B	11. D
3. C	6. D	9. B	12. A

SHORT ANSWER QUESTIONS

1.   Competition, mutualism, commensalism, predation, and parasitism

2.   Altering the characteristics of plants to render them unattractive to
      herbivores prevents the need for pesticides or segregation.

3.   Protective coloration is exemplified by camouflage, Batesian mimicry,
      Mullerian mimicry, and warning coloration.

4.   The successful parasite, usually much smaller than its host, would kill its
      host to keep from depleting necessary resources. Predation results in the
      death of the prey. Predation, usually occurring between a larger predator and
      a smaller prey, is a single action that occurs quickly, while parasitism occurs
      over extended periods of time.

5.   When niches overlap, there is potential competition between organisms.

6.   The habitat is the physical space in which an organism resides, and a niche is
      a functional concept that refers to the characteristics, conditions, and
      behaviors associated with how organisms live.

7.   In Batesian mimicry, what would be a natural prey represents a danger or unpleasant experience to the predator. In Mullerian mimicry, two or more potential prey present unpleasant experiences for their potential predators.

CRITICAL QUESTIONS

1.   In nature quite often a noxious or dangerous organism is conspicuously colored or marked. Predator organisms learn through experience to avoid these prey. If another animal happens to resemble the dangerous one or model, it is protected by virtue of its resemblance. The warning coloration does not afford all the resembling organisms protection, because if the predator species eats them before the model organism, they won't have learned about the negative effects on the model.

2.   Milk is pasteurized to kill any pathogens in the solution, but it does not rid the solution of all living bacteria. As these bacteria feed on the nutrients within the milk, the population and characteristics of the milk solution changes. Lactose, a disaccharide, is broken down into glucose and galactose. Continuation of this process causes lactic acid to by formed and the milk to become acidic and turn sour. The acidic condition in the milk provides an environment in which the proteins can be broken down. Over a sufficient period of time, a different population of bacteria in the milk will favor this changing condition. After several days however, the nutrients become depleted, the bacteria no longer reproduce, and eventually growth of the bacteria stops.

CHAPTER 36

ECOSYSTEMS

KEY CONCEPTS OVERVIEW

1.   Populations, Communities, and Ecosystems

An **ecosystem** is made up of communities of organisms that live within a prescribed area and the nonliving environmental factors with which they interact. There is a hierarchy within nature with regard to the functional units of ecological systems. Sequencing from the less complex to the more complex, the order would be individual organisms, populations, species, communities, ecosystems, biomes, and the biosphere. The functioning unit in nature is the ecosystem. The biotic components of any ecosystem would include consumers, producers, and decomposers. Each of these types of organisms play a significant role within the ecosystem by contributing to the flow of energy and the cycling of nutrients. The abiotic components are the raw materials found within the given area.

2.   Energy Flow Through Ecosystems

All energy originates from the sun. So, the energy that flows through an ecosystem can be traced back to this originating point. The **primary producers** of terrestrial ecosystems are green plants, and they are able to capture radiant energy from the sun and convert it to chemical energy during photosynthesis. Producers are autotrophs, and those organisms that feed directly on them are called **primary consumers,** or herbivores. By eating the producers, these consumers are able to incorporate some of the captured energy into their own metabolic and functional processes. **Secondary consumers**, or carnivores, feed on the primary consumers, reaping energy from the herbivores, producers, and the sun. **Decomposers** are organisms that break down the organic materials of **detritus**, which is the waste material of an ecosystem. They act on producers and consumers when they die, serving as the last link in the flow of energy through an ecosystem and contributing to the recycling of nutrients in the environment.

All the feeding sequences or levels are called trophic levels. Heterotrophs are consumers, and autotrophs are producers. The energy flowing from one organism to the next in an ecosystem is exemplified by a **food chain**. As energy moves through an ecosystem, some is lost as low grade heat and some is used for metabolic processes and growth within the organisms. When an organism is consumed by another, the energy is transferred. Usually the amount transferred is approximately 10% of the energy available.

Food pyramids are a way to accurately represent the concept of food webs. Since organisms do not feed on just one type of food source, nor are they preyed on by only one predator type, the complexity of these interactions makes it difficult to describe. These complex interrelationships can be summarized in one of three ecological pyramids. The **pyramids of biomass** depict the total weight of organisms supported at each trophic level in an ecosystem. The **pyramids of number** depict the total number of organisms at each feeding level. The **pyramids of energy** depict the energy flow through an ecosystem. In each pyramid the base is usually larger than the other levels. Sometimes the pyramids of biomass and numbers can be inverted (larger level on top).

3.   Chemical Cycling Within Ecosystems

There are a number of cycles within nature. Perhaps the best known and easiest to demonstrate is that of the water cycle. Water moves from the atmosphere by condensation and precipitation to the ground. From there it runs off the surface toward the oceans, or it is absorbed into the ground and raises the water table. Water may be stored underground in underground reservoirs surrounded by porous rocks in what is termed an aquifier. Chemicals, pollutants, and contaminants can collect here also. About 2% of the groundwater in the United States is polluted. Water returns to the atmosphere by evaporation and transpiration.

Carbon is fixed on the earth by photosynthesis and transferred to animals by feeding. The processes of respiration or decomposition return carbon to the atmosphere. Carbon dioxide may be stored in the bodies of organisms, the fossil fuels, and in calcium carbonate shells of such animals as mollusks. Such things as deforestation, desertification, and combustion of fossil fuels have caused a recent dramatic increase in atmospheric carbon dioxide.

The nitrogen cycle refers to the movement of nitrogen through the ecosystem. The actions of lightning and nitrogen fixation convert atmospheric nitrogen into nitrates that can be utilized by plants to form proteins. Animals feeding on plants also use the nitrogen to form amino acids and proteins. When the plants and animals die, decomposers break down the nitrogen compounds and produce ammonia. Ammonia is then converted to nitrates by nitrifying bacteria. Denitrifying bacteria release nitrogen back into the atmosphere to maintain the nitrogen component in air.

The phosphorus cycle does not have an atmospheric reservoir. There is also very little stored in the soil. Phosphorus within fertilizers accumulates in aquatic ecosystems due to the amount that is generally applied each year. The increase in phosphates, and perhaps nitrates, causes rivers and streams to develop an overgrowth of algae and other aquatic plants. This process is known as **eutrophication.**

CHAPTER REVIEW ACTIVITIES

CONCEPT MAPPING

Develop a concept map that demonstrates your understanding of ecosystems. Incorporate as many of the key terms as possible and choose the linking terms that will best correlate one term to another. Develop the map with as much detail as you can.

MATCHING

Select the term from the list provided that best relates to the following statements. If you are unsure of the correct answer or chose an incorrect response, review the related concept in Chapter 36.

_____ 1. These are the interwoven and interconnected feeding relationships of an ecosystem.

_____ 2. This term describes an organism that produces its own food by photosynthesis.

_____ 3. This community consists of plants, animals, protists, fungi, and bacteria that interact with one another and with their environment.

_____ 4. This is the refuse or waste material of an ecosystem.

_____ 5. This name is given to various feeding levels within an ecosystem.

_____ 6. These meat eating organisms feed on the herbivores.

_____ 7. These animals eat only plants.

_____ 8. This is the original source of all energy that flows through the biosphere.

_____ 9. As a result of photosynthesis, this molecule is released into the atmosphere as a waste product.

_____ 10. This is the process by which most water is returned to the atmosphere.

_____ 11. This is another term for a secondary consumer.

_____ 12. This process has carbon dioxide as a waste product and allows it to be released into the atmosphere.

_____ 13. This series of organisms, from each trophic level, feed on one another.

_____ 14. These organisms feed directly on green plants.

_____ 15. This is the term for an organism that connot produce its own food.

_____ 16. This special group of consumers obtains nourishment from dead matter, such as fallen leaves or the bodies of dead animals.

_____ 17. These underground reservoirs of water are surrounded by porous rock.

_____ 18. Through this process, plants release water into the atmosphere.

TERMS LIST

a. Food web	j. Ecosystem
b. Primary consumers	k. Carnivore
c. Transpiration	l. Sun
d. Herbivore	m. Autotroph
e, Evaporation	n. Respiration
f. Detritus	o. Aquifer
g. Food chain	p. Oxygen
h. Secondary consumers	q. Trophic levels
i. Heterotrophs	r. Decomposers

LABELING

Label the following diagram. Review the concept that this diagram represents.

Fig.36.1 An ecosystem.

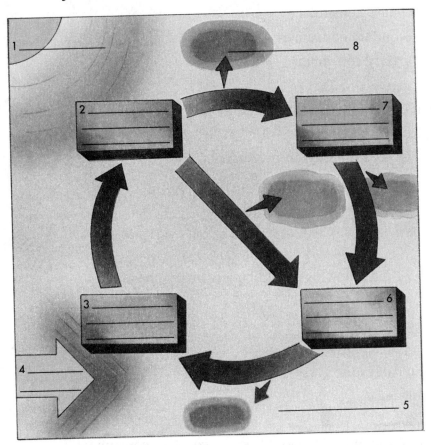

MULTIPLE CHOICE

1. In the following food chain, what is the bird's role?

    Algae ---> aquatic insect ---> fish ---> bird

    a.   Producer

    b.   Decomposer

    c.   Primary consumer

    d.   Secondary consumer

    e.   Tertiary consumer

_____ 2. In the following food chain, what role do the algae play?

Algae ---> aquatic insect ---> fish ---> bird

a. Producer

b. Decomposer

c. Primary consumer

d. Secondary consumer

e. Tertiary consumer

_____ 3. Which of the following statements best relate to secondary consumers?

a. They are carnivores.

b. The feed directly on producers.

c. They are herbivores.

d. They are autotrophs.

e. None of the above

_____ 4. Which of the following would be a primary consumer?

a. Grasshopper

b. Fox

c. A parasite in an antelope

d. St. Augustine grass

e. Algae

_____ 5. Which of the following would be a secondary consumer?

a. Grasshopper

b. Fox

c. A parasite in a fox

d. St. Augustine grass

e. Algae

_____ 6. Which of the following would best be represented by pyramids of biomass?

a. The number of organisms within a particular trophic level

b. The total weight of organisms within a particular trophic level

c. The total amount of energy required by each organism within a Particular trophic level

d. The size of organisms within a particular trophic level

e. None of the above

_____ 7. Which of the following types of pyramids can be inverted?
   a. Number
   b. Biomass
   c. Energy
   d. A and B
   e. A and C

_____ 8. Which of the following would be an autotroph?
   a. Coyote
   b. Fungi
   c. Flea
   d. Cactus
   e. Falcon

_____ 9. What percentage of the evergy available in one trophic level is passed on to the next trophic level?
   a. 2%
   b. 5%
   c. 10%
   d. 15%
   e. 20%

_____ 10. Which of the following elements is not stored in the atmosphere?
   a. Oxygen
   b. Nitrogen
   c. Carbon
   d. Hydrogen
   e. Phosphorus

_____ 11. Which element is found in the greatest concentration in the atmosphere?
   a. Oxygen
   b. Nitrogen
   c. Carbon
   d. Hydrogen
   e. Phosphorus

12. Which of the following is defined as being made up of communities of organisms living within a defined area and the nonliving environmental factors with which they interact?
    a. Biosphere
    b. Biome
    c. Ecosystem
    d. Habitat
    e. Niche

13. Which of the following plants exhibitS symbiotic relationships with nitrogen fixing bacteria ?
    a. Grass
    b. Sunflower
    c. Mint
    d. Legume
    e. Soybean

14. Which occurs first in an aquatic ecosystem?
    a. Massive fish kill
    b. Rapid growth of algae
    c. Leaching of phosphorus from farm lands to rivers
    d. Growth of decomposers
    e. Depletion of available oxygen

15. Which cycle is associated with the process of eutrophication?
    a. Carbon
    b. Phosphorus
    c. Nitrogen
    d. Water
    e. None of the above

SHORT ANSWER QUESTIONS

1. List some of the important biotic components in an ecosystem.

2. What are the ecological pyramids?

3. The chemical components of fertilizers are of interest in monitoring their effects over time. What are the major components of fertilizers?

4. Explain the difference between food webs and food chains.

CRITICAL QUESTIONS

1. The carbon cycle has exhibited a recent modification that is causing environmental concerns. What is this change and what specific effects are we concerned about?

2. Design and describe a method to determine the interrelationships between organisms in a food web.

TELLING THE STORY

Develop a story that demonstrates your understanding of ecosystems. Write as if you were teaching the information to someone. If you have difficulty with any of the associated areas or topics as you write, go back and review that information in Chapter 36.

CHAPTER REVIEW ANSWERS

MATCHING

1. A	7. D	13. G
2. M	8. L	14. B
3. J	9. P	15. I
4. F	10. E	16. R
5. Q	11. K	17. O
6. H	12. N	18. C

MULTIPLE CHOICE QUESTIONS

1. E	4. A	7. D	10. E	13. D
2. A	5. B	8. C	11. B	14. C
3. A	6. B	9. B	12. C	15. B

LABELING
Fig. 36.1

1. Sun
2. Producers
3. Abiotic nutrients
4. Energy nutrients
5. Heat
6. Decomposers
7. Consumers
8. Heat

SHORT ANSWER QUESTIONS

1. There are three large categories of biotic components in an ecosystem: the producers, the consumers, and the decomposers.

2. Pyramid of energy, pyramid of numbers, and pyramid of biomass

3. The many chemical ingredients in fertilizers are: nitrogen, phosphorus, and potassium. There are other less significant elements such as sulfur, calcium, iron, and zinc.

4. Food chains are a way to demonstrate the transfer of energy from one organism to another. This is a very simplified model, because it assumes that an organism has only one source for food and is only consumed by one type of

predator. A food web demonstrates a more complex interaction among organisms and charts the multiple energy flow patterns within an ecosystem.

CRITICAL QUESTIONS

1. Carbon dioxide levels have been dramatically increasing in the atmosphere since the Industrial Revolution. The resulting impact on the environment is referred to as the greenhouse effect, which has been implicated in the recent increase in world temperatures. Some scientists maintain that such increases are part of a very natural variation in temperatures over long periods of time. They suggest that records have not been kept long enough to show the amount of variation that can be expected. The counter argument to this is the correlation between increased temperatures and increased greenhouse gases. The increased levels of carbon dioxide can be attributed to the large amount of fossil fuels being burned and an increase in the rate of forest depletion.

2. Answers may vary. One way would be to use a radioactive isotope as a chemical tracer. By labeling a producer with an isotope, any organism that consumes this plant will demostrate detectable levels of the isotope. These organism will be identified as the primary consumers. Subsequent demonstration of the same radioactive isotope will show up in secondary consumers. The length of time and the amount of radioactivity provide clues as to the level of the organism in the food web or the dominance of that organism within the ecosystem.

CHAPTER 37

BIOMES AND LIFE
ZONES OF THE WORLD

KEY CONCEPTS OVERVIEW

1. Biomes and Climate

As you have no doubt discovered in the previous chapters, organisms vary in many ways. These variations or characteristics are used to differentiate one type of organism from another. Despite the diverse nature of living organisms, no species on Earth lives in total isolation. Each species is found in the context of groupings of plants and animals that interact with each other and their environment. This chapter focuses on the characteristics of diverse locations on Earth and why organisms live where they do.

Biomes are ecosystems consisting of plants and animals that can be found over wide areas of land within specific climatic regions. A biome can be easily distinguished by its appearance, regardless of where it is on the Earth. Rainfall and temperature patterns have a direct influence on the characteristics of a biome. Such patterns result from the relationship between the Earth's structural features and two specific physical factors. These factors are: (1) the amount of heat from the sun that reaches various parts of Earth and the seasonal variations in that heat, and (2) global atmospheric circulation and the resulting patterns of oceanic circulation. Collectively, these factors determine how much and where precipitation will occur.

Not all the surfaces of Earth receive the same amount of energy from the sun. This is due to the spherical nature of planet Earth. The sun's rays are concentrated on the areas nearest the equator and are more spread out at the poles. The greater the distance from the equator (also known as latitude), the colder the climate.

If you've ever paid attention to a globe, you might have observed that the Earth was tilted on its axis. This tilt is responsible for the variations in the seasons. Because of this arrangement, the northern and southern hemispheres do not receive the same amounts of sunlight throughout the year. In the summer, the northern hemisphere is closer to the sun and therefore receives the most direct radiation. Except for the spring and autumn equinoxes, it would be expected that one pole would always be closer to the sun than the other.

The heating process on Earth causes air to rise and move toward the poles. As the earth rotates, the movement causes the air to break into six "coils" of rising and falling air that surround the Earth. As the air moves toward the poles, away from the equator, these air masses cool. It can be stated that for each latitude moving away from the equator, the air is cooler than the previous latitude and warmer than the subsequent latitude. (See Figure 37-3, P. 801 in your text)

Prevailing wind patterns and varying amounts of precipitation result from these moving masses of air that encircle the Earth. As the air is warmed and cooled, the moisture-carrying capacity increases and decreases respectively. Therefore, at or near latitude 60° north and south, the air would be on the rise and cooling. This would account for the presence of temperate forests in these regions of the world. Conversely, at 30° north and south latitude, the air would be sinking and warming. The great deserts would be found in such regions. Although these patterns are predictable, there are other factors which influence climates. Mountain ranges produce climatic variations by forcing warm, moist air upward. As the air cools, it causes precipitation on this side, known as the **windward** side. As the air descends on the **leeward** side, it is warmed and gathers more water. This accounts for the increased moisture and difference in vegetation from that on the windward side. This pattern is known as the **shadow effect.**

On the surface, ocean circulation moves in spiral patterns called **gyrals.** These patterns affect the climate on adjacent land masses. Gyrals that move around the subtropical zones of high pressure between 30° north and 30° south latitude, result in clockwise spirals in the northern hemisphere and counterclockwise spirals in the southern hemisphere. In the northern hemisphere, the western sides of temperate zone continents are generally warmer than their eastern sides. This would be reversed in the southern hemisphere.

2.   Life on Land

There are seven biome categories: tropical rain forests, savannas, deserts, temperate grasslands, temperate deciduous forests, taiga, and tundra. Distribution of these biomes are affected by climatic patterns caused by mountains, shorelines of continents, and sea temperatures. Elevation would also influence climate and vegetation patterns.

Tropical Rain Forests

These forests are found in areas where the rainfall exceeds 80 to 175 inches per year and the average temperature is 77° Fahrenheit. This environment supports the success of giant trees that form an overlapping canopy with their leaves, reducing light intensity by 2% on the forest floor. For this reason, the forest floor has very little vegetation. Some plants grow as epiphytes (air plants) and are nourished by rain caught in modified leaves. Because little nutrient material ends up in the soil, the trees are the "nutrient stores" for this biome. Their removal leaves behind relatively barren soil unable to sustain farming crops for very long.

Savannas

These regions are massive grasslands with scattered trees, which support a large number of grazing herbivores and plant-eating invertebrates. Savannas are situated between tropical forests and deserts, and are found in central and southern Africa, western India, northern Australia, some of the northern and east-central regions of South America, and parts of Malaysia. They are close to the equator but receive less annual rainfall.

Deserts

Within desert biomes, the annual rainfall is less than 10 inches per year. Because of this, the vegetation is sparse. The characteristic organisms are those adapted to dry conditions. Because deserts radiate heat rapidly at night, due to a lack of vegetation, the temperatures can vary some 55° Fahrenheit over a 24 hour period. Desert plants are generally succulents, able to store large amounts of water to assure their survival, have very deep roots, or have seeds that can survive extreme heat and drought conditions. Animals within this type of biome usually limit their activity to short periods of the year when water is more available. They avoid the extreme heat by burrowing deep holes and emerging only at night.

Temperate Grasslands

The temperate grasslands, given many different names around the world, are called **prairies** in North America. These biomes have a yearly rainfall of 10 to 30 inches. They are characterized by large quantities of perennial grasses, burrowing rodents, and large grazing mammals, which prevent woody vegetation from being established. The soil is generally rich and, therefore, often farmed.

Temperate Deciduous Forests

Regions that support the growth and maintenance of **deciduous** trees (those that lose their leaves and remain dormant in winter), are temperate deciduous forests. The climate in these regions is generally warm in the summers, cold in the winters, and the precipitation ranges from 30 to 60 inches per year. Because of the existence of a lower tree canopy and a layer of shrubs, as well as abundant ground vegetation, the deciduous forests differ from the tropical rain forests. These regions support abundant animal life both in the trees and on the ground.

Taiga

The northern coniferous forest is called taiga and is characterized by long, cold winters with very little precipitation. The precipitation occurs during the summer months. Because the days are long, the temperature mild, and the rainfall sufficient, summer allows plants to grow rapidly. These conifer forests house large mammals that are both herbivores and carnivores.

Tundra

The area between the taiga and the permanent ice is the biome called tundra. This enormous area occupies one-fifth of the Earth's surface and is characterized by scattered patches of grasses and sedges, heather, and lichen. The annual precipitation in the tundra is less than 10 inches per year and, what does fall is unavailable to plants for most of the year because of the severe cold. The permafrost, or permanent ice, is impenetrable to water and roots. During the summer, large grazing mammals, birds and waterfowl, and small rodents can be found in the tundra.

3.   Life in Fresh Water

Lakes, ponds, rivers, and streams account for only 2% of the Earth's surface. Within these fresh water biomes, there are three zones: shore zone, open-water zone, and deep zone. Emerging plants such as cattails and water lilies are found in the shallow water near edges of lakes or ponds. These areas are referred to as the **shore zone.** Within the vegetation of these areas, consumers such as frogs, snails, and dragonflies can be found. The **open-water zone** contains phytoplankton, which provide food for the zooplankton. These small animals are then eaten by the fish within the environment. The **deep zone**, or water that is not penetrated by light, is devoid of producers. Only decomposers and some types of animals live in this part of the fresh water bodies. Because ponds are shallower than lakes, they have no deep zone.

Rivers and streams differ from ponds and lakes in that they are open and moving. Because water tumbles over rocks, it mixes with the air and increases the amount of dissolved oxygen. This allows for an abundance of fish and invertebrates to survive in these aquatic environments.

4.   Estuaries

An **estuary** is created when fresh water joins salt water . The shallow area of these environments are often plentiful with grasses , algae, and phytoplankton. The consumers consist of mollusks, crustaceans, fish, and zooplankton. However, the environment has required adaptations on the part of all these organisms in order for them to survive in areas of moving water and changing salinity.

Nutrients are more abundant in the estuaries as compared to the open ocean because of their close proximity to terrestrial ecosystems. These highly productive ecosystems can also be negatively affected by their close association with land. Environmental pollution can significantly impact these successful regions, resulting in a decrease in their fish populations.

5.   Life in the Oceans

Within the marine environment, there are three major life zones: the intertidal zone, the neritic zone, and the open-sea or pelagic zone. The **intertidal zone** is the region between high tide and low tide. This region receives significant light and supports the success of many producers and consumers. The **neritic zone** extends to a depth of 200 meters over the continental shelf. Because light reaches most of this zone, it supports an abundance of plant and animal life. The most notable community within this zone is that of the coral reef. This highly productive structure is constructed by marine animals (coral) that secrete calcium carbonate and form the hardened structures familiar to snorklers and SCUBA divers. The **open-sea zone** is beyond the continental shelf. The distribution of organisms within this zone is controlled by the availability of light and nutrients. The area above 200 meters is the **photic zone** which supports the common diversity displayed in the marine fauna. The fishes and mammals of the sea, called **nekton**, feed on the plentiful plankton and on each other. From 200 to 1000 meters deep, the **mesoplagic zone** exists. Within this region, the temperatures are usually cold, the environment dark, and the pressure severe. These factors directly influence the types of organisms that occupy this zone. Below 1000 meters is the **abyssal zone**, which is populated by very interesting looking creatures called **benthos**. These organisms rely on dead fish falling from the above zones for their food.

CHAPTER REVIEW ACTIVITIES

CONCEPT MAPPING

Develop a concept map that demonstrates your understanding of biomes. Incorporate as many of the key terms as possible and choose the linking terms that will best correlate one term to another. Develop the map with as much detail as you can.

MATCHING

Select the term from the list provided that best relates to the following statements. If you are unsure of the correct answer or chose an incorrect response, review the related concept in Chapter 37.

_____ 1. This is another term for the open ocean.

_____ 2. This margin of land extends out from the intertidal zone, usually 30 to 60 miles, and slopes to a depth of about 650 feet beneath the sea.

_____ 3. These ecosystems of plants and animals occur over wide areas of land within specific climatic regions and are easily recognized by their overall appearance.

_____ 4. These fan-shaped areas of accumulated sediment are deposited by a stream or river as it enters as open body of water such as a lake or the ocean.

_____ 5. This is the area of pond or lake water into which light does not penetrate.

_____ 6. This term is used to describe the concentration of dissolved salts in the water.

_____ 7. In a marine environment, this is the area between the highest tides and the lowest tides.

_____ 8. This is the main body of pond or lake water through which light penetrates.

_____ 9. This term refers to the middle ocean zone (650 feet to 3250 feet deep) where only a little light penetrates the water.

_____ 10. This term describes a place where the fresh water of rivers and streams meets the salt water of the oceans.

_____ 11. This is the shallow water near the edges of a lake or pond in which plants with roots, such as cattails and water lilies, may grow.

_____ 12. These microscopic floating and drifting algae and plant-like organisms are found in aquatic environments.

_____ 13. This debris or decomposing material provides the organic matter for rivers and streams.

_____ 14. These tiny crustaceans are the predominant organisms of sandy shores and often burrow beneath the sand during low tide.

_____ 15. This is an area of the open ocean where light is available to organisms from the water's surface to an approximated depth of 650 feet.

_____ 16. This community of organisms, partially made up of marine animals that secrete calcium carbonate, live on masses of rocks lying at or near the surface of the ocean water.

_____ 17. This is the term for the side of a mountain where warmer air moves up from the ground level and forms precipitation as it cools.

_____ 18. This term is used to refer to distances from the equator northward and southward.

_____ 19. These surface spiral patterns in the oceans affect the climate on adjacent lands.

_____ 20. These floating "animals" feed on the algae and plant-like organisms in an aquatic environment.

_____ 21. This is the side of a mountain from which cooler air descends and increases its moisture-holding capacity.

_____ 22. This is the deepest region of the ocean.

_____ 23. This general term is given to the rather odd-looking organisms that live on the ocean floor and feed on dead or dying organisms that fall from zones above.

_____ 24. This is the general name given to the fishes and mammals of the sea that feed on plant-like organisms and each other.

TERMS LIST

a. Gyrals	i. Detritus	q. Biomes
b. Open-water zone	j. Latitude	r. Nekton
c. Copepods	k. Leeward side	s. Continental shelf
d. Coral reef	l. Windward side	t. Zooplankton
e. Pelagic zone	m. Estuary	u. Benthos
f. Deltas	n. Deep zone	v. Photic zone
g. Shore zone	o. Mesopelagic zone	w. Salinity
h. Abyssal zone	p. Photoplankton	x. Intertidal zone

DIAGRAMMING AND LABELING

1. Label the zones of the ocean floor in the illustration below.

Fig. 37.1 Zones of the ocean.

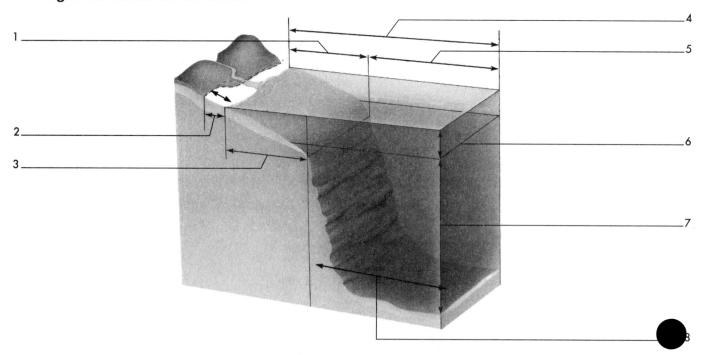

2. Illustrate the rain shadow effect by drawing a diagram depicting the direction of the wind, the location of the rain, and the resulting desert locations. Beneath your illustration, write a brief description of this effect using the western United States as your example.

MULTIPLE CHOICE QUESTIONS

_____ 1. In the tropical rain forest, the most likely limiting factor for ground level vegetation would be which of the following?
   a. Light
   b. Nutrients
   c. Moisture
   d. A and C
   e. A and B

_____ 2. The Earth's atmospheric coils can be explained by which of the following events?
   a. Spin of the Earth
   b. Effects of the moon
   c. Heating and cooling of the atmosphere
   d. A and C
   e. B and C

_____ 3. Which of the following terms best describe the accumulation of seasonal weather events over a long period of time?
a. Coils
b. Climate
c. Systems
d. Tradewinds
e. Rain shadow effect

_____ 4. The Earth's atmosphere is divided into how many coils of rising and falling air masses?
a. 2
b. 4
c. 6
d. 8
e. This number changes as the seasons change.

_____ 5. At which of the following latitudes would precipitation most likely occur?
a. 60
b. 30
c. 20
d. The equator
e. Latitude does not influence the amount of precipitation.

_____ 6. At which latitude would the deserts most likely be found?
a. 60
b. 40
c. 30
d. 20
e. A and D equally

_____ 7. Which biome is found on the top of a tall mountain in the tropics?
a. Temperate deciduous forest
b. Taiga
c. Tundra
d. Tropical rain forest
e. Savannas

_____ 8. Which of the following would include open grasslands with scattered trees?

   a. Temperate deciduous forest
   b. Taiga
   c. Tundra
   d. Tropical rain forest
   e. Savannas

_____ 9. In which of the following biomes would you expect to see the greatest fluctuation in daily temperature?

   a. Temperate deciduous forest
   b. Desert
   c. Prairies
   d. Tropical rain forest
   e. Savannas

_____ 10. Which of the following biomes is characterized by cone-bearing evergreen trees?

   a. Temperate deciduous forest
   b. Prairies
   c. Taiga
   d. Tropical rain forest
   e. Savannas

_____ 11. Which biome would be characterized by dwarf trees, lichens, grasses, herbaceous plants, and permafrost?

   a. Temperate deciduous forest
   b. Temperate grasslands
   c. Taiga
   d. Tundra
   e. Savannas

_____ 12. In which biome would you most likely expect to find lemmings, caribou, and wolves?
  a. Tundra
  b. Temperate grasslands
  c. Taiga
  d. Temperate deciduous forest
  e. Savannas

_____ 13. In terms of biomass, which of the following ecosystems would be considered the most productive?
  a. Estuaries
  b. Coral reefs
  c. Open oceans
  d. Savannas
  e. Tundra

_____ 14. Which of the following organisms would be found in the deepest regions of the intertidal zone?
  a. Benthos
  b. Barnacles
  c. Red and brown algae
  d. Sea cucumbers
  e. Oysters

_____ 15. What percent of the Earth's surface is covered with freshwater?
  a. 12%
  b. 10%
  c. 8%
  d. 5%
  e. 2%

_____ 16. In which of the following would air be the most dense?
  a. When it flows over the oceans
  b. When it reaches land from the oceans
  c. At the base of a mountain
  d. As it nears the top of a mountain
  e. Over a desert

BIOMES

Match the following biomes with the characteristics below.

Key:

a. Tropical rain forest	d. Temperate grassland	g. Temperate deciduous forest
b. Savanna	e. Tundra	
c. Desert	f. Taiga	

_____ 1. This treeless region is often subjected to fires and supports rapidly growing plants.

_____ 2. This area contains short trees and grasses.

_____ 3. Orchids are common here.

_____ 4. This region contains permafrost.

_____ 5. Succulent plants are most common here.

_____ 6. Pampas, steppes, and veldt are found here.

_____ 7. Cold, treeless, plains are found here.

_____ 8. Kangaroos and dingoes would be found living here.

_____ 9. Very tall trees with broad leaves forming a thick canopy are common here.

_____ 10. Long, cold, winters, with very little precipitation and very little daylight, occur here.

SHORT ANSWER QUESTIONS

1. Describe some of the factors that could be associated with formation, locations, and characteristics of deserts?

2. What controls the distribution of biomes?

3. What are the different kinds of freshwater habitats and how do they differ?

4. What makes the abyssal zone such a unique place to find living organisms?

CRITICAL QUESTIONS

1. What do you think the central grasslands of North America were like prior to its agricultural use?

2. If 2/3 of the Earth's surface is covered with water, why are we so concerned about water supplies throughout the world?

TELLING THE STORY

Develop a story that demonstrates your understanding of biomes. Write as if you were teaching the information to someone. If you have difficulty with any of the associated areas or topics as you write, go back and review that information in Chapter 37.

CHAPTER REVIEW ANSWERS

MATCHING

1. E	5. N	9. O	13. I	17. L	21. K
2. S	6. W	10. M	14. C	18. J	22. H
3. Q	7. X	11. G	15. V	19. A	23. U
4. F	8. B	12. P	16. D	20. T	24. R

DIAGRAMMING AND LABELING

1.   Fig. 37.1.

1. Neritic	5. Oceanic
2. Intertidal	6. Photic
3. Continental shelf	7. Aphotic
4. Pelagic	8. Abyssal

2.   Please refer to Fig. 37.4 on page 802 in the text for the answer.

MULTIPLE CHOICE QUESTIONS

1. E	5. A	9. B	13. B
2. D	6. A	10. C	14. C
3. B	7. C	11. D	15. E
4. C	8. E	12. A	16. D

BIOMES

1. D	3. A	5. C	7. E	9. A
2. B	4. E	6. D	8. B	10. F

SHORT ANSWER QUESTIONS

1.   The majority of deserts are found at 30° north and south latitude, where   the air descends and warms. As it warms, it increases its moisture capacity and therefore renders the land regions dry. Some deserts are located within the interior of continents and result because of the long distance between these locations and a source of moisture.

2.	The primary controlling factor over the distribution of biomes is climate. Two important variables are latitude and altitude. The proximity to large bodies of water and the presence of warm or cold air and water currents affect temperature. The presence of mountain ranges influences the precipitation patterns.

3.	The freshwater body types are rivers, streams, ponds, and lakes. Rivers differ from streams in their size. Lakes are generally larger and deeper than ponds. The water in lakes and ponds does not move and, therefore, they do not have as much dissolved oxygen as would a river or stream. Rivers and streams have very few producers and are strongly tied to the terrestrial ecosystems surrounding them.

4.	The abyssal zone is characterized by increasing pressure, the closer to the ocean floor you go. There is no light, and extremely cold temperatures. The animals that live in this region must rely on dead animals falling from above for food. Because of these very harsh conditions, only recently have we been able to study these organisms and observe their living patterns and morphological characteristics.

CRITICAL QUESTIONS

1.	Prior to agricultural use, the temperate grasslands were very different. This biome covered regions in the rain shadow of the Rocky Mountains south of the northern conifer forests. The rainfall was moderate and ideal for hundreds of species of perennial sod grasses, which grew from the Rockies to the Mississippi River. These grasses are better able than bunchgrasses to withstand grazing and trampling from large herbivores. Very strong, warm winds, known as Chinooks, blow leeward in the winter, causing snow to evaporate and expose the grasses. Because of this, the grasslands are excellent for winter grazing.

2.	The water that covers the majority of the Earth's surface is saliniated and, therefore, unsuitable for drinking. The freshwater bodies on the Earth only make up 2% of the total surface and make freshwater a very important commodity in the world.

CHAPTER 38

THE BIOSPHERE:
TODAY AND TOMORROW

KEY CONCEPTS OVERVIEW

1. The Biosphere

We often have such a limited view of our world, because most of our attention is given to the small area we inhabit. Given that your text has devoted numerous chapters to introducing you to the many diverse aspects of the living world, it offers you a different perspective about your world. The global ecosystem in which all other ecosystems exist is referred to as the **biosphere**. This region in which all living things are confined extends from approximately 30,000 feet above sea level to about 36,000 feet below sea level.

2. The Land

Because humans have such a dramatic impact on the land, it is critical that the consciousness be raised in every citizen of the world regarding how, why, and when our actions take a costly toll. While some attention has been paid to educating people about these issues, our dependency on finite resources has changed very little.

The land is very rich with natural resources. Because many of these are **nonrenewable**, meaning that they form at a slower rate than they are consumed, their supply and their uses must be carefully monitored. Many of the fuels we use today are fossil fuels, formed from decomposed carbon compounds of organisms that died millions of years ago. Our dependency on these fuels has raised serious concerns about their supply and their environmental effects. While nuclear energy is dependent on a finite supply of uranium, the supply is abundant. However, the concerns regarding waste disposal and safety issues have limited the use of nuclear power. Another resource upon which we rely for fuel is that of plant material, primarily wood from trees. The use of living plants to produce energy is referred to as **bioenergy**. The high demand for wood in the world has placed this resource at serious risk of depletion. **Reforestation** simply cannot replenish the forests at the rate we are destroying them.

Renewable resources are those produced by natural systems that replace themselves quickly enough to keep pace with consumption. While the supply of these resources may not be an issue, their uses should be. The primary sources of power are available from the sun, wind, water, geothermal energy, and bioenergy. Solar power can be captured and used for numerous purposes. Water, or **hydropower**, has been used for many years to produce electricity. In fact, 20% of the world's electricity is currently produced using water. Wave power and tidal power may also be used to generate electricity. High-tech windmills are designed and located in windy regions of the country to generate electricity. Obviously, this would not be a practical means of energy production in all locations. **Geothermal** energy refers to the use of heat deep within the Earth and can be manipulated to produce heat that can be converted to electricity. In many of the developing countries, microorganisms are used to enhance decomposition of excrement and organism matter to produce methane gas, which is used to fuel stoves, lamps, and to produce electricity. Newer methods of bioenergy involve the use of corn and sugar cane to produce ethanol from a fermentation process. This process has raised questions about the use of crops for fuel production.

Mineral resources are inorganic substances that occur naturally in the Earth's crust. Because these materials are not evenly distributed throughout the world, their value is often dependent on availability and usage. The minerals most commonly used by humans are zinc, lead, copper, aluminum, and iron. These minerals are usually mined or extracted, used and reused, or recycled.

Deforestation is one of the major concerns in the world today. The use of forest products for paper, lumber, and fuel has placed a significant drain on the present supply. The tropical rain forests have demonstrated a critical depletion of trees because they are being cleared for agricultural purposes. As you have learned in the last chapters, the soil in this biome is nutrient poor and not able to sustain crop production for long. The method of clearing is usually to cut and burn the trees. The resulting release of carbon dioxide into the atmosphere has been linked to the greenhouse effect and global warming. Additionally, the loss of plants and animals that inhabit these areas has a significant impact on biodiversity and organism populations.

While habitat destruction has led to the decline and extinction of many species of organisms, direct human exploitation may also result in the elimination of some endangered species. Poaching, smuggling, and trading in products such as ivory or succulent plants can reduce such species to nonrenewable levels. Species extinction profoundly effects the diversity of living organisms and ultimately impacts the very fiber of Earth's delicate balance. Zoos and seed banks have been established in an effort to preserve the genetic diversity and replenish depleted populations of organisms.

The average American produces about 19 pounds of garbage and trash per week! The amount of **solid waste** discarded in the United States alone is staggering, and the space for disposing of this waste is quickly diminishing. Landfills are covered each day, rather than being burned. However, while there has been a reduction in the amount of carbon dioxide and toxic fumes released by these burns, the amount of soluble materials that drain into water sources has increased. Additionally, increases in methane gas from the decomposition of organic matter has occurred. Smarter packaging of consumer products, recycling, and mulching are ways to reduce the waste being disposed of in the landfills.

3.   The Water

The **hydrosphere** of this planet lies mainly in the oceans, but also consists of the freshwater reservoirs, atmospheric water vapor, and groundwater. This water is continually being cycled from the land to the air. Because this process can involve mixing water with pollutants from the land and air, water resources can become contaminated and be harmful to living and nonliving things.

Water pollution may originate from a single point source or from dispersed areas that carry pollution into surface runoff water. Organic materials that are added to water will undergo bacterial decomposition and deplete the dissolved oxygen available in the water. The effect on aquatic organisms is devastating. The addition of inorganic nutrients, such as nitrates and phosphates, leads to eutrophication which triggers an algae bloom. As these algae die, they too will be decomposed and usurp oxygen from the water.

Since most toxic pollutants that enter the surface waters of the world do not degrade, they accumulate over periods of time and begin to contaminate the organisms within these aquatic environments. This process is known as **biological concentration.** Then, as other organisms within the environment consume contaminated organisms, they too become contaminated, a process referred to as **biological magnification.**

Acid rain refers to precipitation that has a pH less than 5.7. The two primary causes for acidic precipitation are nitrogen oxides produced by automobile emissions and sulfur dioxide produced from burning high sulfur coal in power plants across the world. These two gases combine with atmospheric water vapor to form nitric acid and sulfuric acid. The resulting rain lowers the pH of lakes and kills many of the inhabitants. The lower pH makes the toxic heavy metals more mobile in the ecosystem and may result in increased incidence of growth abnormalities, cancer, and death in aquatic organisms.

4.   The Atmosphere

Earth's atmosphere actually extends beyond that included in the biosphere. The troposphere extends approximately 36,000 feet into the atmosphere and slopes downward toward the equator. The cycling of cooled and warmed air is descriptive of the constant turnover in the air. The **troposphere** is the portion of the atmosphere most affected by pollution. Particulate matter can be a serious problem in industrial areas or in places where the combustion of fuels is high. Carbon monoxide, hydrocarbons, and nitrogen oxides all contribute to the production of **smog** which hovers in the air. Once the hydrocarbons and nitrogen oxides undergo photochemical reactions, new secondary pollutants are formed. The most well known of these is **ozone**, a chemical that affects the eyes and upper respiratory tract.

It is ironic that we produce ozone in our environment and destroy it in the **stratosphere** were it is needed. This atmospheric layer just above the biosphere forms ozone when sunlight reacts with oxygen and produces a protective shield against the harmful ultraviolet rays of the sun. Chlorofluorocarbons (CFCs), which are used in the manufacturing of Styrofoam and foam insulation, split up ozone faster than it can be made. A reduction in the ozone layer in the stratosphere can increase the risk of skin cancer.

CHAPTER REVIEW ACTIVITIES

CONCEPT MAPPING

Develop a concept map that demonstrates your understanding of the biosphere. Incorporate as many of the key terms as possible and choose the linking terms that will best correlate one term to another. Develop the map with as much detail as you can.

MATCHING

Select the term from the list provided that best relates to the following statements. If you are unsure of the correct answer or chose an incorrect response, review the related concept in Chapter 38.

_____ 1. These substances are formed over time from the undercomposed carbon compounds of organisms that died millions of years ago.

_____ 2. This term discribes the use of heat deep within the Earth for heating purposes or as part of a process to produce electricity.

_____ 3. This chemical air pollutant is formed as a secondary pollutant when hydrocarbons and nitrogen oxides undergo photochemical reactions.

_____ 4. This is the use of the sun for heating or electricity production.

_____ 5. This term describes the richness and variety of species on Earth.

_____ 6. This is the blocking of outward heat radiation from the Earth by carbon dioxide in the atmosphere.

_____ 7. This is the loss of the plants and animals that live in the tropical rain forest through cut-and-burn agriculture and logging.

_____ 8. This is the process by which toxins in organisms form in high concentrations when they consume tainted organisms lower on the food chain.

_____ 9. Precipitation having a low pH is produced when sulfur dioxide and nitrogen dioxide pollutants combine with water vapor in the atmosphere.

_____ 10. This type of air pollution results from the burning of fossil fuels.

_____ 11. This is the use of living plants to produce energy.

_____ 12. This is the global ecosystem of life on Earth extends from the tops of the tallest mountains to the depths of the deepest seas.

_____ 13. This describes a worldwide temperature increase that could result from the greenhouse effect.

_____ 14. These are resources produced by natural systems that replace themselves quickly enough to keep pace with consumption.

_____ 15. This type of air pollution is caused by reactions between hydrocarbons and nitrogen oxides taking place in the presence of sunlight.

_____ 16. This is a general term for the biosphere, encompassing the land, air, water, and every living thing on Earth.

_____ 17. The accumulation of inorganic nutrients in a lake stimulates plant growth.

_____ 18. A change in water temperature occurs when industries release heated water back into rivers, after using the water for cooling purposes.

_____ 19. This is a process by which some organisms accumulate certain harmful or deadly chemicals within their bodies which are present in their environments or in the food they eat.

_____ 20. These resources are formed at a rate much slower than their consumption and are therefore in finite supply.

_____ 21. An energy source derives its power from the splitting apart of the nuclei of large atoms or from the combining of the nuclei of certain small atoms.

_____ 22. This is another term for garbage and trash.

TERMS LIST

a. Global warming	l. Fossil fuels
b. Ozone	m. Smog
c. Deforestation	n. Biological concentration
d. Nuclear power	o. Biodiversity
e. Environment	p. Biosphere
f. Eutrophication	q. Acid rain
g. Solar power	r. Photochemical smog
h. Nonrenewable resources	s. Geothermal energy
i. Solid waste	t. Thermal pollution
j. Biological magnification	u. Bioenergy
k. Greenhouse effect	v. Renewable resources

MULTIPLE CHOICE QUESTIONS

_____ 1. By what year is it predicted that the world's population will double?
   a. 2010
   b. 2020
   c. 2030
   d. 2040
   e. 2050

_____ 2. Which of the following mineral resources would last the longest if the current consumption rate continues?
   a. Aluminum
   b. Iron
   c. Mercury
   d. Lead
   e. Copper

_____ 3. Which of the following mineral resources would be used up first, at the current rate of consumption?
   a. Platinum
   b. Nickel
   c. Phosphorus
   d. Gold
   e. Copper

_____ 4. What percent of the garbage produced by a typical American household is paper?
   a. 10%
   b. 20%
   c. 30%
   d. 40%
   e. 50%

_____ 5. At the current rate of destruction, how many years will it take to deplete the rain forest?
   a. 25 years
   b. 30 years
   c. 50 years
   d. 100 years
   e. 175 years

_____ 6. Approximately how many different species exist on Earth?
   a. 4 million
   b. 40 million
   c. 12 million
   d. 20 million
   e. 32 million

_____ 7. What is the most common cause of extinction?
   a. Man's direct elimination of species
   b. The use of pesticides
   c. Climatic changes
   d. Man's destruction of habitats
   e. Desertification

8. Which of the numbers below represents the pounds of garbage produced by the average American in one week?
   a. 4 pounds
   b. 8 pounds
   c. 10 pounds
   d. 15 pounds
   e. 19 pounds

9. Which of the following are used in a fermentation process to develop methanol for fuel?
   a. Peanuts
   b. Corn
   c. Potatoes
   d. Cabbage
   e. All of the above

10. Most acid rain is formed by dissolving which of the following in atmospheric water vapor?
    a. Chromic acid
    b. Sulfuric acid
    c. Carboxylic acid
    d. Hydrochloric acid
    e. Acetic acid

11. Which of the following would NOT be a predicted effect of global warming?
    a. Melting of the polar ice cap
    b. Alteration of rainfall patterns
    c. Coastal area flooding
    d. Increased crop yield
    e. Increased temperature

SHORT ANSWER QUESTIONS

1. What are the three primary fossil fuels?

2. What are the sustainable renewable resources?

3. List some strategies for averting a mineral shortage.

4. How does the current extinction compare to massive extinction in the geological past?

5. What might account for the decline in genetic diversity in agricultural crops, and what danger does this pose?

CRITICAL QUESTIONS

1. Explain how genetic engineering is being used to improve production of food crops. Describe some of the potential problems in using such technology.

2. How might humans disrupt or interfere with the nitrogen cycle?

3. Discuss the basic flaw in the concept that: "The solution to pollution is dilution."

TELLING THE STORY

Develop a story that demonstrates your understanding of the biosphere. Write as if you were teaching the information to someone. If you have difficulty with any of the associated areas or topics as you write, go back and review that information in Chapter 38.

CHAPTER REVIEW ANSWERS

MATCHING

1. L	5. O	9. Q	13. A	17. F	21. D
2. S	6. K	10. M	14. V	18. T	22. I
3. B	7. C	11. U	15. R	19. N	
4. G	8. J	12. P	16. E	20. H	

MULTIPLE CHOICE QUESTIONS

1. C	5. C	9. B
2. A	6. B	10. B
3. D	7. D	11. D
4. D	8. E	

SHORT ANSWER QUESTIONS

1. The three primary fossil fuels are coal, natural gas, and oil.

2. Sustainable renewable resources include the sun, wind, moving water, geothermal energy, and bioenergy.

3. A shortage or depletion of minerals could be averted by decreased population growth, reduction in the demand for the mineral, reusing the mineral for the same or other purposes, recycling the mineral, or by developing another substance or product that could replace the use of the mineral.

4. The extinctions of today are occurring much more rapidly and are the result of man's actions. Those of the geological past were caused by climatic events or changes or geophysical factors.

5. Scientists have utilized selective breeding processes to produce varieties of plants which have specific desirable traits. The seeds of these plants are then sold to the exclusion of other varieties that do not have the desired traits. Varieties with little diversity are more vulnerable to disease, pests, and climatic changes.

CRITICAL QUESTIONS

1.   Transferring genes from one plant to another in order to code for such things as nitrogen fixation, reduction in water needs, herbicide resistance, high yields, and deterrents to herbivores, have increased the popularity of genetically altered varieties of crop plants. The technology of DNA recombination makes it quite possible to develop strains that could eliminate the need for nitrogen fertilizers and result in a significant decrease in its demand. A reduction in the use of such fertilizers will eliminate the buildup of nitrogen byproducts in groundwater, rivers, and streams. The down side of this technology is the potential depletion of plant varieties. Variation in plant types reduceS the risk of extinction.

2.   Many of the activities of humans have a profound effect on the nitrogen cycle. Humans convert a massive amount of nitrogen (N_2) to nitrates yearly in the commercial production of fertilizers. When this fertilizer is applied to the soil where agricultural crops are grown, the distribution of nitrogen on the Earth is altered. Without the addition of this nitrogen, crops would have to be rotated with legumes or let the fields go without being planted every third or fourth year so that the ammonia from nitrogen-fixing bacteria could replenish nitrogen supplies. When the crops that have been fertilized with commercial nitrogen products are harvested, they become food for humans and animals. Their waste will contain much of the nitrogen from the plants. This nitrogen waste builds up in the sewage and feedlots and eventually ends up in our rivers and streams or in the groundwater. Any nitrogenous compounds that make their way to lakes often result in algae bloom which causes eutrophication and depletes the dissolved oxygen in the water. Nitrogen is also introduced into the atmosphere by burning fossil fuels and automobile emissions containing nitrogen oxide. These have been linked to the depletion of the ozone in the stratosphere.

3.   Dilution is a physical process resulting in dispersion of toxins throughout nature. The biological side of nature will absorb the toxins and concentrate them in a process known as biological magnification. The concentration, and ultimately the effects, of the toxins increase as it passes through the food chain, counteracting the dilution factor.